The Biology, Structure, and Systematics of the Cycadales

Proceedings of the Symposium CYCAD 87, Beaulieu-sur-Mer, France, April 17–22, 1987.

Dennis Wm. Stevenson,
Editor

NYBG

The New York Botanical Garden
Bronx, New York 10458, U.S.A.
Issued: 7 March 1990

MEMOIRS OF THE NEW YORK BOTANICAL GARDEN VOLUME 57

Cover: *Chigua restrepoi* megasporangiate and microsporangiate strobili and two views of a megasporophyll. Priscilla Fawcett, del.

Published by
The New York Botanical Garden
Bronx, New York 10458

International Standard Serial Number 0071-5794

Library of Congress Cataloging-in-Publication Data

The Biology, structure, and systematics of the Cycadales : proceedings of the symposium CYCAD 87, Beaulieu-sur-Mer, France, April 17–27, 1987. / Dennis Wm. Stevenson, editor.

p. cm. — (Memoirs of the New York Botanical Garden ; v. 57)
Includes bibliographical references.
ISBN 0-89327-350-3
1. Cycadales—Congresses. I. Stevenson, Dennis Wm. (Dennis William), 1942- . II. Series.
QK1.N525 vol. 57
[QK494]
581 s—dc20
[585′.9] 89-13570
CIP

International Standard Book Number 0-89327-350-3

CONTRIBUTORS

Paul Bamps, National Botanic Garden of Belgium, Meise B 1860, Belgium.

Kabita Das, Department of Botany, University of Allahabad, Allahabad, 211002 India.

Yves Delange, Museum National d'Histoire, Paris Cedex 05, France.

Paolo De Luca, Orto Botanico, 223 via Foria, 80139 Naples, Italy.

Cynthia Giddy, Giddy's Nursery, P.O. Box 45, Umlaas Road, 3730 Natal, South Africa.

Maria Grilli Caiola, Department of Biology, The University of Rome, via O. Raimondo 00173, Rome, Italy.

John Hendricks, 110 Brookmeade Avenue, Statesville, North Carolina 28677, USA.

Peter Lindblad, Institute of Physiological Botany, University of Uppsala, Box 540, S-751 21 Uppsala, Sweden.

Stanislas Lisowski, Department of Geobotany, A. Mickiewicz University, PL 61713, Poznan, Poland.

François Malaisse, Laboratory of Ecology, Phytogeography, and Phytosociology, Faculty of Agronomic Sciences, University of Belgium, 5800 Gembleux, Belgium.

Aldo Moretti, Department of Plant Biology, University of Naples, 223 via Foria, 80139 Naples, Italy.

Knut Norstog, Fairchild Tropical Garden, 11935 Old Cutler Road, Miami, Florida 33156, USA.

Roy Osborne, Department of Chemistry and Applied Chemistry, University of Natal, King George V Avenue, Durban 4001, South Africa.

Divya Darshan Pant, Department of Botany, University of Allahabad, Allahabad, 211002 India.

P. Mick Richardson, New York Botanical Garden, Bronx, New York 10458, USA.

Sergio Sabato, Department of Plant Biology, University of Naples, 233 via Foria, 80139 Naples, Italy.

Jean-Pierre Sclavo, Villa "La Finca," Plateau du Mont Boron, 06300 Nice, France.

Gesualdo Siniscalco Gigliano, Department of Plant Biology, University of Naples, 223 via Foria, 80139 Naples, Italy.

Dennis Stevenson, New York Botanical Garden, Bronx, New York 10458, USA.

Ian Turner, Springs Farm, P.O. Box 2162, Harare, Zimbabwe.

Mario Vazquez Torres, Center of Biological Investigations, University of Veracruz, Apartado Postal 294, Xalapa, Veracruz 91000, Mexico.

Siyuan Yang, Dukou Horticulture Department, Dukou City, Sichuan Province, Peoples Republic of China.

Lin Zhou, Dukou Horticulture Department, Dukou City, Sichuan Province, Peoples Republic of China.

Zheng Zhou, Dukou Horticulture Department, Dukou City, Sichuan Province, Peoples Republic of China.

In Memoriam

Padre Sergio Restrepo

Martyred, June 1989

Introduction

The contents of this volume represent a series of papers presented by the authors at a symposium entitled CYCAD 87 that was held at Beaulieu-sur-Mer in France from April 17 through April 22, 1987. The theme of Cycad 87 was to emphasize current research trends on the Cycadales. The papers, revised for publication, have been loosely organized into several categories as indicated in the Table of Contents. I would like to take the opportunity to make a few brief comments on these areas with respect to where we are and where we need to go.

At the onset it should be understood that cycads, as non-flowering seed plants like the conifers and particularly the gnetopsids, have received considerably less research attention because they are a small group that are generally perceived as esoteric taxonomic relicts. Other contributing factors include (a) the lack of available material for study due to the rarity of cycads in the wild, in cultivation, and in herbaria; (b) only a very rudimentary understanding of cycad reproductive biology; (c) virtually no research on cycad population biology; and (d) a poor understanding of the phytogeography of cycads. In addition, although the cycads have a long history, their fossil record remains incomplete, impeding the resolution of major evolutionary and morphological trends. Perhaps the best example reflecting the paucity of work on the cycads is the fact that a recent publication entitled "Bibliography of the Living Cycads (Annotated)" by Robert Read and Marie Solt (Lyonia **2**: 33–200. 1986) contains only 118 pages of citations. Moreover, the citations include short annotations on each publication! Surely, one would find it quite difficult to find an order within the flowering plants, containing as many species as the Cycadales, that had as few or fewer published works. I would point out that the evolutionary history of the cycads far exceeds that of the flowering plants which by comparison have yet to prove their mettle.

For the preceding reasons, our understanding of the biology of the Cycadales is still in its infancy. In fact the basic taxonomy is still incomplete as evidenced by the numerous new taxa that have been described in the past few years in the neotropical genera and currently in the paleotropical genera, *Cycas* and *Encephalartos*. How could it be otherwise, with little complete information concerning basic morphology, phytochemistry, cytology, etc.? The works contained in the present volume represent renewed interest in various aspects of cycad biology. Consequently, many of the articles' authors try to pull together past work to be synthesized with their current research efforts. Upon completion, the newly obtained information can and will be brought to bear on a phylogenetic classification of the order.

Not only has there been renewed interest in the cycads in recent years, but this interest has been quite diverse, as evidenced by the articles presented in this volume and in their bibliographies. For example, the papers by Linblad and Grilli Caiola on the physiological and ultrastructural aspects of the symbiotic relationship of cycads with cyanobacteria represent long-needed studies. Similarly, the papers on reproductive biology, propagation via tissue culture, and conservation by Norstog, by Osborne, and by Giddy, respectively, also represent a new attempt to understand the reproduction in cycads and how this information can be used in preventing their extinction through overcollecting and habitat destruction.

There is little published information on the ecology of cycads, particularly the genus *Cycas*. The paper by Zhou, Yang, and Zhou certainly helps in our understanding of the habitat requirements for the species of *Cycas* occurring in

China. The paper by De Lange makes a similar long needed contribution to our knowledge of *Encephalartos* in central Africa. Certainly, the contribution by Sabato on the phytogeography of Mesoamerican and Caribbean cycads is useful not only to those interested in cycads but also for those interested in the biogeography of these regions.

The consolidation of information and continued research into the phytochemistry of the Cycadales by Richardson and by Siniscalco Gigliano mark important steps in the acquisition of much needed information that is being and will continue to be used by systematists working on the order. Moreover, this information is important to the medical community who are studying the toxicity effects of cycads on people who use these plants as a supplemental food source. In fact, most of our ethnobotanical information on cycads has in the past focused on this aspect. Thus, the contribution by Vazquez Torres on other uses of the cycads in Mesoamerica will, I hope, initiate a continuing trend. Based on my field experience in the Neotropics, I know that indigenous peoples attach special significance to these plants which they are reluctant to discuss.

The bulk of the cycad literature, other than species descriptions, is on anatomy and morphology. Yet here too our information is not complete, as we find basic observations presented in the papers by Stevenson and by Pant and Das. It is interesting that the latter authors find evidence for the presence of vessel elements in the roots of *Cycas*. Given the well known trend of vessel elements first occurring phylogenetically in the roots of monocotyledons, it is very interesting that such a trend may well occur in another group of plants such as the cycads. This aspect of xylem structure certainly needs a more thorough examination in the Cycadales as a whole.

The preceding comments are also applicable to the karyology of the cycads in the sense that cytological variability is just now being thoroughly examined and understood by Moretti. While this work has a long way to go, it is well on its way. The reproductive problems associated with populations that have an odd diploid number (e.g., $2n = 27$) is not only interesting but also perplexing.

If one were to ask most systematists their opinion on the status of cycad systematics, one would in most instances receive the answer that they are a small group that is completely known. The range of contributions at all taxonomic levels in the present volume certainly refutes such an idea. New species are presented here and in many other recent publications dealing with the Neotropics and Africa. Even a new genus, previously unknown at any level (in other words not segregated from an existing genus), is described. It should be clear that generic and even family relationships have yet to be resolved. However, the addition of new data from the diverse areas presented in this volume and continuing research will contribute to a more comprehensive taxonomy.

The production of this volume could not have been undertaken without the generous assistance of the organizers and sponsors of the symposium. Principally, I extend my deep gratitude to the Conseil Général des Alpes-Maritimes and the officials of Nice, France for their generous financial support and hospitality. While many people were important to the success of the symposium, it is impossible to thank all individually. However, the herculean efforts of M. Jean-Pierre Sclavo of Nice, France, Marianne Aardenburg and Dott. Aldo Moretti of Naples, Italy should certainly be credited. Simply stated, without the editorial efforts of H. David Hammond, this volume would not have been produced.

Finally, this volume is dedicated to the memory of Padre Sergio Restrepo of Colombia. Padre Restrepo was an avid amateur botanist who guided Knut Norstog and me to the locality of the new genus *Chigua. Chigua restrepoi* was named in his honor when I first wrote the manuscript for this volume. In June 1989, Padre Restrepo was assassinated by an unknown person(s). This is a tragic loss of a truly remarkable man who was highly respected and loved by his local community. His love of botany was responsible for our opportunity to search for cycads in an area of considerable political and armed danger; a danger which I at the time did not take too seriously, but which obviously was real and led to his murder. The frontispiece photograph of Sergio Restrepo was taken during our trip together in March 1987.

DENNIS WM. STEVENSON
New York Botanical Garden

MARCH 1990 MEMOIRS OF THE NEW YORK BOTANICAL GARDEN 57: 1–210

Contents

A Historical Perspective on Cycads from Antiquity to the Present

PAOLO DE LUCA

Abstract

DE LUCA, P. (Dipartimento di Biologia vegetale and Orto Botanico, Universita di Napoli, via Foria 223, 80139 Napoli, Italia). A historical perspective on cycads from antiquity to the present. Memoirs of the New York Botanical Garden **57**: 1–7. 1990.—A brief survey of the history of cycads in various cultures is given beginning with the Etruscans. Their use in the religous practices of several cultures is also reviewed. The early exploration for cycads by naturalists and horticulturalists is discussed with particular reference to the discoveries and expeditions that led to the descriptions of the cycad genera. The historical aspects of the acquisitions of the more important large living collections in botanical gardens are outlined.

Key words: Cycads; living collections; exploration; cultural uses.

Historical Survey

CLASSICAL TIMES AND PLACES

In Western classical literature cycads are not reported because these plants were extinct before man arrived in the Mediterranean and surrounding regions. Thus, cycads are not mentioned by Theophrastus, the Greek botanical student of Aristotle; nor by Dioscorides, the Greek botanist from Cilicia in the 1st century A.D.; nor by Pliny, the noted Roman botanist also of the 1st century A.D.

It is understandable that in Greek and Roman times African cycads were not known, because the most northern species of this group occur only to the south of the areas known at the time as "*Hic sunt leones.*" However, it is strange that Asian species (viz., *Cycas* spp.), were not known, as most of them are distributed in the Indian sub-continent, which Alexander the Great had already arrived at, and from which, in the next century, were brought several exotic plants to Europe for the gardens of Patricians. Among these exotic Asians we should note citrus plants, of which a painting is still present in a Pompeian villa covered in the Vesuvian eruption of 79 A.D. and discovered at the end of the last century.

Some authors have recognized a plant described by Theophrastus and named χύχαs (xuxas), as a species of cycad referable to the genus *Cycas.* However, a particular passage from Theophrastus leads us to understand that this plant was an Ethiopian palm; probably *Hyphaene thebaica* (χοι̂ξ).

EXTRA-EUROPEAN CULTURES

So, while the European people did not have contact with cycads, the same is not the case for people of other continents, where cycads are present. People from America, central and southern Africa, Asia, and Oceania have had a good knowledge of these plants and often used their seeds and/or stems to prepare a flour for bread making. Some natives of South Africa, for example, named a local species of cycad "bread tree."

But the interest of man for this plant has been often stimulated by their particularly primitive features and perhaps by the difference in appearance which they show, in contrast to the other plants amongst which they live.

In many localities, cycads have even acquired a sacred significance. Indian people, before conversion to Christianity, used to decorate their temples with fronds of *Cycas* for festivals; this tradition continued after their conversion and *Cycas* became known as the "Church palm." The phenomenon of sacredness of cycads from non-Christian cultures is also present in Mexico, where *Dioon* fronds are still used to decorate churches, to the point that campesinos named *Dioon* "Palma de la Virgen."

The austere appearance of cycads is probably the reason for which these plants were considered the symbol of veneration of the dead in many parts of the world. People of Java used to plant *Cycas* around tombs. *Dioon* plants in Mexico, besides being the "Palma de la Virgen," are also known as "Palma de Dolores," because of their use at funerals. A species of this genus, *Dioon merolae,* was reported for the first time from plants cultivated near a funeral chapel in the state of Chiapas in southern Mexico (De Luca et al., 1981). An African *Encephalartos, E. hildebrandtii* var. *dentatus,* was found for the first time planted near the Muslim cemetery in Dar es Salaam, Tanzania (Melville, 1957; Heenan, 1977).

In Europe, as we have already said, cycads are not present: even though we know of a case of interest by ancient Italian people in a group related to these plants, the Cycadeoidales. More than four thousand years ago, ancient Etruscans found a fossil trunk of a member of the Cycadeoidales, previously used by Neolithic people as a sharpening stone. Probably they understood that they had found something of interest, and they put this trunk in a tomb of the Marzabotto necropolis. This fossil, that has the longest known history of any fossil, is the specimen from which the species *Cycadeoidea etrusca* was established and is today in the Museo Cappelli in Bologna (Wieland, 1916).

The above, not strictly pertinent to the study of cycads, pertains rather to that series of approaches that man has made to plant study and represents the beginning of that scientific interest which became botany.

Beginnings of Modern Knowledge

The first new report of Cycadales to the Western world was of the genus *Cycas,* and was due to two Arabian naturalists of the 9th century who, upon returning from travels to India, referred to some plants of this genus as being utilized as a source of flour. Later, we have reports by Antonio Pigafetta and Fernão Lopez de Castanheda, in the 16th century, with further particulars by Francis Drake, who found *Cycas* plants in the Moluccas, where Indians used to eat the seeds. With regard to American cycads, certainly the first report in the West is that of Giovanni Lerio, whose report of his trip to Brazil in 1576 contains his observations of a plant named Ayrius, which certainly refers to the genus *Zamia.*

With the beginning of exploration of our world, many naturalist voyagers made notes in their diaries of news and paintings of various cycads which they came across. They introduced living plants to European botanical gardens and, for the first time, took care to collect dried specimens which they deposited in various herbaria, from that moment leaving material help for botanists who study the systematics of cycads. There are a notable number of explorers, but I would like to confine myself to explorers of the Americas.

The first of these were Humboldt and Bonpland, who collected valuable material in the course of their explorations in tropical America from 1799–1804. Later, Poeppig, in the 1830's, collected cycad samples in Peru and Bolivia.

Of particular interest is the work of Liebmann, a Danish naturalist explorer, who in 1840 collected material in Mexico which later was to be attributed to the new genera *Dioon* and *Ceratozamia.*

Karwinski, a Russian baron, explored Mexico for a long time in the first half of the last century and transported to the St. Petersburg (now Leningrad) Botanical Gardens live material on which a new species was later described.

At the same time, C. d'Orbigny explored, on behalf of the Paris Science Museum, in the Mato

Grosso of Bolivia, collecting samples from which a new species was described.

Von Warsewicz, a Polish traveller, in the middle of the last century, explored Colombia and Central America, and Seemann, an English traveller of the same period, explored the Chóco in Colombia, both finding new species of cycads.

In the second half of the last century Roezl and Wallis travelled to South America, finding new species of cycads described later by botanists, who named two of these species after the two discoverers.

All these naturalist travellers increased the collected material of both botanical gardens and private collections, as seen in the very long list published by Wendland (1854).

Early Private Collections

Among the private collections, we note that of James Yates in England. An amateur botanist, he had a deep knowledge of cycads, and had continuous scientific relationships with the botanists of the time. His herbarium material was donated to the British Musem in London.

Numerous other amateur students of cycads, at great personal sacrifice, built up very important living collections. To many of them we owe our thanks, because they donated their collections to botanical gardens, offering researchers precious study material.

While not wishing to present another list, I must mention Garbari, a physician in Trento in North Italy, whose cycad collection was donated to the botanical garden of Florence in 1907. Nor can I omit to mention Professor Luigi Califano, an internationally famous pathologist and a large scale collector and scholar of cycads, who bequeathed his collection, upon his death in 1976, to the Naples Botanical Garden.

Scientific History of Cycads

Descriptions of New Genera

It is now opportune to move to the scientific history of cycads. The first botanical description of a plant belonging to the order Cycadales is that of the pre-Linnean botanist Rheede Tot Draakestein, who described a cycad by the name of *Todda panna* in 1682. Linneaus, in 1753, using the illustrations of this botanist, named the genus *Cycas,* the first cycad genus of modern botany. Linneaus probably thought that the plant described by Rheede and classified by him was the same one described by Theophrastus as χύχας (xuxas). On the basis of this erroneous interpretation, he named the genus, *Cycas,* a Latinization of the word used by Theophrastus.

The second genus of cycads described by botanists is American. Again Linneaus described it, in 1763, and called it *Zamia.* The origin of this name is obscure and its derivation from the writings of Pliny, as claimed by Schuster (1932), is erroneous.

The third genus described (by Lehmann, 1834) is African, with the name *Encephalartos.* As stated above, the natives produced flour from the seeds of the strobili which are at the top of the trunk of this plant. Lehmann formed the genus name from the Greek words en-cephali-artos (ἐνκέφαλοςά-τος) which mean bread on the head.

In 1842, Miquel described the first Australian genus, which he called *Macrozamia,* by adding the prefix macro (large) to Linneaus' name *Zamia.* The following year Lindley described a new North American genus which he named *Dioon* from the two Greek words δίς (dis) and ὠόν (oon), which loosely means twice ovulate with reference to the two seeds on each megasporophyll. In 1846, Brongniart described a new North American genus which he named *Ceratozamia,* again by prefixing to the name *Zamia* a word, χέρας (xeras = horn), to note the presence of two small horns on the sporophylls.

In 1853, Moore described a small South African genus which he named *Stangeria* in honor of Dr. Stanger, general superintendant of Natal, who in 1851 had sent the plant to him in England from Port Natal. The genus *Stangeria* was originally attributed by Kunze to the fern genus *Lomaria* and only later, with Moore's description of the strobili, was it correctly attributed to the Cycadales.

In 1857, Regel described a new Australian genus to which he gave the name *Lepidozamia,* by prefixing to the name *Zamia* the Greek name λεπίς (lepis = scale), because the leaf scars on the trunk of this plant resemble the scales of a snake.

Later, in 1863, Joseph Hooker described a new Australian genus, *Bowenia* in honor of Sir George F. Bowen, a botanist and the first governor of Queensland.

De Candolle (1868) described the genus *Microcycas,* a Cuban endemic, by adding the word micro (=small) to the name *Cycas,* although this is certainly a misnomer, because *Microcycas* is often larger in size than *Cycas.*

As exploration for new cycads continues today, it perhaps is not surprising that recently Dennis Stevenson and Knut Norstog have discovered a new and very locally endemic genus in Northern South America, which is described in this volume. This genus, named *Chigua* after the name for cycads used by the indigenous people of this region, is the only cycad genus endemic to South America.

Above, in my description of the eleven genera of Cycadales, while it was superfluous to mention the new species attributed to known genera, I cannot forbear mentioning that many botanists have described new species on the basis of a single herbarium specimen or a single living plant, without a sound knowledge of previously described species. Consequently, several species already described were redescribed as new, and this has led to great confusion in the systematics and nomenclature of the Cycadales.

Systematic Revisions

Of great help in rectifying this inconvenience were the large systematic revisions of the order, the first by Miquel, the Dutch botanist mentioned earlier. His first major cycad work, published in 1842, is exemplary for its scientific rigor and quantity of information. In 1861, Miquel published the *Prodromus Systematis Cycadearum* and, in 1868 and 1870, "Nouveaux materiaux pour servir a la connaissance des Cycadées," a work in which he reported everything then known on the biology and systematics of these plants.

Of equal importance is the "Cycadearum generum specierumque revisio" by Regel (1876), a work done with an acute systematic sense, the crowning glory of the research activity of that Russian botanist.

Another excellent monograph on cycads was published by de Candolle (his *Prodromus Systematis Cycadearum* of 1868). Also important is the 1884 monograph of American cycads published by Thiselton-Dyer, then Director of Kew Gardens, in "Biologia centrali-americana."

Among others who have worked in the present century, of particularly important relevance was C. J. Chamberlain, noted American botanist, both for the quantity of data and the novelty of his approach to the study of cycads. He was the first to understand the importance of field studies with the purpose of evaluating the extreme morphological variability in the distinctive characters of these plants during the course of their life; and therefore, he travelled for 15 years in America, Africa, and Australia to observe cycads in their habitat. In 1919, he published *The living cycads,* a flowing volume, rich in data, still current for its synthesis of the taxonomy, morphology, and reproductive biology of these plants, containing most of the data from his original research.

In 1932, the German paleobotanist Schuster published the volume "Cycadaceae" as part of the important work *Das Pflanzenreich* edited by Engler. This monograph was intended to be a complete synthesis of what was known up to that time and is valuable for the richness of the bibliographic citations. Unfortunately, Schuster had no familiarity with living cycads, consulted few herbaria (among them Berlin, later destroyed in the Second World War), and was superficial with regards to nomenclature. Moreover, he did not understand the nomenclatural type concept in the modern sense and never annotated the specimens cited as types in his work, creating great confusion and instability in the nomenclature of cycads. All these deficiencies constrain and confound the contemporary botanist, complicating nomenclatural revisions.

At the beginning of the 1940's, Chamberlain wrote an excellent monographic work on Cycadales in which he reported the mistakes and confusions of all earlier studies of this group. For example, he strongly criticizes the practice of describing new species from a single herbarium specimen or living plant of dubious origin without understanding the great variability of these plants during their life cycle. Unfortunately, this work was never published, probably due to his death, and thus botanists could not use this fundamental contribution to the knowledge of cycads. Copies of the Chamberlain manuscript are stored at the Fairchild Tropical Garden of Miami, the New York Botanical Garden, and the Botanical Garden of Naples. From his writings, it is possible to see the very high number of correct intuitions by Chamberlain.

There are no other complete works on the cycads, even though they are needed. Useful publications, even though partial, are the monograph on Australian Zamiaceae by Johnson (1959); the monograph on South African cycads by Dyer (1965); the volume on *Cycas and the Cycadales* by Pant (1973), already in a second edition and with a third in preparation. Also, there is the popular science volume *The Cycads of South Africa* by Cynthia Giddy.[1]

In the last decade there has been a growing interest in the biosystematics, chemosystematics, comparative morphology, ecology, physiology, karyology, and reproductive biology of Cycadales, as reported in this volume on the present state of research in these fields by the most qualified researchers in the world. It is not possible to discuss them all individually, but I must mention Professor Knut Norstog, the leader in the study of the reproductive biology of this group.

Living Collections in Botanical Gardens

All the research reported in this volume could only take place thanks to the existence of collections of cycads in botanical gardens. In the field, these plants are distributed in areas very distant from one another, and often they are found in areas of difficult access. Only in botanical gardens is it possible to obtain an overall view and have access to the material to undertake comparative research among families and genera.

As an example, let me relate the history of the cycad collection of the Naples Botanical Garden, with which I have some familiarity. The Botanical Garden of Naples has a priceless collection of Cycadales, notable for both the number of specimens and the number of species. All ten known genera are represented. This collection was started in 1809, two years after the foundation of the Botanical Garden, with the acquisition of a specimen *Encephalartos altensteinii.* In that year, a Neapolitan noble (and plant collector), gave to the Botanical Garden a specimen of this species started at the end of the 18th century from seed collected in the wild. This plant, still present in the collection today, became reproductive for the first time in 1864 and formed two magnificent male cones.

The first donation was followed by a second, in 1813, given by the queen of Naples, Maria Carolina Bonaparte, the wife of King Giacchino Murat. She imported from Japan two large specimens of *Cycas revoluta* and gave one to the Naples Botanical Garden; this plant lived until 1905. The following years saw new acquisitions, but they were not recorded.

Michele Tenore's 1845 catalogue of plants cultivated in the botanical garden shows that, at that time, there were grown outside specimens of Cycadales belonging to three genera (*Cycas, Encephalartos,* and *Zamia*), with a total of nine species.

In the 1867 catalogue of the "Royal Botanical Garden" of Naples, prepared by Giuseppe Antonio Pasquale, the list of cycads was brought up to date. It included a specimen of *Dioon edule.* There is no further mention of the collection for about fifty years.

In 1913, Fridiano Cavara in a note of the Royal Botanical Garden writes: "in (cultivation) in open ground are specimens of Cycadales, including *Dioon edule, Cycas revoluta, Encephalartos altensteinii, E. horridus, Zamia muricata. . . .*" These plants were grown outside and during the winters they were protected by simple reed-matting. They were transplanted from the open ground into pots a little after 1929.

In 1934, the collection was enlarged with a pseudobulb of a male *Cycas revoluta* from the Botanical Garden of Pisa, in Central Italy, which periodically produced fertile pollen, used for pollination and production of seeds.

In 1966, the Cycadales were exhibited in the area occupied by Gymnosperms and at this time an inventory listed the following specimens: *Ceratozamia mexicana* male and female, *Cycas circinalis* female, *C. revoluta* male and female, *Dioon edule* female, *Encephalartos altensteinii* male, *E. horridus* male, *E. longifolius* female and *Macrozamia communis.*

At the end of the 1960's interest in the Cycadales underwent a notable expansion and the collection was notably increased, with the idea of protecting, conserving, multiplying and spreading these plants. The Naples Botanical Garden has been a leader in this cultural operation. Our collection has increased in three different ways: acquisition of specimens, donations and field collections.

Between 1966 and the present day, we have acquired from South Africa 193 specimens of 32 species of *Encephalartos* and 15 specimens of *Stangeria eriopus* collected from their original localities. From 1974 to the present day, we have acquired from Australia 83 specimens of all the known species of *Macrozamia,* 25 specimens of the two species of *Bowenia,* 12 specimens of the two species of *Lepidozamia* and also 50 specimens of the nine species of Australian *Cycas.* We also acquired cultivated specimens of unknown origin. From 1966 to the present day in this manner, we have obtained 34 specimens belonging to the genera *Cycas, Dioon, Encephalartos,* and *Macrozamia.*

The collection of Cycadales of the Botanical Garden has also notably increased thanks to donations from private collections and scientific institutions.

In 1973, a Catholic missionary sent two specimens of *Encephalartos laurentianus* collected in the wild on the border of Zaire and Angola. In 1974, the Singapore Botanical Garden donated plants of *Cycas rumphii.* In 1976, Professor Muniz of the Havana Botanical Garden (Cuba) visited Naples, bringing as a gift a magnificent specimen of *Microcycas calocoma.*

On the 14th of January 1976, the death of the earlier mentioned Professor Luigi Califano occurred. His family, respecting the wishes of the deceased, gave to the Botanical Garden his entire collection of plants, which consisted of a precious and useful group of Cycadales. In this way we had the honor and the fortune to inherit 102 specimens of 59 species belonging to 8 genera.

In 1977 and 1978 Professor Viviani of the Botanical Garden of Puerto Rico sent specimens of *Zamia latifoliolata, Z. media, Z. portoricensis,* and *Z. pumila.*

In the years 1978–1980, our collection was enriched by some specimens of *Ceratozamia, Stangeria,* and *Macrozamia* donated by the estate of Professor Angelo Verga, a lawyer by profession and an enlightened collector of plants. The donations from Professor Verga included a priceless specimen of *Encephalartos woodii.* Two specimens of *Cycas thouarsii,* collected in Madagascar were donated by him in 1980 and 1981.

The collection has been enriched with interesting specimens by exchange with the Fairchild Tropical Garden in Miami. From 1980 to this day Fairchild Tropical Garden has sent specimens of *Zamia floridana, Z. chiqua, Z. muricata, Z. pseudoparasitica* and *Microcycas calocoma.*

From 1969 onwards, there was a notable growth of the collection thanks to a number of botanical expeditions. The first expeditions took place in 1969, 1971, and 1974 and were organized on behalf of the Accademia Nazionale dei Lincei by Professors Luigi Califano, Aldo Merola (the late director of the Naples Botanical Garden) and Ruggero Tomaselli (the late director of the Botanical Garden of Pavia). Professor Sergio Sabato and myself also participated in these expeditions.

In the course of these botanical expeditions we collected numerous specimens of *Dioon edule, D. mejiae, D. spinulosum* and specimens of an unknown species of *Dioon,* numerous species of *Zamia* and *Ceratozamia,* as well as seeds of many species.

In 1975, on a research trip in the east with Professor Aldo Merola, I collected specimens of *Cycas rumphii* in Indonesia and *C. siamensis* in Thailand.

From 1980 to the present day, there have been seven botanical expeditions to Latin America. These expeditions, led by Professor Sergio Sabato with the help of Professors Aldo Moretti and Gesualdo Siniscalco Gigliano of the Department of Plant Biology in Naples, led to the acquisition of numerous specimens of *Dioon, Ceratozamia,* and *Zamia.* Most of these expeditions were made in collaboration with Professors Mario Vazquez Torres, Dennis Wm. Stevenson, and Knut Norstog. This material has been used for phytochemical, systematic, embryological and other studies. In this regard I would like to mention our contribution to the knowledge of the systematics of the genus *Dioon.* On the basis of the specimens collected in the wild and cultivated in the botanical garden, six new species and one new variety were described, adding to the four species known in 1980.

The collection at the Botanical Garden of the University of Naples is, therefore, one of the most important in the world today, consisting of 570 adult specimens and numerous seedlings belonging to 111 species of all 10 known genera. Of these specimens 95% are classified with certainty and about 80% are of known collection locality.

This collection is of notable scientific importance, not only for the number of specimens, but because of the number of species and genera represented and for the presence of the frequently reproductive male and female specimens. Other important living collections for both scientific and conservation purposes are to be found at Kew Gardens; Fairchild Tropical Garden, which has the most extensive collection of cycads with respect to both the number of plants and number of taxa; New York Botanical Garden; Hortus Amsterdam; and Leiden Botanical Garden. The latter two are notable because they have living plants which are the types of many of the species described by Miquel in his 36 years of work on cycad systematics.

To conclude my introduction, I hope that this volume will lead to further scientific meetings on cycads with the aim of the exchange and comparison of both ideas and plant material. Such living material is invaluable for studies on each field of cycad research.

Literature Cited

Brongniart, A. 1846. Note sur un nouveau genre de Cycadacée du Mexique. Ann. Sci. Nat. Bot., sér. 3, **5:** 5–9.

Candolle, A. de. 1868. Cycadaceae. Prodromus systematis naturalis **16(2):** 522–548.

Chamberlain, C. J. 1919. The living cycads. University of Chicago Press, Chicago.

De Luca, P., S. Sabato & M. Vazquez Torres. 1981. *Dioon merolae* (Zamiaceae), a new species from Mexico. Brittonia **33:** 179–185.

Dyer, R. 1965. The cycads of Southern Africa. Bothalia **8:** 405–518.

Giddy, C. 1984. Cycads of South Africa. 2nd ed. C. Struik (Pty.) Ltd., Cape Town.

Heenan, D. 1977. Some observations on the cycads of Central Africa. J. Linn. Soc., Bot. **74:** 279–288.

Hooker, J. D. 1863. *Bowenia spectabilis* Australian *Bowenia.* Bot. Mag. **89:** t. 5398 & text.

Johnson, L. A. S. 1959. The families of cycads and the Zamiaceae of Australia. Proc. Linn. Soc. New South Wales **84:** 64–117.

Lehmann, F. 1834. Novarum et minus cognitarum stirpium pugillus **6(3):** 3–14.

Linneaus, C. 1753. Species plantarum. Page 1188.

———. 1763. Species plantarum. Ed. 2, **2:** 1659.

Melville, R. 1957. *Encephalartos* in Central Africa. Kew Bull. **12:** 237–257.

Miquel, F. A. W. 1842. Monographia cycadearum. R. Natan, Utrecht.

———. 1861. Prodromus systematis cycadearum. C. v. D. Post Amsterdam, C. G. v. D. Post, Jr., Utrecht.

———. 1868. Nouveaux matériaux pour servir a la connaissance des Cycadées. Première-troisième parties. Arch. neerl. Sci. exactes nat. **3(3):** 193–254. Quatrième-cinquième parties. Arch. neerl. Sci. exactes nat. **3(5):** 403–427.

———. 1870. Nouveaux matériaux pour servir a la connaissance des Cycadées. Sixième partie. Arch. neerl. Sci. exactes nat. **5(1):** 74–88.

Moore, T. 1853. List of Mr. Plant's Natal ferns. Hooker's J. Bot. Kew Gard. Misc. **5:** 228–229.

Pant, D. 1973. *Cycas* and the Cycadales. 2nd ed. Central Book Depot, Allahabad.

Regel, E. 1857. Zwei neue Cycadeen. Bull. Soc. Imp. Naturalistes Moscou **30:** 163–191.

———. 1876. Cycadearum generum specierumque revisio. Acta Horti Petrop. **4:** 273–320.

Schuster, J. 1932. Cycadaceae. *In* A. Engler (ed.), Das Pflanzenreich **99(IV. 1):** 1–168.

Thiselton-Dyer, W. 1884. *In* W. B. Hemsley, Biologia cent.-americ., Bot. **3(16):** 190–195.

Wendland, H. 1854. Index Palmarum, Cyclanthearum, Pandanearum, and Cycadearum. Hahnii, Hanover.

Wieland, G. 1916. American fossil cycads. Vol. II. Taxonomy. Carnegie Institution of Washington.

MARCH 1990 MEMOIRS OF THE NEW YORK BOTANICAL GARDEN 57: 8–55

Morphology and Systematics of the Cycadales

Dennis Wm. Stevenson

Table of Contents

Abstract

Stevenson, D. W. (New York Botanical Garden, Bronx, New York 10458, USA). Morphology and systematics of the Cycadales. Memoirs of the New York Botanical Garden **57:** 8–55. 1990.—The morphology, anatomy, and systematics of the Cycadales are reviewed and discussed with emphasis upon new information published since the last comprehensive work on Gymnosperms by Chamberlain, fifty-five years ago. The morphology and anatomy of vegetative and reproductive structures are discussed in terms of the systematic distribution of the character state at the generic level and gaps in our information concerning these features are pointed out. Vegetative topics include growth habit, secondary growth with emphasis on polyxyly and extrafascicular vasculature, leaf morphology and petiole vasculature, and root morphology. Reproductive topics include ovule morphology and vasculature, mega- and microsporophyll morphology and vasculature, and reproductive branching. The historical aspects of classification, at the generic level and above, are briefly reviewed. Cladistic analyses are provided with discussions on character distributions and the interpretations of the cladograms. A provisional classification is presented based on the cladistic analyses.

Key words: Cladistics; leaves; megasporophylls; microsporophylls; ovules; roots; stems.

Introduction

Compared to other groups of seed plants, the cycads have received little attention from morphologists and systematists since their last major treatment by Chamberlain (1935). It would be fair to say that research on cycads or the lack thereof has not been the result of lack of interest but rather the lack of availability of suitable research material. Renewed interest in the group has, however, been developing in the past 10–15 years. In general, the plants are rare and usually encountered as prized specimens in botanical gardens. It comes as no surprise then that a plant anatomist or morphologist is not going to be able to destroy such specimens simply to look at the shoot apex. The generally large size of adult plants has presented technical problems for those wishing to study growth patterns or the lack of predictable branching patterns or strobilus production. Also, the space requirement for individual plants has generally mitigated against having large populations of any given species in a glasshouse in temperate areas, and the individual plants usually constitute an important part of living teaching specimens. Another problem facing researchers has been the lack of information concerning phenology. It would appear that most cycads have very long life cycles. The time from the germination of a seed until that seedling grows into a reproductive plant is unknown for most species and would appear to be a very long time, perhaps in some cases exceeding our life span. Thus, systematics, for example, has been hampered by the inability to observe successive generations of putative hybrids. In fact, meiosis has not even been described, in any detail, in the cycad literature.

Another problem facing systematists is the paucity of herbarium specimens, particularly of neotropical and Asian species. Often only a single sterile specimen is available for study and in most cases there is no representative sampling of the variability either between or within populations. Another exasperating situation is that there are no type specimens for many of the species described in the last century. The last complete monograph of the order was done by Schuster (1932), and he based this monograph mostly on specimens in Berlin Herbarium. Unfortunately as a result of the vicissitudes of World War II, all cycad specimens in Berlin and Vienna were destroyed with the exception of the few in the Willdenow Herbarium in Berlin.

Having pointed out the major impediments to cycad research, the fact remains that a considerable body of recent research has been published. In this paper, I will confine my review to a limited portion of this research. With respect to anatomy and morphology in the broad sense, I will exclude work on cytology, reproductive biology, and morphogenesis because these topics are well covered in other papers in this volume. Also, I will not discuss systematics at the species level because my goal here is to consider generic and familial relationships and useful characters that are now being investigated at these levels.

Vegetative Morphology

Growth Habit and Trunk Anatomy

Cycads vary from large columnar, arborescent forms to subterranean forms. All cycads are pachycaulous, although some species appear more so than others. That is, they are generally sparingly branched and more or less succulent, having a massive parenchymatous pith and cortex and relatively little secondary, manoxylic wood. Arborescent forms may or may not have a persistent armor of leaf bases. Thus, for instance, the arborescent species of *Cycas* and *Dioon* have a persistent armor of old leaf bases covering the trunk, whereas *Microcycas* and the arborescent species of *Zamia* have smooth trunks without persistent leaf bases.

Those genera with few species (less than 3) have only one growth form. Thus, *Stangeria* (monotypic) and *Bowenia* and *Chigua* (two species each) have only subterranean axes. In contrast, *Lepidozamia* (two species) and *Microcycas* (monotypic) are arborescent. The other cycad genera, *Ceratozamia, Cycas, Dioon, Encephalartos, Macrozamia,* and *Zamia,* all contain ten or more species. Each of these genera has growth forms ranging from completely subterranean species through species with short, bulbous trunks, to tall arborescent species with trunks greater than 3 meters.

Some genera have polyxylic stems while others have monoxylic stems. Polyxylic anatomy consists of concentric rings of secondary vascular tissue in which each ring is composed of a xylem cylinder surrounded externally by a phloem cylinder. Thus, the polyxylic anatomy consists of pith surrounded by alternating bands of xylem and phloem. These concentric rings of vascular tissue appear to be the result of successive cambia. Apparently, the successive cambia originate in the phloem produced by the preceding cambium, but the actual development is unknown. Also, the nature of cambial activity is not known in the sense that is there only one of these cambia active at a given time, or are several of them active simultaneously? Monoxylic stems have just one ring of vascular tissue composed of xylem surrounded by phloem. The genera with polyxylic stems include *Cycas, Encephalartos, Lepidozamia,* and *Macrozamia* (Greguss, 1968). However, not all species in these genera have this character and its expression appears to be related to the age of the individual and perhaps its stem diameter. Polyxyly appears to be more developed in arborescent taxa than in subterranean taxa although it does occur in species with the latter growth form. *Dioon, Zamia, Ceratozamia, Microcycas, Stangeria,* and *Bowenia* are monoxylic. The distribution of this character is among cycad genera is given under 1.18 in Table I.

Another unusual, for extant seed plants, feature of the vascular architecture that occurs in cycads is the presence of extrafascicular vascular tissue. Little is known about this aspect of the vasculature other than the earlier descriptive literature concerning its occurrence (Matte, 1904; Worsdell, 1896, 1900, 1901). Moreover, the terminology used to describe this condition and that of polyxyly is confusing in this early literature, although the two features are distinct and apparently not related causally. Extrafascicular vasculature is restricted here to those bundles or groups of bundles that occur in the cortex of the stems of some cycads. Single bundles may be concentric and amphicribral (i.e., primary xylem surrounded by primary phloem) or nearly so. Again the descriptive terminology in the earlier literature is confusing because concentric bundle is used thusly and also for collateral bundles with centripetal development of metaxylem (i.e., the protoxylem is next to the phloem). These amphicribral bundles, which look like small protosteles, have a vascular cambium. In other regions of the cortex, in the same transection, there may be a group of open collateral bundles. Eventually, a continuous vascular cambium is formed between the bundles and secondary xylem and phloem are produced by it. Thus, the stem has a central stele (ring of vascular tissue) surrounded by peripheral smaller vascular cylinders. The resulting anatomy appears surprisingly similar to that of medullosans. Worsdell (1901) was particularly impressed by the similarity of this feature of vascular organization in cycads and that of medullosans, although for some reason, this aspect received no comment by Chamberlain (1935). According to Worsdell (1901), the occurrence of this type of vascular architecture is found only in *Cycas, Encephalartos, Lepidozamia, Macrozamia,* and *Bowenia.* Later, Matte (1907) and Dorety (1909) reported on the occurrence of extrafascicular cambia in seedlings of *Ceratozamia,* but no vascular tissue was noted as having been produced by these cambia. Worsdell (1901) explicitly stated that extrafascicular organization was not present in older stems of *Ceratozamia.* I have never seen either a polyxylic system or cortical vascular cylinders in *Ceratozamia.* I would point out that the extrafascicular system is not related to the presence of the cone domes (see discussion of these structures in the section on reproductive morphology) associated with reproductive branching, as this type of vascular organization has never been attributed to *Dioon* or *Zamia,* both of which have conspicuous cone domes. The distribution of this character is given under 1.29 in Table I.

Any conclusions concerning polyxyly and extrafascicular vasculature in the cycads must remain tentative at this time because several aspects of both remain to be thoroughly investigated. These include the developmental anatomy of the polyxylic vascular cylinders (this has not been investigated for any taxon). The relationship between polyxyly and size, age, and growth form is still ambiguous. The same is true with respect to extrafascicular vasculature. Moreover, the three-dimensional organization of extrafascicular vasculature and its relationship, if any, with that of polyxyly needs attention, and the occurrence of extrafascicular cambia needs to be substantiated for *Ceratozamia.* The possibility that extrafascicular vasculature is related to adventitious branches should also be investigated.

Encephalartos, Lepidozamia, and *Macrozamia* are the only cycad genera with medullary bundles. While the occurrence and descriptive anatomy of these bundles was thoroughly inves-

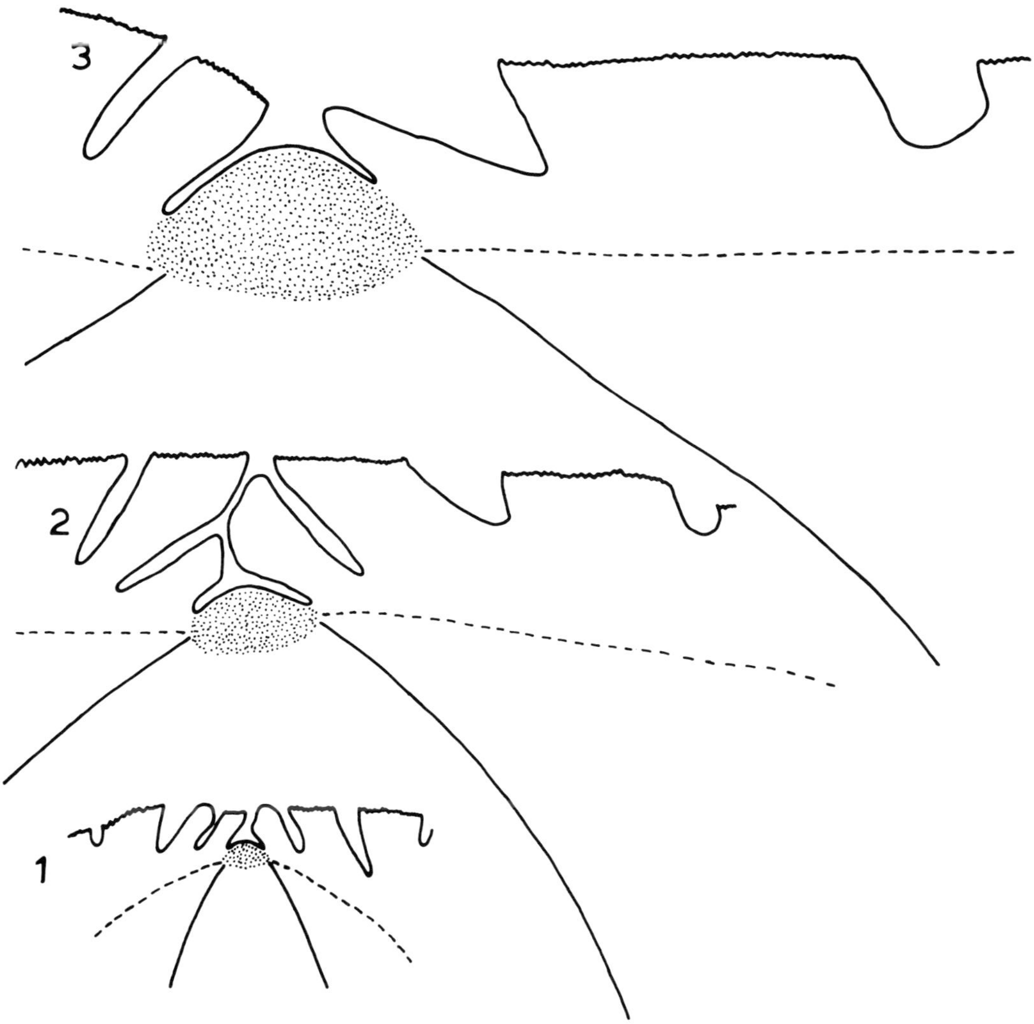

Figs. 1–3. Schematic drawing of successively older shoots based on *Zamia pumila.* Dashed lines represent outer limits of tissue derived from the primary thickening meristem (solid line). Stippled areas represent the apical meristem. From Stevenson (1980a). **1.** An approximately five year old shoot. **2.** An approximately 25 year old shoot. **3.** An approximately 50 year old shoot.

tigated by Worsdell (1901), the three-dimensional architecture and development of medullary bundles is still unknown. However, as with extrafascicular vasculature, medullary bundles are not associated with cone domes.

The overall pachycaulous habit of cycads results from a primary thickening meristem (Figs. 1–3) that is continuous with the apical meristem (Stevenson, 1980b), a massive apical meristem (Stevenson, 1980b and literature cited therein), and a very active and extensive pith-rib meristem (Stevenson, 1980b). The massive parenchymatous cortex results from the activity of the primary thickening meristem, whereas the vast parenchymatous pith results from the activity of the extensive pith-rib meristem and its derivatives (Figs. 1–3). Many other extant groups with pachycaulous forms such as the Marattiales, Filicales, and arborescent and bulbous Liliopsida [e.g., Arecaceae (palms), Agavaceae, Iridaceae, Liliaceae, etc.] also have a primary thickening meristem (Rudall, 1989; Stevenson, 1980a). However, that of the cycads is unique because the primary thickening meristem lies next to the pith and produces most of its derivatives centrifugally (Stevenson, 1980b). In contrast, the primary thickening meristem of ferns (Stevenson, 1976, 1978) and monocots (Stevenson, 1980a) lies just under the leaf bases and produces most of its derivatives centripetally. In addition, the cycads have the largest shoot apices in vascular plants, some (e.g., *Cycas*), exceeding 3 mm in diameter (Stevenson, 1980b and literature cited therein), and the most extensive pith-rib meristem in vascular plants that accounts for the formation of the large central pith (Stevenson, 1980b).

Another feature unique to the cycads is the mechanism by which the vertical stem remains underground. In those taxa which have a subterranean stem but yet produce up to fifty new leaves in a growth flush, the stem has to increase in length to at least the equivalent of the number of new nodes. However, the shoot apex is always buried under the substrate surface. When the height of leaf bases is analysed from the shoot apex to the base of the stem, one finds that the height of these leaf bases decreases up to 60% towards the base of the stem while at the same time the width of the scars remains more or less constant (Stevenson, 1980c). This is because cycads with subterranean axes have contractile stems. These contractile stems show the same developmental anatomy as contractile tap roots in the cycads and other vascular plants (Stevenson, 1980c). These contractile stems are known in vascular plants only in the cycads.

Several recent papers on fossil cycads (Delevoryas, 1982; Delevoryas & Hope, 1971, 1976) have suggested that the original growth form of the cycads was not pachycaulous but rather leptocaulous. This conclusion rests on two observations about the trunk of *Leptocycas gracilis*; namely, that this fossil taxon consisted of plants with slender trunks and the trunks do not have an armor of persistent leaf bases. The conclusion that these plants are not pachycaulous is not supported by extant cycad morphology. For example, *Zamia roezli* has the typical massive trunk (i.e., greater than 10 cm in diameter) of arborescent cycads, whereas *Z. obliqua* has a slender trunk (i.e., 5–8 cm diameter) up to 4 m tall. Neither species has an armor of persistent leaf bases covering their stems. Both of these species are, however, basically unbranched, with a small amount of secondary xylem relative to the parenchyma of the pith and cortex and both have a primary thickening meristem of the cycad type that gives rise to this parenchyma. Thus, even species such as *Z. obliqua* are pachycaulous because the criteria are not how massive the stem appears and whether or not there is a persistent armor of leaf bases, but rather the growth pattern of the stem with respect to branching, the development of primary parenchyma via a primary thickening meristem and extensive pith-rib meristem, and the paucity of secondary xylem in relation to primary parenchyma. The fact that some fossil taxa appear to have slender trunks without a persistent armor of leaf bases is not relevant to the argument because these plants are at the same time sparingly branched. They appear similar to *Z. obliqua* in growth habit. Unfortunately, the anatomical preservation of these fossils is not good enough to use those critera. Nevertheless, I maintain that there are no data inconsistent with the conclusion that the early cycads were indeed pachycaulous.

BRANCHING

Three types of vegetative branching have been described for the cycads (Stevenson, 1988). These are the normal dichotomous branching of the apex, adventitious branching in intact and uninjured plants, and adventitious branching as a result of injury or decapitation under natural conditions. Branching from axillary buds does not occur in the cycads because axillary buds do not occur (Stevenson, 1980a, 1988). Therefore, the cycads are limited to these types of branching.

Dichotomous branching of vegetative shoots occurs both in natural habitats and in cultivated plants, but is rare. These dichotomies are isotomous with successive isotomies occurring at right angles to preceding isotomies. The developmental evidence for this is meager as only a few apices in the early stages of isotomous branching have been observed (Stevenson, 1988). However, these observations are supported by vascular anatomy where it can be shown that the vascular cylinder is divided into two equal cylinders corresponding to the two isotomous branches.

Adventitious branching is common in the cycads. It commonly occurs in the production of bulbils along the stem as in *Cycas* or as suckers or bulbils arising near the base of older trunks as is common, for example, in *Ceratozamia* and *Encephalartos.* These bulbils arise as adventitious buds produced from callus arising as internal derivatives of the phellogen involved in the abcission of foliage leaf bases (Stevenson 1988). These bulbils, which emerge by growing through the remnants of the leaf bases, remain dormant until the trunk has increased in length. Adventitious buds are also produced as a wound response, particularly in decapitated plants (Coulter & Chrysler, 1904; Stevenson, 1988). In these cases, adventitious buds may arise from the vascular cambium or the phellogen or both. In contrast to the equal division of the vascular cylinder that is associated with isotomous branching, the vascular cylinder of the adventitious branch remains diminutive with respect to that of the parent axis (Stevenson, 1988). It should be noted that adventitious branching in the cycads does not result in the dissemination of asexual propagules, although it does lead to the production of multiaxial individuals.

Leaves

In general, the vegetative shoots of cycads produce two types of leaves, scale leaves or cataphylls, and foliage leaves (Figs. 4–6). The exceptions to this are *Stangeria* and *Bowenia* (Stevenson, 1981) which are discussed later. Other than in *Bowenia,* cycad leaves consist of a pinnately compound lamina without a terminal leaflet, a swollen base either with or without stipules, and a petiole that may be pronounced, as in *Zamia,* or obscure, as in *Dioon* and *Encephalartos. Bowenia* differs in having a bipinnate lamina with a terminal leaflet. The broadest leaflets are usually found in the central portion of the rachis with a series of leaflets of decreasing length occurring distally and proximally, giving the leaves an oval to oblong or lanceolate outline. Only *Microcycas* differs significantly in this respect because its terminal leaflets are more or less the same length as its median leaflets. Thus, the leaves of *Microcycas* appear truncated, as if someone had cut off their ends with a machete.

Bowenia and *Stangeria* do not produce cataphylls on vegetative shoots. Instead, the petiole and blade portions of the leaves abscise distally to the stipulate leaf (Figs. 7, 10), and the leaf base and enclosing stipules function as a cataphyll (Stevenson, 1981). Cataphylls (Figs. 8, 9) are produced on reproductive branches in *Bowenia* (Stevenson, 1981). In 1981, I thought that no cataphylls were produced on the reproductive axes of *Stangeria.* However, Worsdell (1898) had reported their association with microsporangiate strobili, but made no mention of their occurrence with megasporangiate strobili. I have recently been able to confirm this, but have yet to make similar observations concerning megasporangiate strobili. It is interesting to note that decurrent cataphylls are present on the cone peduncles in *Dioon, Encephalartos, Lepidozamia,* and *Macrozamia* but not in the other genera, although I have found an occasional diminutive and apparently teratological cataphyll on a few peduncles in *Zamia integrifolia.* I would draw attention to the fact that the ubiquitousness of peduncular cataphylls in both sexes of and all species of the aforementioned genera needs to be surveyed. Also, because the peduncles of male cones in *Cycas* are very short, it may appear, superficially, that *Cycas* has decurrent cataphylls on its cone peduncles, but these are actually borne on the vegetative stem. A summary of the distribution of cataphylls among the genera appears under characters 1.20 and 1.30 in Table I.

The form of young emerging foliage leaves, called ptyxis, was extensively studied by Stevenson (1981). Contrary to previous descriptions, *Cycas* does not have circinnate ptyxis of the whole leaf but rather only of the individual leaflets. Circinnate ptyxis of whole leaves is found only in the bipinnate leaves of *Bowenia.* The restriction of circinnate ptyxis to *Bowenia* among the cycads appears to represent a difference in the developmental pattern in the leaves of *Bowenia.* Although extensive developmental studies of leaf ontogeny in the cycads have yet to be published, it appears that the sequence of leaflet initiation is basipetal in all genera except *Bowenia.* This basipetal sequence precludes the occurrence of circinnate ptyxis because the leaf apex does not remain meristematic but ceases apical growth early in development. Extended apical growth occurs only in *Bowenia.*

The other major difference, erect versus inflexed, in the pytxis of whole cycad leaves appears to be related to growth habit. The ptyxis of the whole leaf is erect in those taxa with aerial trunks and in contrast inflexed in those taxa with subterranean stems. Thus, in any given genus with both arborescent and subterranean species,

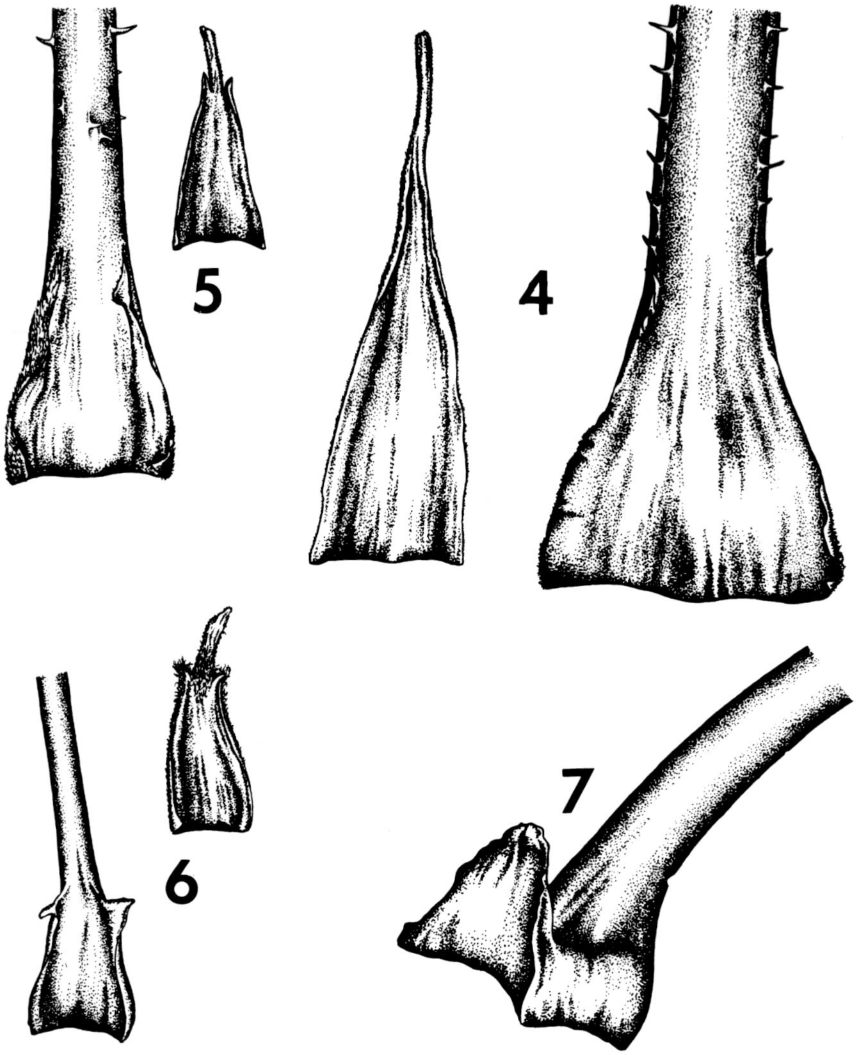

Figs. 4–7. Leaf bases and cataphylls of cycads. From Stevenson (1981). **4.** *Cycas circinalis.* Note lateral rows of spines on petiole base. **5.** *Ceratozamia mexicana.* Note small prickles on petiole base. **6.** *Zamia pumila.* **7.** *Stangeria eriopus.* Note cowl-like stipule on adaxial side of leaf base.

such as in *Zamia* or *Encephalartos,* one finds erect ptyxis in the arborescent species and inflexed ptyxis in the subterranean species. In fact, the ptyxis changes from inflexed to erect in a given individual plant as it progresses from its subterranean seedling stage to its arborescent adult form. The association of inflexed ptyxis with a subterranean growth habit appears to be functional in that the inflexed form of the emerging leaves functions in a protective manner similar to the hypocotylar hook of many hypogeous dicotyledonous seedlings found in the flowering plants.

The petiole and rachis of the leaves of the cycads exhibit some interesting variations which have some systematic significance. Often there are spines or prickles on the petiole and rachis. These occur in two forms, reduced leaflets which are true spines (Fig. 4) and epidermal projections which are prickles (Fig. 5). Spines, found only on petioles, occur in two parallel rows continuous distally with the leaflets (Fig. 4). In general,

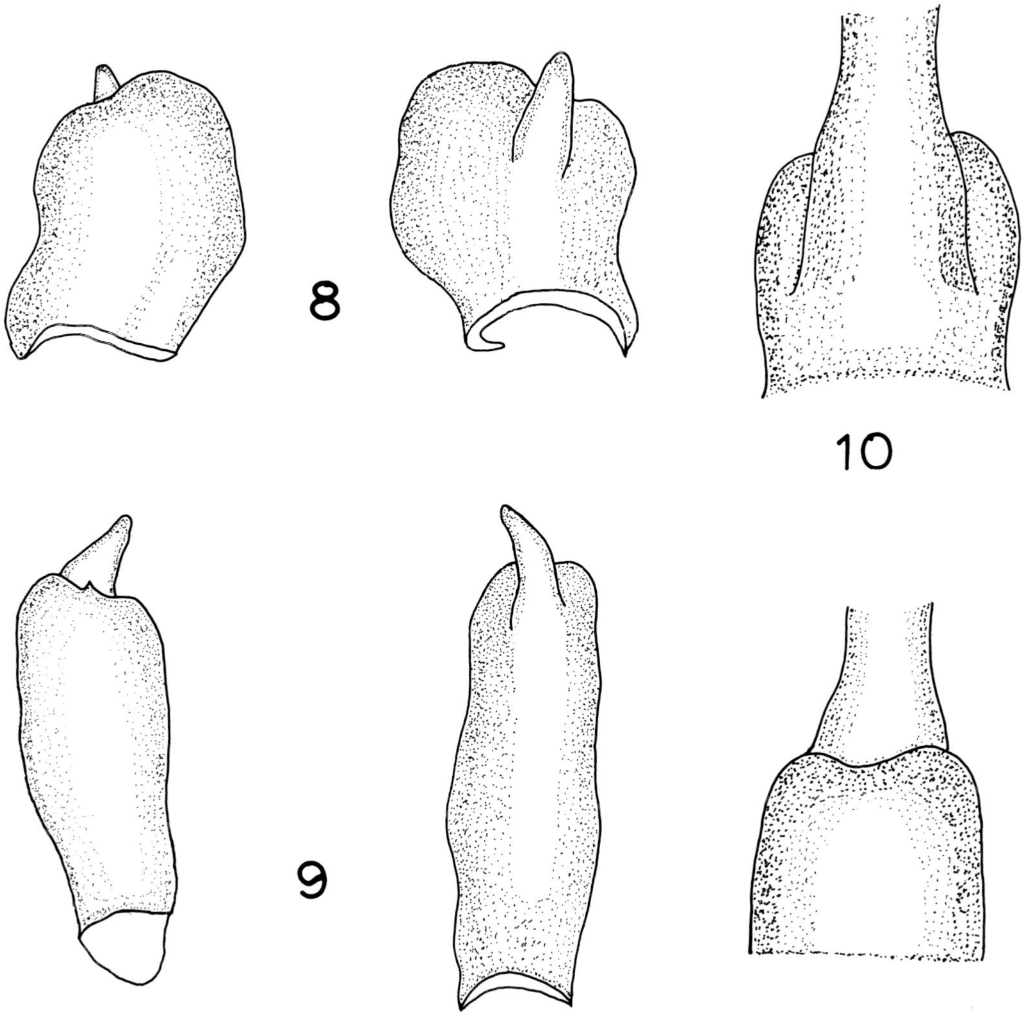

Figs. 8–10. Leaf bases and cataphylls of a branch bearing a megasporangiate strobilus of *Bowenia spectabilis*. **8.** Ad- and abaxial views of a cataphyll with an expanded form. **9.** Ad- and abaxial views of a cataphyll with a linear form. **10.** Ad- and abaxial views of a leaf base with stipules. Note the similarity in form to those of *Stangeria* in Figure 7.

the spines exhibit a gradual transitional series between leaflets, true leaflets, and spines. The spines occur on leaves of adult plants in *Cycas* (Fig. 4), *Dioon, Encephalartos, Lepidozamia,* and *Macrozamia,* but are absent from seedling leaves. In contrast, prickles (Fig. 5), which are randomly located on the petioles and sometimes the rachii of *Ceratozamia* and *Zamia,* are not reduced leaflets and exhibit no morphological transition as reduced leaflets. Some genera have neither spines nor prickles. This includes *Bowenia, Stangeria,* and *Microcycas.* On the other hand, *Zamia* has some species with prickles and others without, for example, the West Indian species and *Z. inermis.* In other species, for example, *Z. polymorpha* and *Z. cunaria,* some leaves of an individual plant may have prickles and other leaves may be without prickles. In *Ceratozamia,* all species have leaves with prickles although they may be absent in individual leaves and plants. For instance, *C. keusteriana* is generally thought not to have armed petioles but small prickles can be found on some leaves of some plants although this condition is not common. Conversely, *C.*

hildae and *C. latifolia* have an occasional leaf or individual plant in which prickles are absent. A summary of the distribution of prickles and reduced spinose leaflets among the genera is given under characters 1.24 and 1.25 in Table I.

The position and mode of leaflet attachment to the rachis exhibit some variability. There are two basic patterns of leaflet attachment, decurrent (Figs. 11–17) and articulate (Figs. 18–23). In the latter, which occurs in *Ceratozamia* (Fig. 18), *Chigua, Microcycas* (Fig. 19), and *Zamia* (Figs. 20–23), the leaflets tend to abscise before the leaf, thus leaving only the rachis and petiole attached to the stem. Associated with the articulation is a small callous rim at constricted leaflet base (Figs. 18–23). Decurrent leaflet bases occur in *Cycas* (Fig. 11), *Dioon* (Fig. 12), *Encephalartos, Lepidozamia* (Fig. 15), *Macrozamia* (Figs. 16, 17), and *Stangeria* (Fig. 13). In *Bowenia,* both the pinnae and pinnules (Fig. 14) are decurrent. This character is obvious in all the species of *Cycas, Dioon, Lepidozamia,* and *Stangeria.* In some species of *Encephalartos* and *Macrozamia* this character appears obscure and the leaflets appear constricted at their bases as in articulated species (compare Fig. 16 with Fig. 18). However, the leaflets do not abscise and a groove, however slight, can be observed on the rachis between leaflets. This groove or depression demarcates the decurrent leaflet base. *Macrozamia* is unusual in that the leaflets have a swollen callous area near their base that is yellow in striking contrast to the green leaflets (Figs. 16, 17). This also appears to occur in *Lepidozamia* but is not clear (Fig. 15). The function of this swollen yellow callous area is unknown and is morphologically and anatomically dissimilar to that described for *Zamia manicata.* All genera, except *Lepidozamia,* have their leaflets inserted laterally on the rachis. In *Lepidozamia,* the leaflets are inserted adaxially and medianly on the rachis (Fig. 15). There is a tendency for adaxial leaflet insertion in some species of *Macrozamia* (compare Fig. 16 with Fig. 17), but this is not nearly as pronounced as in *Lepidozamia.*

A petiolule (Figs. 22, 23) is present at the base of the lamina of the leaflets in only two species of *Zamia.* In *Z. wallisii,* this petiolule is from 4–8 cm long and is furrowed adaxially. In *Z. manicata* (=*Z. madida*), the petiolule is from 0.5–2 cm long and has a prominent glandular collar at the point of attachment of the lamina to the petiolule (Figs. 22, 23). This glandular collar is semicircular, extending across the abaxial surface from margin to margin (Figs. 22, 23). *Zamia manicata* exhibits an interesting transition series from seedlings to adult plants with respect to the development of the petiolule and the glandular collar. Seedling leaflets lack the petiolule and the collar. As the plants grow and produce larger leaves and leaflets, a petiolule develops but without a glandular collar. Later, as more leaves are produced, they develop both the petiolule and glandular collar. Generally, both of these structures are constant features of adult reproductive plants. However, in damaged plants or plants that have been transplanted, the first new leaves produced, which are often diminutive, lack the petiolule and collar. Both structures eventually develop on successive leaves, as described above.

Leaf bases of cycads are often swollen to the point of appearing bulbous particularly in *Cycas, Dioon, Lepidozamia,* and some species of *Encephalartos* and *Macrozamia.* Stipules are present on the foliage leaf bases and cataphylls (where present) in several genera. The stipules of cycads are of two types. The first type, which is present only in *Stangeria* (Fig. 7) and *Bowenia* (Fig. 10), are somewhat fleshy, well developed, and vascularized. In *Stangeria,* the stipules appear to be fused laterally (Fig. 7) and form a cowl-like structure on the adaxial side of the leaf base (Stevenson, 1981). In contrast, the stipules present on both foliage leaves and cataphylls of *Ceratozamia* (Fig. 5), *Chigua, Microcycas,* and *Zamia* (Fig. 6) are diminutive, inconspicuous, and unvascularized. In *Microcycas,* stipules are present on the cataphylls and foliage leaves in seedlings and juvenile plants but not on adult plants. Presumably all of these stipules are of the same developmental origin, but this remains to be demonstrated.

The vascular architecture of cycad leaves with respect to the vascular supply entering the pinnae is remarkably similar (Matte, 1904). However, the arrangement of vascular bundles in the petiole of *Bowenia* and *Stangeria* is quite distinct from that of the other genera. In the former two genera, the vascular bundles are arranged in a circular pattern, with *Stangeria* having a single circle and *Bowenia* having two or more concentric circles. All other genera have their petiolar bundles arranged in an omega pattern (e.g., *Dioon*)

Figs. 11–14. Decurrent leaflet bases in cycads. **11.** *Cycas revoluta.* **12.** *Dioon spinulosum.* **13.** *Stangeria eriopus.* **14.** *Bowenia spectabilis,* pinnule bases.

or a modified omega pattern as in *Encephalartos*. *Encephalartos* is interesting because the petiole bundles in any given transection appear to be scattered so that a defined omega pattern is not discernible. However, by following serial sections, Matte (1904) was able to determine that the basic pattern is an omega, with many small, numerous anastomosing branches that give rise

Figs. 15–19. Types of leaflet base attachment in cycads. **15.** Decurrent leaflets attached adaxially in *Lepidozamia peroffskyana.* **16.** *Macrozamia spiralis* with pseudo-median leaflet attachment and swollen callous area, which appears white, point of leaflet attachment to rachis. **17.** *Macrozamia riedlei* with lateral attachment of leaflets, white areas near leaflet bases are the small swollen callous regions. **18.** Articulated leaflet bases of *Ceratozamia matudae.* The small rim-like articulation is less pronounced than that in *Microcycas.* **19.** Articulated leaflet bases in *Microcycas calocoma.*

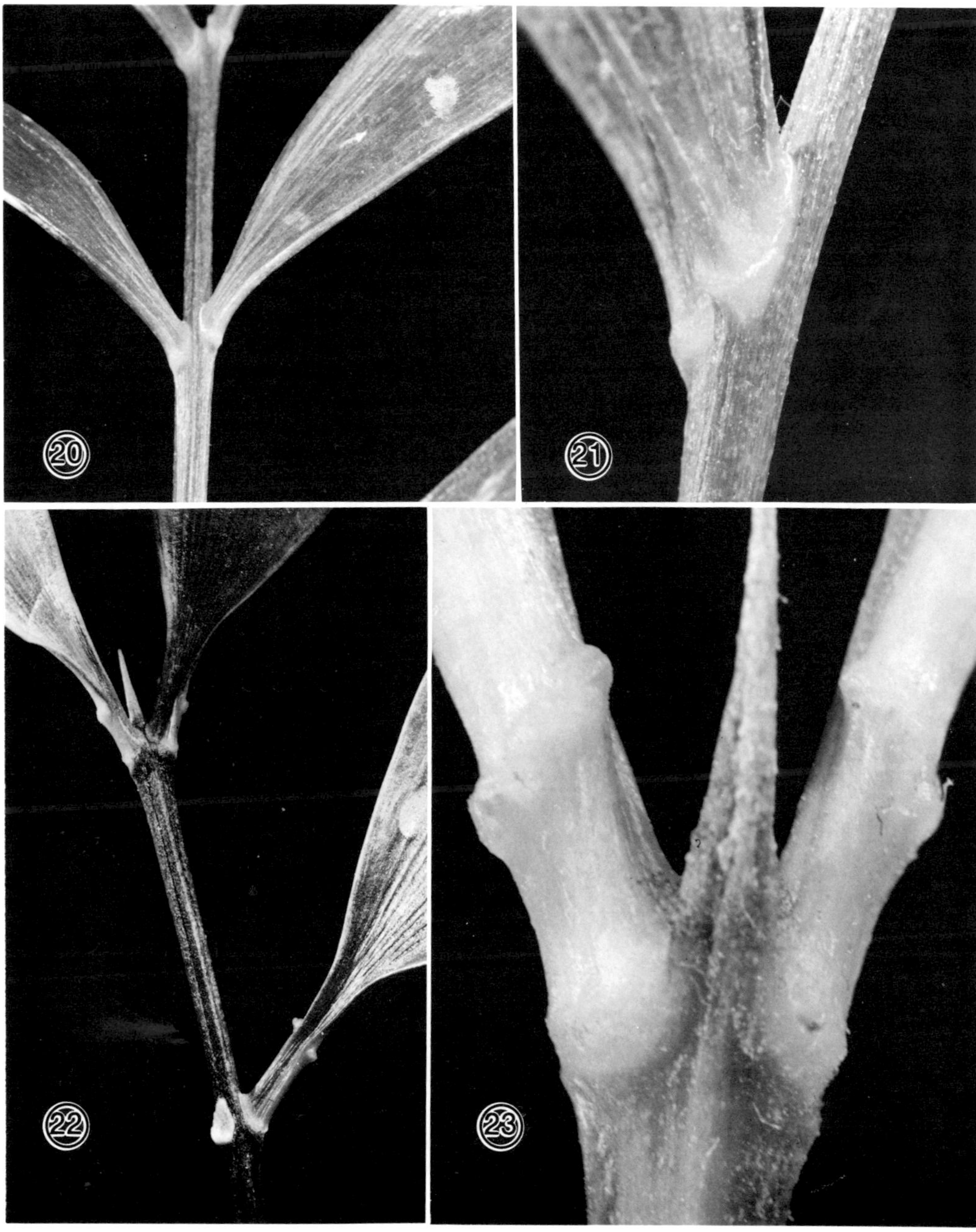

Figs. 20–23. Leaflet bases in two species of *Zamia*. **20, 21.** *Zamia pumila*. **20.** Articulated leaflet bases. **21.** Greater magnification of leaflet base in which the rim-like articulation is well developed. **22, 23.** *Zamia manicata*. **22.** Leaflet bases with petiolules and glandular collars at the points of attachment of the lamina to the petiolules. Note that the collar does not extend across the adaxial side of the petiolules and the leaf is not terminated by a leaflet. **23.** Close-up of abaxial side of the terminal portion of the leaf in Figure 22. Note glandular collars and swollen articulated area.

to the scattered appearance obscuring the omega pattern.

In terms of leaf morphology and anatomy, *Bowenia* and *Stangeria* share several unique characters (see characters 1.13, 1.14, and 1.30 in Table I) for petiole vasculature, large fleshy stipules with vasculature, and cataphylls only in association with reproductive axes. In contrast, the other genera have an omega pattern of petiole vasculature, cataphylls on vegetative axes, and (when present) unvascularized rudimentary stipules.

ROOTS

Cycads can be characterized as having a taproot system with diminutive lateral roots. Some of the lateral roots become ageotrophic. These in turn develop into the "coralloid roots," containing nitrogen-fixing Cyanobacteria, so characteristic of cycads. In addition, *Cycas* generally produces aerial adventitious roots. Adventitious roots are also frequently found associated with bulbils. The tap roots of all cycad seedlings are contractile (Stevenson, 1980c), as well as older lateral roots. In general, the primary tap root is present throughout the life of the individual. In those taxa with subterranean stems, there is usually only a single tap root (although there may be several storage tap roots in some species of *Zamia*). In arborescent taxa, several of the first formed lateral roots may also develop into tap roots so that the plant then has several tap roots. This seems to aid in anchoring the massive pachycaulous stems.

The presence of a coleorhiza surrounding the embryonic tap root has been claimed for cycads by Dorety (1908), Chamberlain (1919, 1935), Hooft (1970), Bierhorst (1971), Pant (1973), and Gifford and Foster (1989). Within seed plants a coleorhiza thus would be unique to only grasses and cycads. The evidence for a coleorhiza in cycads is based upon Dorety's (1908) interpretations, reiterated by Hooft (1970). Their evidence rests upon the occurrence of a discolored root tip and papery fragments adhering to the sides of the emerging tap root during germination. However, the critical evidence should rest in the comparative embryology and a careful examination of the origin of the papery fragments. One only has to compare the organization of the embryos of representatives of other gymnosperms (Figs. 25, 26, 29) and grasses (Figs. 27, 28) and other monocotyledons (Fig. 30) with that of cycads (Figs. 24, 31) to see that grasses are unique. Cycads appear identical to the other monocotyledons, in this case *Beaucarnea* (Fig. 30), and gymnospermous representatives, *Ginkgo* (Fig. 29), *Picea* (Fig. 25), and *Ephedra* (Fig. 26; also, see Singh, 1978 for illustrations of many other gymnosperms). The discolored tissue associated with the root cap is composed of the remnants of the suspensor and the papery fragments adhering to the emerging root are merely pieces of the inner papery layer of the integument near the micropyle that become attached as the root grows through the seed coat (Figs. 24a–c, 44, 45). Moreover, there is no collar surrounding the young root of cycads as in, for example, *Zea* (Figs. 27, 28). In addition, there is no tissue disruption in the cycads as there is in the grasses (compare Fig. 24d–g with Fig. 28). There may occasionally be a swollen area or damaged tissue at the root base as a result of contact with the sclerotesta of the seed coat which can give a similar appearance, but this region has no equivalency. Thus, the cycads lack a true coleorhiza. For those who require a term, perhaps they can call this structure in the cycads a "pseudocoleorhiza."

Reproductive Morphology

GROWTH HABIT

All cycads are dioecious and annually or sometimes biennially produce reproductive structures. The reproductive structures are produced on specialized, determinate reproductive axes termed strobili (with the exception of megasporangiate plants of *Cycas*, which might be said to have an indeterminate strobilus). Occasional teratisms occur, particularly in some species of *Encephalartos* (Stevenson, 1988), in which a diminutive and abortive vegetative shoot terminates a strobilus. The reproductive axes consist either of a series of pollen-bearing microsporophylls or a series of ovule-bearing megasporophylls, but never both on the same axis. The strobili are simple; that is, they consist only of sporophylls without any other associated structures, such as the compound ovulate cones of the conifers.

Historically, the nature and ontogeny of strobili in the cycads have been surrounded by controversy (Stevenson, 1988). This controversy has centered around two mutually exclusive concepts regarding the ontogeny of strobili—the sympodial concept and the monopodial concept (for a complete review of the historical arguments see Stevenson, 1988 and literature cited therein). One of the long noted anatomical features of most strobilar-producing cycads has been the presence of "cone domes" or medullary vascular bundles that are continuous with the vascular system of the strobili. The presence or absence of cone domes was fundamental to whether or not a taxon was considered to be sympodial or monopodial. However, given that axillary buds do not occur in the cycads (Stevenson, 1980a), the monopodial concept is not viable and the sympodial concept becomes unduly complicated in those plants producing multiple strobili (Stevenson, 1988). The interpretation of the ontogeny of strobili, which is in accord with the known available evidence, is that the production of a single strobilus results from anisotomous branching, with one branch developing into a strobilus and the other a vegetative shoot. In producing multiple strobili, the vegetative apex forms a series of anisotomous reproductive branches (Stevenson, 1988). In the latter case, the strobili may continue their development and emerge from the center of the crown or they may become dormant to emerge in non-axillary positions among the old leaf bases on the trunk the following year (Stevenson, 1988).

SPOROPHYLLS

The general description of micro- and megasporophylls of the cycad genera are unnecessary here as they can be found in almost any textbook on general botany. Instead, I will restrict my comments to those aspects that have been overlooked in earlier literature, the results of a survey I have made on the vascular organization of sporophylls, the interpretation of the megasporophyll of *Cycas*, and the aggregation of microsporangia.

Megasporophylls

From a morphological point of view, megasporangiate plants of *Cycas* are enigmatic. The classical and most common interpretation is that each ovule-bearing organ is a megasporophyll. The evidence for this is quite striking in that many species, such as *C. revoluta*, have a pinnatifid sterile tip (Fig. 32) on each megasporophyll and the megasporophylls are produced in a loose crown surrounding the vegetative apex. Moreover, in the phenology of *Cycas*, a flush of megasporophylls is followed sequentially in the growing season by a flush of cataphylls, foliage leaves, and cataphylls. The vascular anatomy of the megasporophyll petiole, rachis, and sterile tip is the same as that of a foliage leaf with the exception of the presence of diminutive adaxial bundles which are discussed below. This classic concept of the homology of the ovule-bearing structure with a fertile leaf, and thus the term megasporophyll, has been questioned by Boureau (1946) and particularly by Meeuse (1963, 1986). Boureau observed that the main vascular bundles of the ovulate reproductive structures are essentially the same as those of a vegetative leaf, with the addition of a set of diminutive, inverted adaxial bundles. He noted that the presence of these inverted bundles was similar to that seen in the ovuliferous scale of conifers, which is the result of fusion of a subtending bract and a fertile axillary shoot. He pointed out that the inverted bundles in *Cycas* had an anomalous developmental pattern, with the primary xylem developing before the primary phloem. Thus, Boureau (1946) considered this anatomical feature to be a convergence in organization. Meeuse (1963) went even further with this argument and described the ovule-bearing organ of *Cycas* as a "gonoclad," a shoot bearing ovule. He noted the absence of a bract subtending this structure but considered it to be completely fused to the ovule-bearing axis. The reason for this conclusion about bract fusion was that Meeuse considered the strobili of cycads to be subtended by a bract and therefore the gonoclad of *Cycas* is the equivalent of a strobilus and a fused bract. There are several reasons for rejecting Meeuse's interpretation, some of them being the result of observations published since his 1963 work. The strobili of cycads are not produced from axillary buds and thus are not subtended by a bract (Stevenson, 1980a, 1981, 1988). Therefore, the fact that there is no subtending bract associated with the megasporophyll in *Cycas* is not surprising because even if the megasporophyll were a gonoclad there

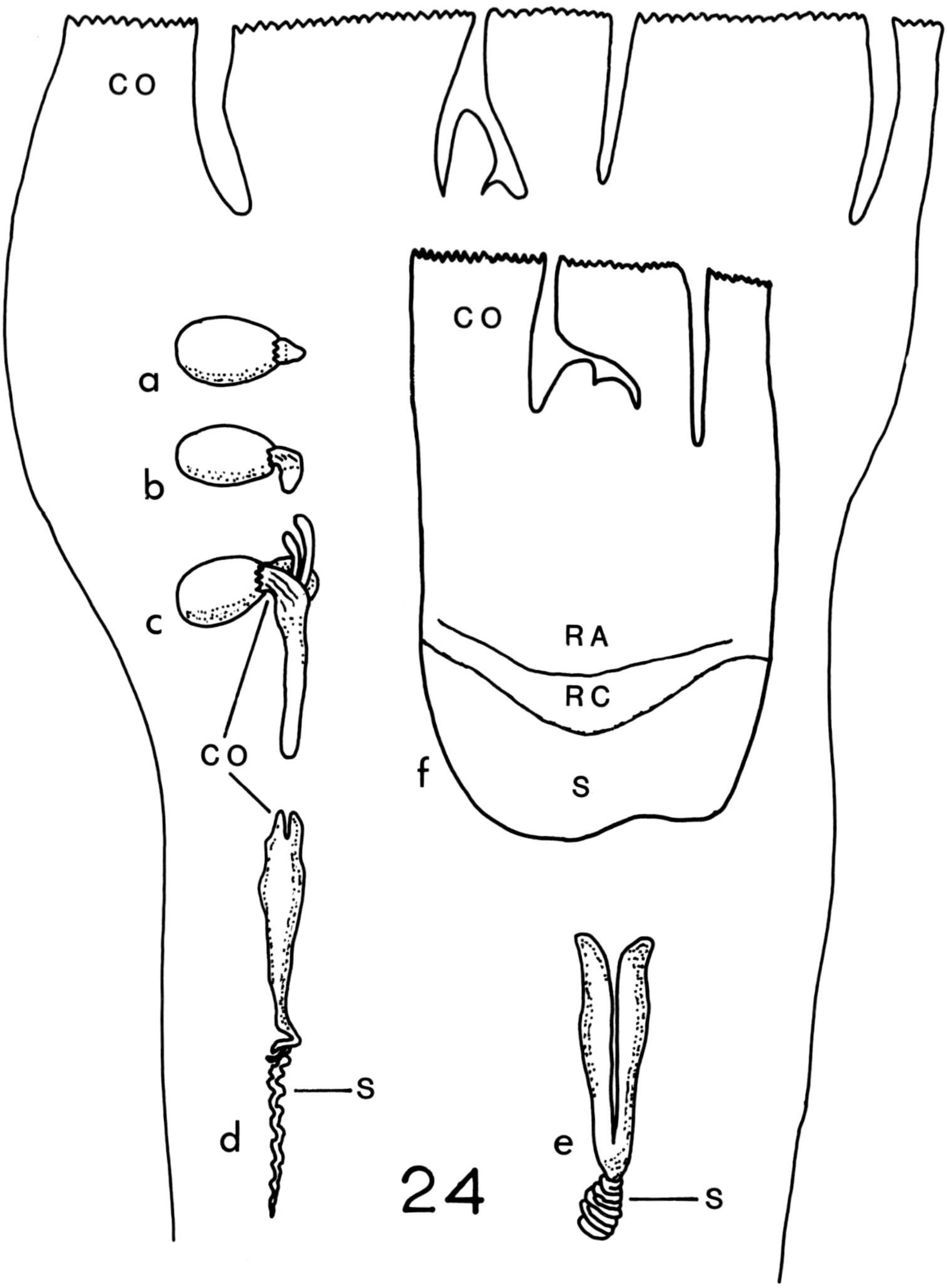

Figs. 24–26. Embryos and seedlings of selected gymnosperms. **24.** *Zamia furfuracea.* **a–c.** Stages in seed germination. **d, e.** Embryos dissected fron seeds. **f, g.** Drawings of median longisections of germinating embryos in a and b, respectively. Note lack of tissue disruption, similarity to Figures 25, 26, and dissimilarity to Figures 27, 28. **25.** Drawings of median longisections of stages in embryo germination of *Ephedra nevadensis.* **26.** Drawings of median longisections of stages in embryo germination of *Picea alba.*

Key to Labelling: CO, cotyledons; RA, root apex; RC, root cap; S, suspensor.

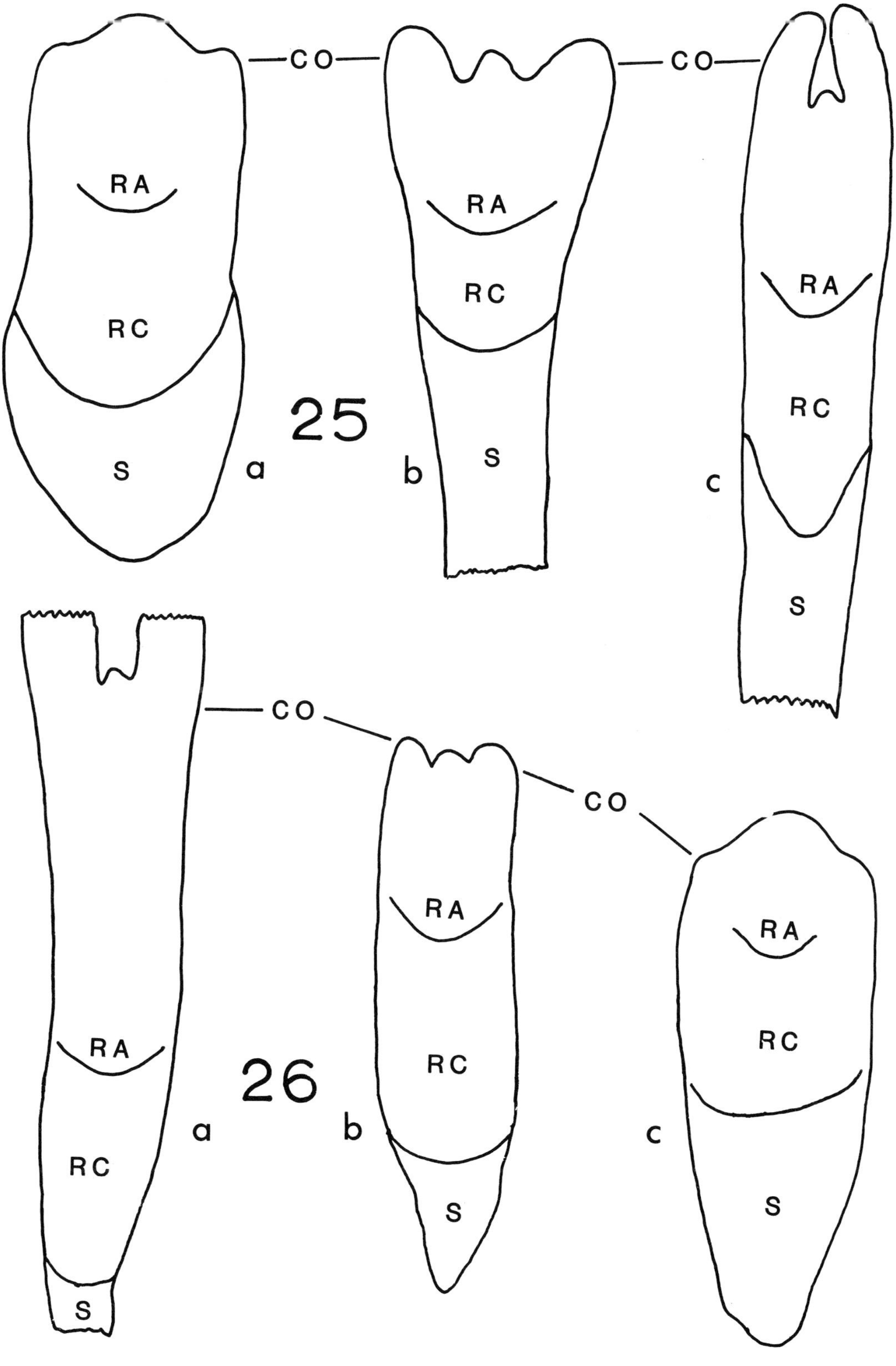
CO
CO
RA
RC
S
a
25
b
RA
RC
S
c
RA
RC
S
CO
CO
RA
RC
S
a
26
b
RA
RC
S
c
RA
RC
S

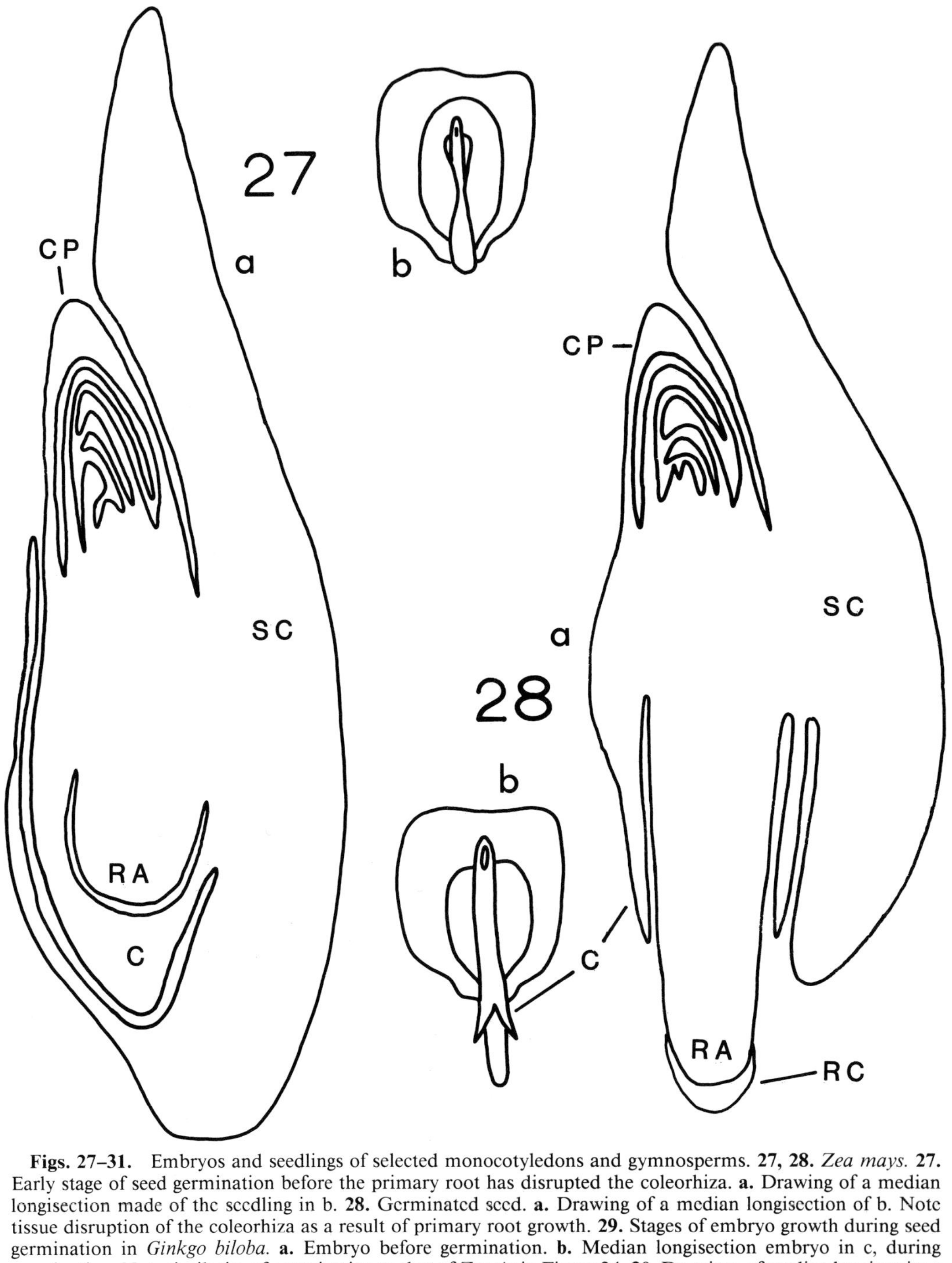

Figs. 27–31. Embryos and seedlings of selected monocotyledons and gymnosperms. **27, 28.** *Zea mays.* **27.** Early stage of seed germination before the primary root has disrupted the coleorhiza. **a.** Drawing of a median longisection made of the seedling in b. **28.** Germinated seed. **a.** Drawing of a median longisection of b. Note tissue disruption of the coleorhiza as a result of primary root growth. **29.** Stages of embryo growth during seed germination in *Ginkgo biloba.* **a.** Embryo before germination. **b.** Median longisection embryo in c, during germination. Note similarity of organization to that of *Zamia* in Figure 24. **30.** Drawings of median longisections of *Beaucarnea recurvata* (Agavaceae). **a.** Embryo before germination. **b.** Embryo after germination. Note dissimilarity to that of Figures 27a and 28a. **31.** Drawing of longisection of an embryo of *Cycas revoluta.*

Key to Labelling: C, coleorhiza; CO, cotyledons; CP, coleoptile; RA, root apex; RC, root cap; S, suspensor; SC, scutellum.

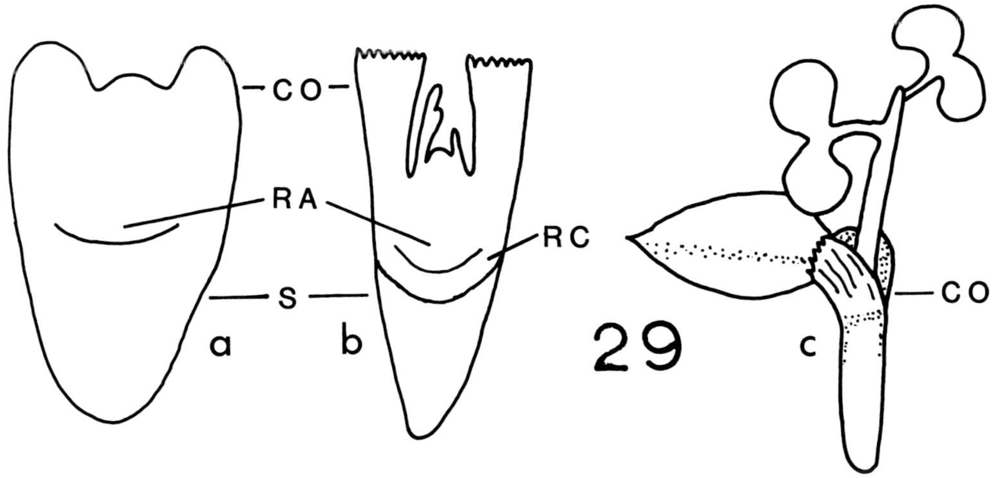
CO
RA
RC
S
a
b
29
c
CO

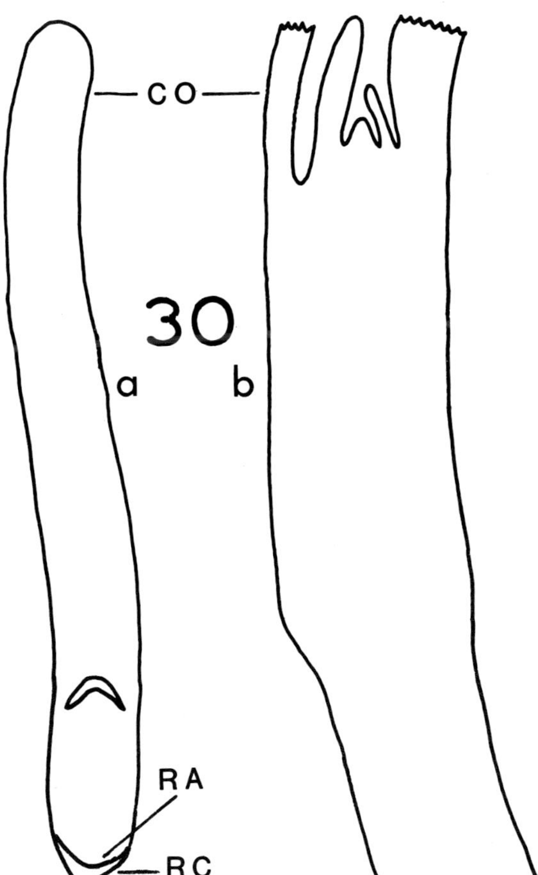
CO
30
a
b
RA
RC

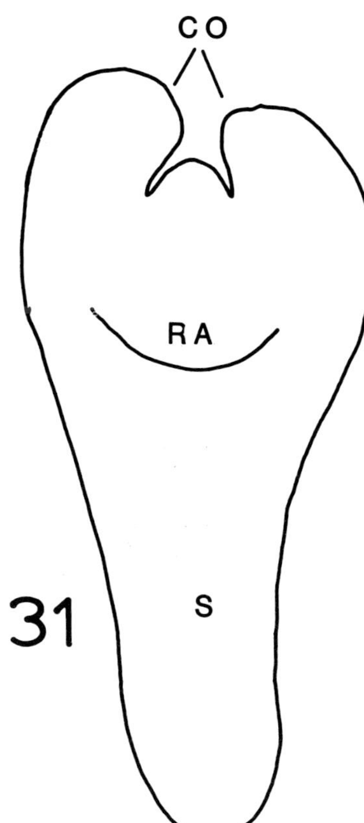
CO
RA
31
S

would be no subtending bract. Thus, to try and explain the absence of the bract as a complete fusion has no morphological basis. However, this in itself does not refute Meeuse's interpretation because it only demonstrates that the megasporophyll and strobilus both lack subtending bracts (in fact, one could use this feature to support Meeuse). The presence of adaxial inverted bundles still needs discussion. The diminutive adaxial bundles differentiate late in the life of the megasporophyll, after the seed is in its final development phase. Thus, their occurrence was not noted by Matte (1904), who used young sporophylls in his study on vasculature, but these bundles were thoroughly described by Worsdell (1898), who used only mature sporophylls and also noted their sporadic occurrence in the megasporophyll stalks of many other cycad genera. I have been able to confirm that these adaxial inverted bundles are present only in mature sporophylls and, furthermore, at maturity they have considerably more primary and secondary phloem than primary and secondary xylem. Thus, it appears that these adaxial bundles differentiate in response to the nutritional sink created by the development of the massive starch-rich megagametophyte.

Another, until now, unexplored aspect of the vasculature of *Cycas* has been the origin of the megasporophyll traces with respect to the axis on which they are borne. All cycads have foliage leaves with girdling traces (see e.g., Bierhorst, 1971 or Gifford & Foster, 1989 and literature cited therein). Thus, the homology of foliage leaves with megasporophylls would be more strongly supported if the megasporophylls of *Cycas* had girdling traces. In fact, the traces from the stem that are continuous into the megasporophylls are not girdling. Rather, a single trace departs from the stem vasculature and then undergoes a series of dichotomies near the base of the megasporophyll so that two to four traces enter the meagsporophyll base. In most instances, after the first dichotomy, only one of the two resultant shanks undergoes another immediate dichotomy before the traces enter the megasporophyll so that only three traces appear in the base. In fact, these dichotomies occur so near to each other that the single original trace appears to trifurcate. This is exactly like the vascular pattern of the peduncle and megasporophylls of the ovulate cones of *Dioon*. The trunk region of a megasporangiate plant of *Cycas* that bears the megasporophylls has a vascular organization virtually identical to that of the peduncle and megasporophyll bearing axis of *Dioon* and is quite similar, except for details, to that of all other cycads. This evidence supports the interpretation that megasporangiate plants of *Cycas* have an indeterminate strobilus.

Therefore, in my opinion, the ovule-bearing structures of *Cycas* are just what they look like, megasporophylls. Moreover, the real question is whether or not the region of the trunk that bears them is a step on the way to the evolution of a true strobilus or rather represents the dissolution of a strobilus in response to the rather different pollination mechanism proposed for *Cycas* (Niklas & Norstog, 1984).

Typologically, the megasporophylls of the other cycad genera can be divided into two forms: the peltate form found in *Bowenia, Ceratozamia* (Fig. 40), *Chigua, Encephalartos* (Fig. 38), *Lepidozamia, Macrozamia* (Fig. 39), *Microcycas* (Figs. 42, 43), and *Zamia*; and the expanded laminal tip form found in *Dioon* (Figs. 34, 35), and *Stangeria* (Fig. 36). This latter form appears in many ways similar to that of *Cycas* (Fig. 32), and has a vascular organization similar to *Cycas*. The vascular pattern of the laminar sterile tip in *Cycas* (Fig. 46e–l), *Dioon*, and *Stangeria* (Fig. 48) consists of an open dichotomous venation pattern with the dichotomies all occurring in one plane (Figs. 46, 48). In contrast, the vascular pattern of the sterile tip in the peltate form consists of an open dichotomous system with successive dichotomies occurring at right angles to the previous dichotomy (Figs. 47, 49, 50). Thus, the vascular system of the planar sterile tip is planar whereas that of the peltate sterile tip is three-dimensional. The degree of three-dimensional branching appears to be related to the size of the peltate sterile tip and the larger the three-dimensional volume of the sterile tip the more elaborated is the vasculature (compare Figs. 47, 49, 50). For example, *Bowenia* (Fig. 50), which has the largest peltate sterile region, has each successive dichotomy at a right angle to the previous dichotomy. In contrast, the other seven genera have the second dichotomy at a right angle to the first dichotomy and then all successive dichotomies in the same plane, more or less. The same is true for the laminar sterile tip. The large sterile tip of *Dioon* has a more elaborated two-

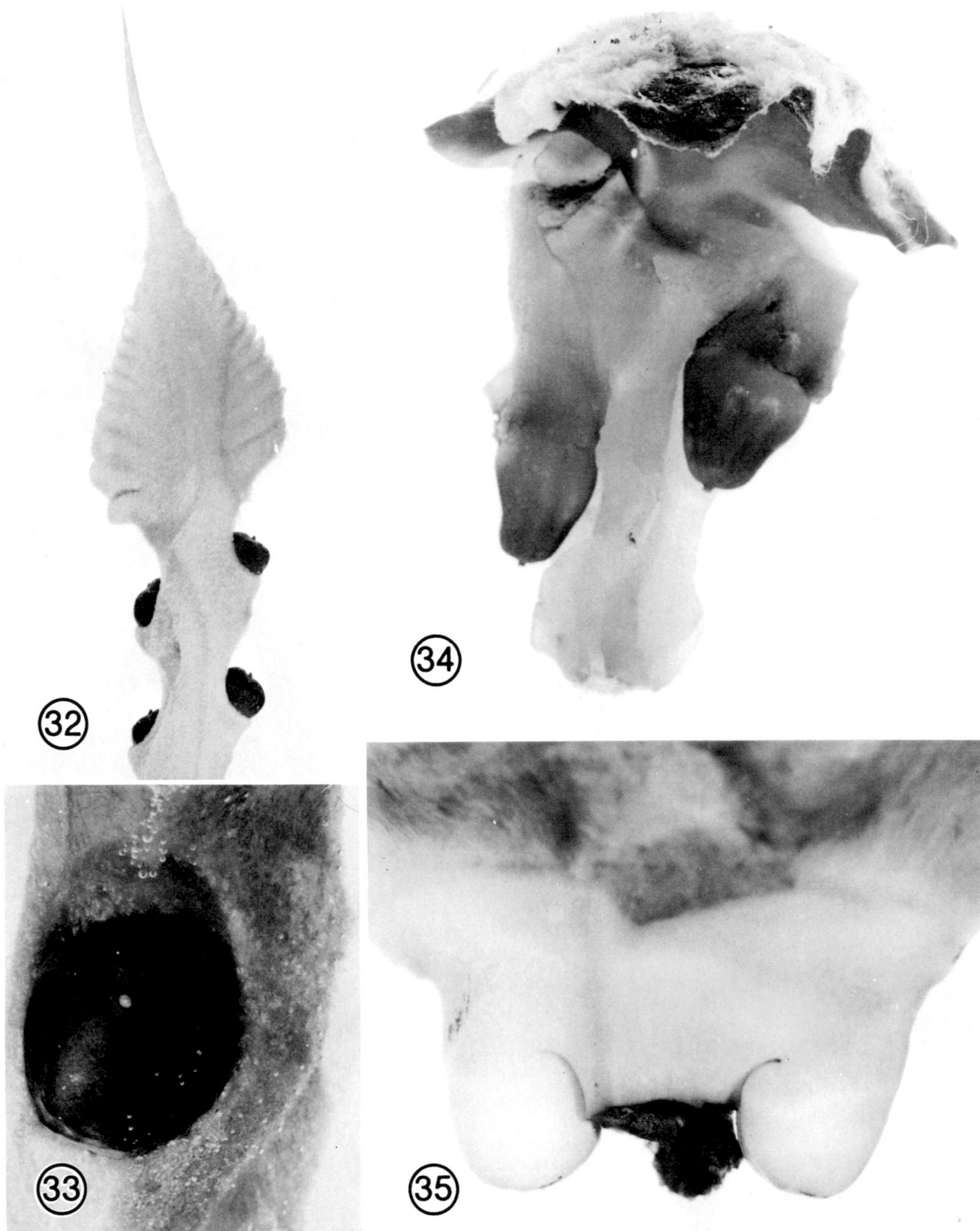

Figs. 32–35. Megasporophylls and ovule attachment in cycads. **32, 33.** *Cycas media.* **32.** Megasporophyll with young platyspermic ovules pointing obliquely upwards. **33.** Close-up of young ovule with platyspermic morphology. Note that the ovule seems to be slightly sunken into the sporophyll. **34, 35.** *Dioon spinulosum.* **34.** Stalked ovules on megasporophyll. Ovule exhibits slight constriction at juncture with the stalk and the laminar portion of the sterile megasporophyll tip is pointing towards the viewer. **35.** Young megasporophyll with two ovules each with a short stalk.

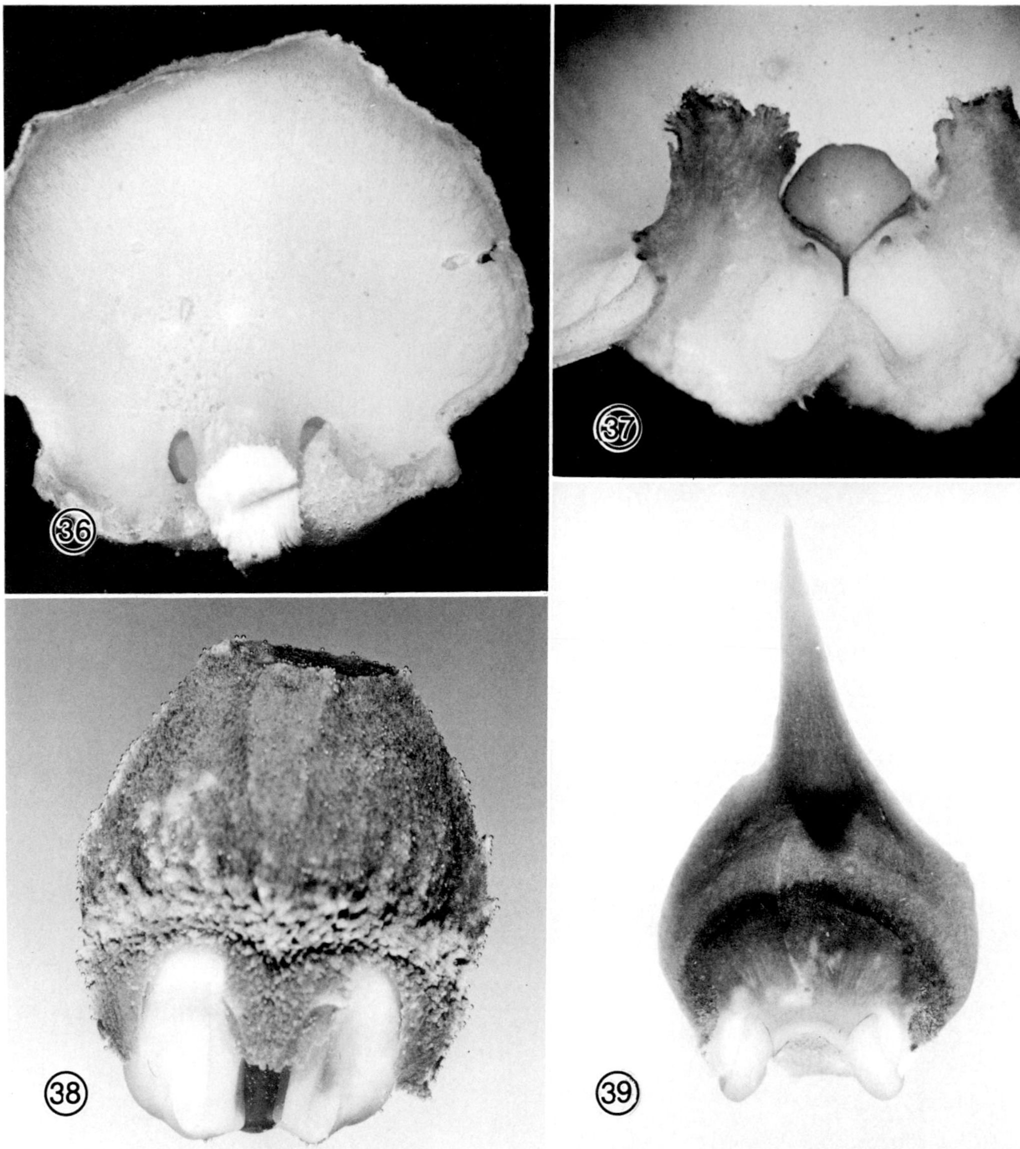

Figs. 36–39. Megasporophylls of cycads. **36, 37.** *Stangeria eriopus.* **36.** Adaxaial view of a megasporophyll with young enclosed ovules. **37.** Sterile enclosing tissue removed to expose young ovules. **38.** Abaxial view of a young megasporophyll of *Encephalartos gratus* with a median abaxial lobe and two lateral lobes. **39.** Adaxial view of a young megasporophyll of *Macrozamia communis* with thickened area in the middle region of the sterile tip (below the terminal spinose region) and two small lateral lobes.

dimensional vascular pattern than does the smaller sterile tip found in *Stangeria.*

The genera of the cycads are easily recognizable by their megasporophyll morphology and the most obvious features have been described many times in most morphology text books (e.g., see Bierhorst, 1971; Gifford & Foster, 1989; etc.). I would like to discuss here the less obvious aspects of megasporophyll morphology receiving little or no attention in the past. For example, although the megasporophylls of *Dioon* and *Stangeria* appear similar, they have quite dis-

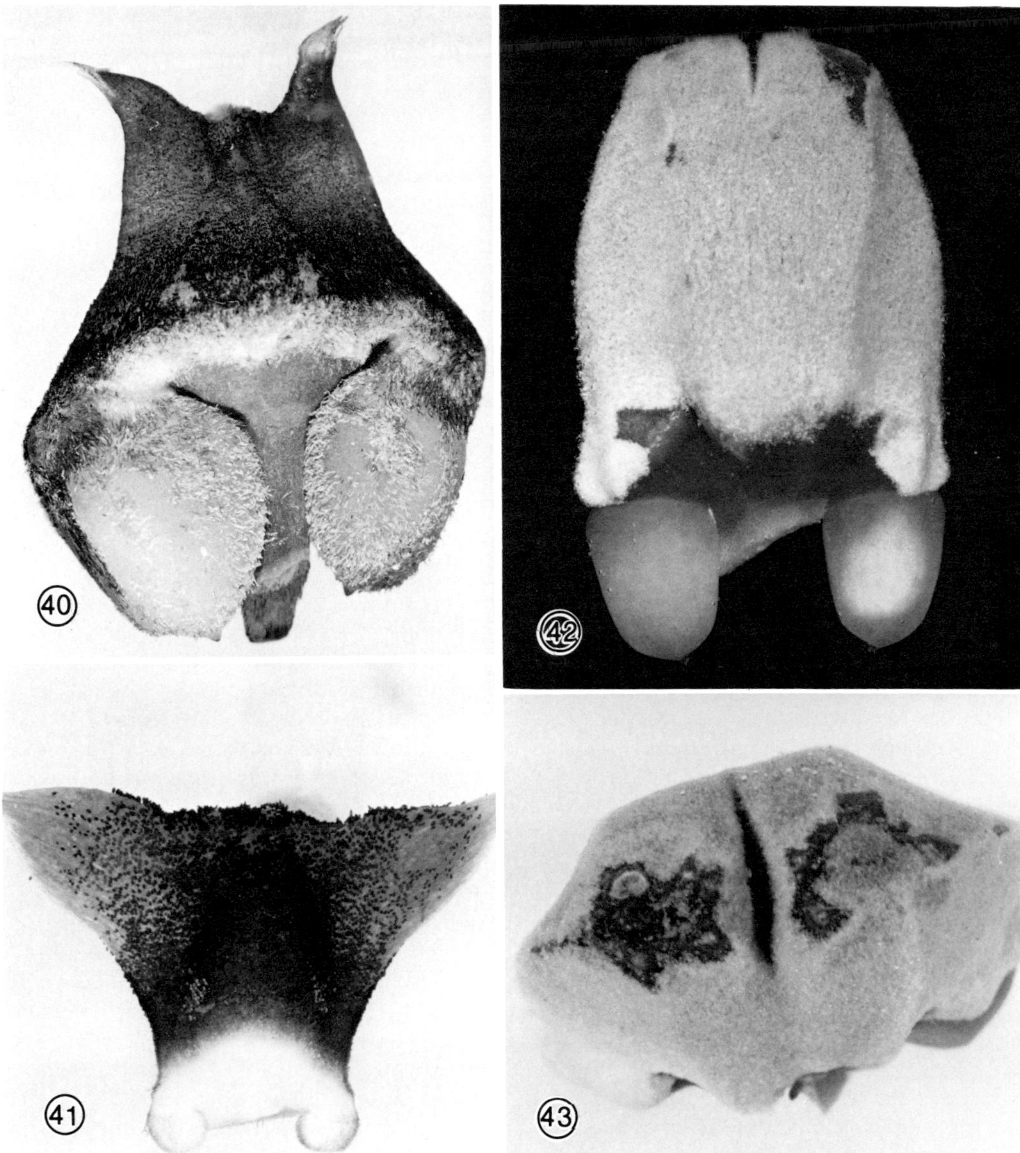

Figs. 40–43. Megasporophylls of cycads. **40, 41.** *Ceratozamia robusta.* **40.** Abaxial view of a nearly mature megasporophyll with the two spinose tips pointing obliquely vertical. **41.** Adaxial view of a young megasporophyll with the two spinose tips pointing laterally. **42, 43.** *Microcycas calocoma.* **42.** Adaxial view of a nearly mature megasporophyll with a distinct cleft or bifurcation at the apex of the sterile tip. **43.** End view of the sterile tip of a megasporophyll with two blunt lobes.

tinctive autapomorphic features. *Dioon* has its ovules attached to the megasporophylls on a distinct stalk (Figs. 34, 35). This stalk is present during the early ontogeny of the ovule (Fig. 35), but it is not clear yet whether the stalk is formed before ovule initiation or after. *Stangeria,* in contrast, has its ovules enclosed by sterile tissue of the megasporophyll (Figs. 36, 37). When these ovules are young they are almost completely enclosed (Figs. 36, 37); but as the ovule enlarges to

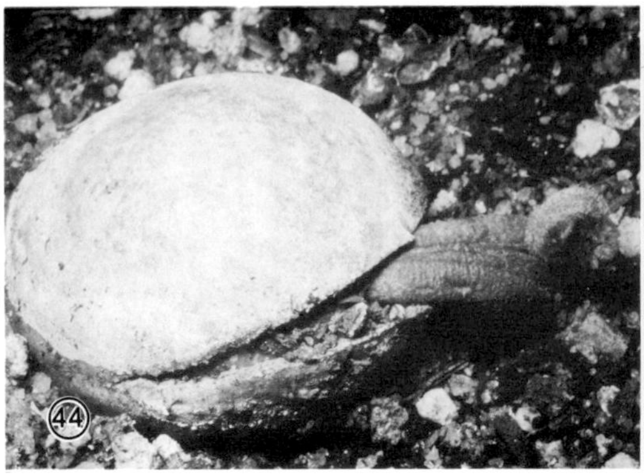

Figs. 44, 45. Germinated seeds of cycads. **44.** Germinated platyspermic seed of *Cycas revoluta.* Note the splitting of the sclerotesta at the microplyar end into two halves. **45.** Germinated radiospermic seed of *Dioon spinulosum.* Note the emergence of the seedling axis through a pore at the micropylar end.

become a mature seed, the enclosing tissue becomes relatively obscure as it does not enlarge with the ovule. In *Encephalartos,* there are two lateral lobes at the base of each sterile sporophyll tip, one extending parallel to and along the outside of each ovule (Fig. 38). These lobes are diminutive in some species (e.g., in *E. voiensis* about 2 mm long) and well developed in others (e.g., in *E. gratus* to 2 cm long, Fig. 38). *Encephalartos* also has a median adaxial lobe extending distally from the base of the sterile tip (Fig. 38). This median lobe is diminutive in many species (e.g., *E. voiensis*) or well developed in others (e.g., *E. gratus*). When both the lateral and median lobes are well developed the ovule has the appearance of being enclosed, but as can be seen by comparing Figure 36 with Figure 38, these are really quite different conditions. Both *Macrozamia* and *Lepidozamia* have lateral lobes that are similar to those described for *Encephalartos,* but in those two genera the lateral lobes are generally small and more like the small lateral lobes found in certain species of *Encephalartos.* Occasionally, the lateral lobes may be fused with the ovule in *Macrozamia* (Fig. 49d). A median lobe is not present in *Macrozamia* and appears not to be present in *Lepidozamia.* This needs to be examined in more detail in *L. hopei* because Schuster (1932) has illustrated such a lobe in his figure 13L, but Johnson (1959), in the most recent monograph, does not mention this feature in his description of the genus or this species. I have been unable to confirm or reject this as I have not seen megasporophylls of this species but there is no median lobe associated with the megasporophylls of *L. peroffskyana,* the only other species in the genus. The presence of lateral lobes on the megasporophylls provides a synapomorphy for these three genera (see character 1.26 in Table I). Interestingly, if it is unequivocally demonstrated that a median lobe is present in *Lepidozamia,* this would provide a synapomorphy for *Lepidozamia* and *Encephalartos.* This would certainly change the current thinking about the close relationship between *Lepidozamia* and *Macrozamia* (Johnson, 1959 and also the systematics section of this paper).

Previous interpretations of the megasporophyll of *Macrozamia* and *Lepidozamia* (e.g., Chamberlain, 1935 and Pant, 1973) as a reduced form of a *Dioon* type are, in in my opinion, an oversimplification. This concept is based on the

---

→

Figs. 46, 47. Drawings from representative serial transections of megasporophylls. **46.** *Cycas revoluta.* **a–l.** Acropetal series from base, a, to apex, l. **a.** Three traces that enter the megasporophyll. **b, c.** Traces undergo random dichotomies and anastomoses. **d.** Departure of ovular trace on left. **e–l.** Basal to apical region of sterile tip. Decrease in number of veins reflects departures to rudimentary pinnae on sterile tip (see Fig. 32). **47.** *Ceratozamia mexicana.* **a–d.** Acropetal series beginning with a. **a.** Basal region after the first dichotomy of the two entering traces. **b.** Further dichotomies in the basal region of the sterile tip, two lateral traces are the ovular supplies. **c.** Middle region of sterile tip. Note diminutive dichotomous branching at right angles. **d.** Apical region just below the two spinose tips (see Fig. 40). Lateral traces will enter the tips.

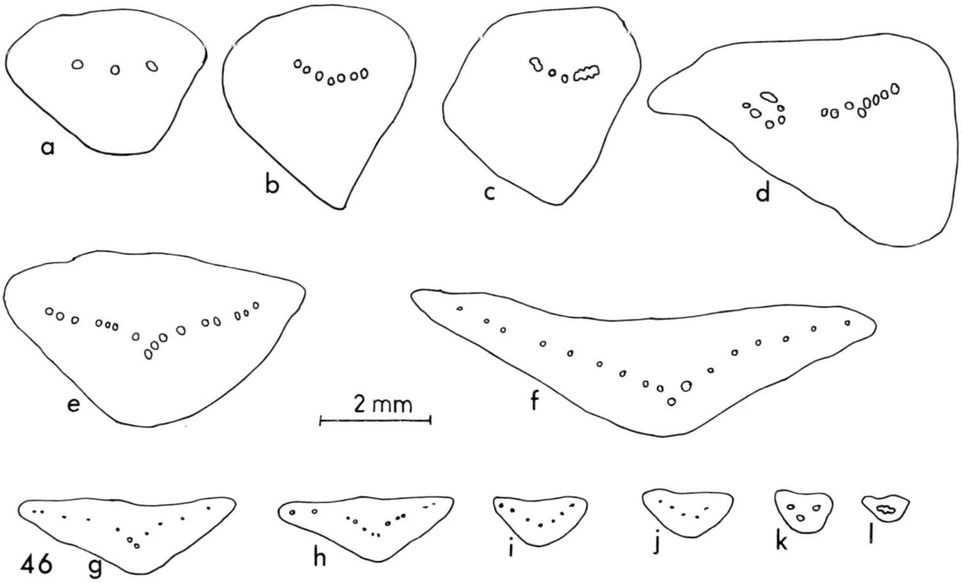

46

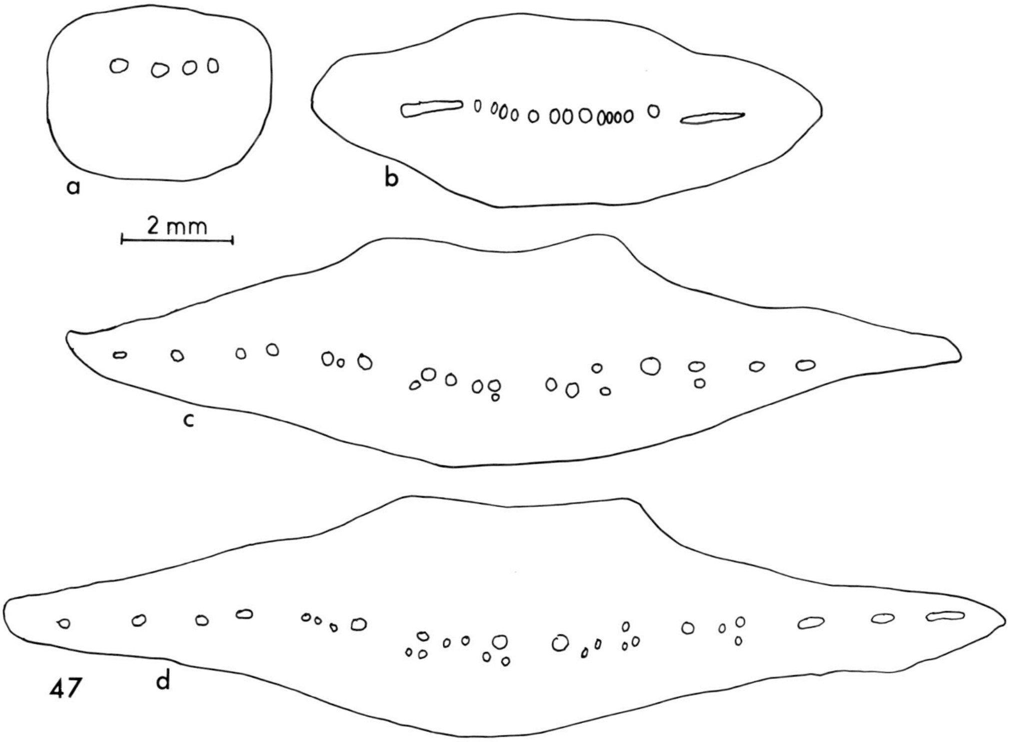

47

48

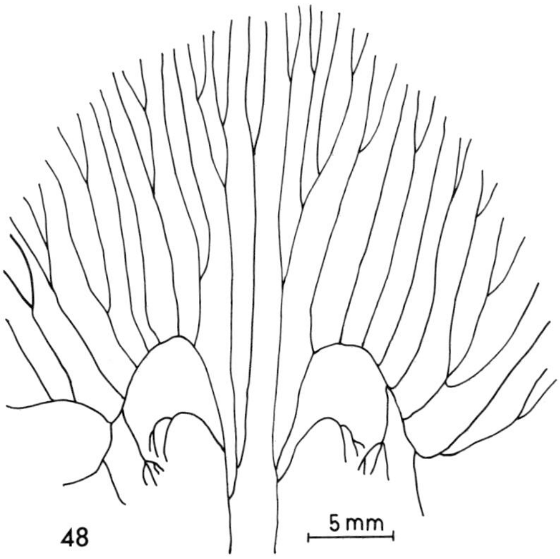

Fig. 48. Vascular reconstruction of megasporophyll of *Stangeria eriopus*. Pattern is similar to that of *Cycas* (Fig. 46) and *Dioon*.

idea of the reduction of the laminar sterile tip to a smaller spinose sterile tip. However, examination of the sterile tip of *Macrozamia* and *Lepidozamia* shows that there is a thickened region at the base that is quite pronounced in some species of *Macrozamia* (Fig. 39). Moreover, this adaxially thickened region has the same elaborated three-dimensional vascular system (Fig. 49e–g), in conformity with the discussion given above on megasporophyll vasculature. In fact, in the region below the terminal spine, the vascular pattern is identical to that of truncate megasporophylls of *Encephalartos*. While a definitive interpretation of the megasporophylls of *Macrozamia* and *Lepidozamia* may not be possible until we have comparative developmental data on cycad megasporophylls, it is clear that this feature of these two genera is most likely a synapomorphy.

The megasporophylls of *Ceratozamia* have a distinctive apex composed of two prominent horns or spines (Figs. 40, 41). This feature in *Ceratozamia* is distinctive in its degree of expression, but a similar morphology also occurs in most, but not all, of the megasporophylls of *Microcycas*. The apices of the megasporophylls of *Microcycas* are shallowly bifurcated (Fig. 42), with each shank of the bifurcation having a blunt ending (Fig. 43). Unlike *Ceratozamia*, these tips are neither spinose nor obvious. Also, in *Ceratozamia*, the two projections extend laterally in the young megasporophyll (Fig. 41) and then become oriented apically as the sporophyll and ovules mature (Fig. 40). The interpretation of this character in *Microcycas* is enigmatic, but its common occurrence does not appear to be teratological. A more thorough comparative study of the megasporophylls of these two genera with those of *Zamia* and *Chigua* is needed to put this character in context.

It should be obvious that there are new unexploited megasporophyll characters for use in discerning relationships among cycad genera. For systematics, some of these, such as the stalked ovules in *Dioon* and the enclosed ovules in *Stangeria*, are autapomorphic for those genera and do not help in elucidating relationships. Other characters are, however, synapomorphic, the lateral lobes found in *Encephalartos, Macrozamia,* and *Lepidozamia,* for example. For other characters, such as the megasporophyll tips in *Ceratozamia* and *Microcycas,* it is not yet clear how they can be interpreted.

Ovules and Seeds

Cycads have two distinct types of ovule morphology. The platyspermic ovule is distinctly flattened (Figs. 32, 33, 44) and thus bilaterally symmetrical. This type, in extant cycads, is restricted to *Cycas*. All other cycads have radiospermic ovules. Even in early developmental stages, the platyspermic form of *Cycas* ovules can be clearly seen because they are distinctly flattened and two lobed with an apical depression (Figs. 32, 33). In contrast, the young radiospermic ovules are radiospermic without any signs of bilateral symmetry (Figs. 35, 37–42).

Associated with platyspermic morphology is a distinct germination mechanism in which the seed splits open into two distinct halves near the micropyle (Fig. 44). This is a strikingly different pattern to that seen in the germination of the radiospermic seeds of the other cycads, where the embryo, apparently by partial digestion of the woody layer, forces its way through the seed coat by rupturing it in the micropylar region (Figs. 24a–c, 45).

In general, platyspermic ovules have been considered derived with respect to radiospermic ovules, but Crane (1988) has recently questioned this and come to the opposite conclusion. This conclusion is based upon the occurrence of pu-

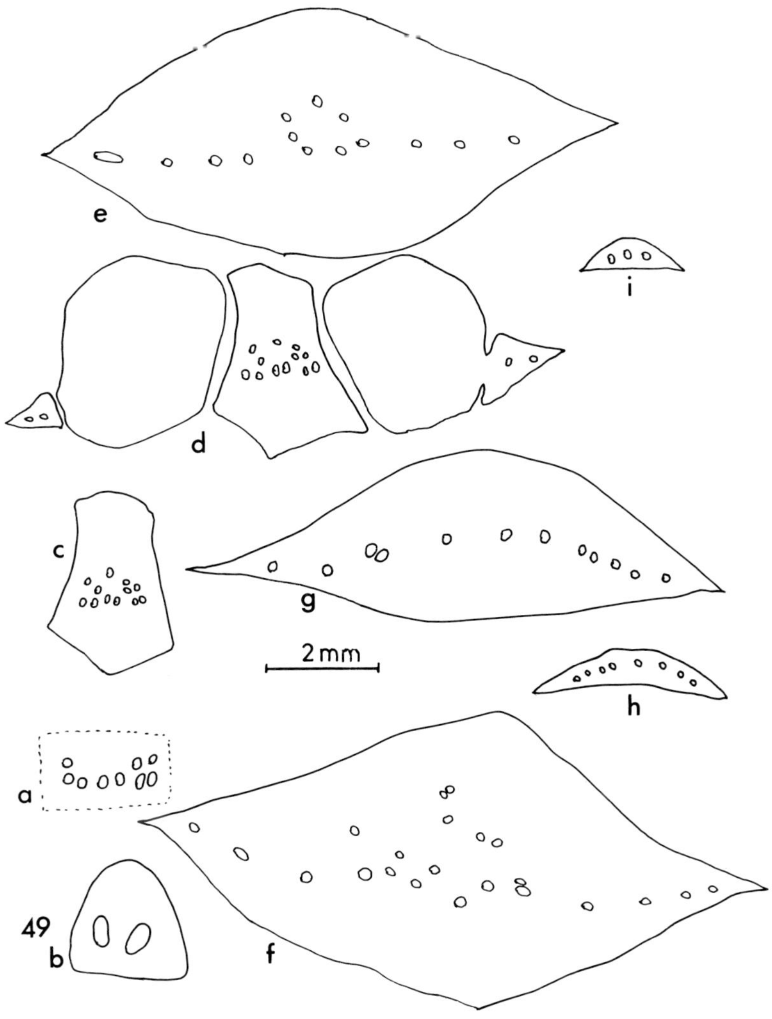

Fig. 49. Drawings of megasporophyll vasculature based on serial transections. *Macrozamia communis.* **a–i.** Representative acropetal serial transections from attachment to strobilus axis, a, through sterile tip. **a, b.** Several of the same traces originating as dichotomies of two original traces anastomose to form two traces that enter the base of the megasporophyll. **c, d.** Repeated random dichotomies of the two traces give rise to a three-dimensional vascular system in the sporophyll stalk. **d.** Lateral lobes of the megasporophyll are vascularized by branches originating from a common supply with the ovules. Note fusion of lateral lobe on the right with ovule. **e, f.** Elaborated vasculature of adaxially thickened region of the sterile tip. **g–i.** Planar vasculature of the spinose tip (see Fig. 39).

tative *Cycas*-like platyspermic ovules in Lower Permian fossils from China (Crane, 1988 and literature cited therein) and the bilateral nature of the vasculature and vascular supply of all extant cycad ovules. The platyspermic morphology of fossil cycad and putative cycad ovules from the Permian and early Paleozoic seems to be controversial and unresolved (see Crane, 1988 for a thorough discussion). Regardless of the interpretation of these fossils, they still only give us minimum ages for this type of ovule and do not resolve the question of which is plesiomorphic in the extant Cycadales.

The bilateral vascular organization of all extant cycad ovules is used by Crane (1988), in conjunction with his interpretation of the early fossils, to support the concept that the radiospermic morphology is derived. In other words, this aspect of the vasculature is thought to be relatively unchanged. However, given that the primary vasculature of all cycad stems, leaves, and sporophylls exhibits dichotomous branching, then it is not surprising that the vasculature of the ovules is also dichotomous. In my opinion, neither the nature of the vascular architecture nor the ultimate origin of the ovule vasculature can be used to support the apomorphic status of either the platyspermic or radiospermic conditions in extant cycads because they remain the same regardless of changes in ovule morphology.

The origin of the two bundles that enter the ovule of all extant cycads is always from one common bundle by two anisotomous branches of that bundle (Worsdell, 1898; Matte, 1904). The level at which these branchings may occur varies from the base of the ovule to a more distal region of the megasporophyll, but the pattern is the same. The fact that both bundles entering the ovule contribute branches to both the outer and inner vasculature of the ovule is quite distinct from that of medullosans, where the lower level branch gives rise to the outer ovular vasculature and the upper branch the inner, as pointed out by Crane (1988). The lack of homology between these two types of dual systems appears to be a correct interpretation of their vascular architecture.

Finally, it should be noted that although most species of *Cycas* have decidedly platyspermic ovules, *Cycas wadei* has very radiospermic seeds. I have not been able to observe either young ovules or germinating seeds of this species, but it certainly raises interesting questions and possibilities regarding the platyspermic condition in *Cycas.*

Microsporophylls

Except for *Cycas,* the microsporophylls of each genus and species in genera closely resemble (*Stangeria*) or are identical to the megasporophylls in general form. This may very well be correlated with insect pollination (Norstog et al., 1986; Tang, 1987). As with the discussion of the megasporophylls, I will refrain from comments on the well-known features (see Bierhorst, 1971; Pant, 1973; Gifford & Foster, 1989) and restrict myself to more recent and unpublished observations.

Like the megasporophylls, the vascular pattern of microsporophylls is related to the overall form (Figs. 51–56). Thus, those genera with planar microsporophylls, *Cycas* (Fig. 52), *Dioon* (Fig. 54), and *Stangeria* (Fig. 53), have a basic two-dimensional vascular system (exclusive of the vascular supply to the sporangia). The other genera with peltate microsporophyll tips have a basic three-dimensional vascular pattern (Figs. 51, 55, 56).

The short stalked microsporangia occur in groups of from 2–5 and, thus, these groups have often been referred to as sori. Bierhorst (1971), in his limited study on cycad microsporophylls, noted not only this clustering of microsporangia, but also that they can be fused, particularly in the stalk region. He interpreted these fusions as synangia. My broader survey of all the genera that included the vascular supply to the microsporangia supports Bierhorst's conclusion. The degree of synangial development is different for different genera. Thus, in *Cycas* (Fig. 52), fusions are uncommon; in *Dioon,* the stalks are always fused (Fig. 54); and in *Macrozamia,* the stalks are fused and at least some of the sporangia are always fused, with at least one purilocular structure occurring in a sporangial cluster (Fig. 56). The vascular supply follows suit. Those taxa, *Cycas, Bowenia,* and *Stangeria,* with individual microsporangia in a cluster have a common bundle that terminates in as many branches as there are sporangia with each branch entering the stalk

---

→

Figs. 50, 51. Drawings of the vasculature, based on representative serial transections, of the sporophylls of *Bowenia spectabilis.* **50.** Acropetal series, beginning with a, of a megasporophyll. **a.** Two traces in the base of the megasporophyll stalk. **b, c.** Elaboration of the vasculature by random dichotomies in the stalk. **d.** Base of the sterile tip just above ovule attachment. Lateral traces, as seen longitudinally, supply the ovules. **e–h.** Serial sections from the expanded sterile tip with highly elaborated vasculature. **51.** Acropetal series, beginning with a, of a microsporophyll. **a.** First dichotomies of the two traces that enter the base. **b, c.** Elaboration of the vasculature by random dichotomies in the sterile base. **d, e.** Representative sections of the vascular supply to the sporangial groups. **d.** Stalks and sporangia separate on left and fused on right. Note branches from a common supply to the sporangial groups. **f.** Both fused and separate sporangial stalks, each group with a common vascular supply, and vascular elaboration with increasing transectional area. **f, g.** Vasculature of the expanded sterile tip.

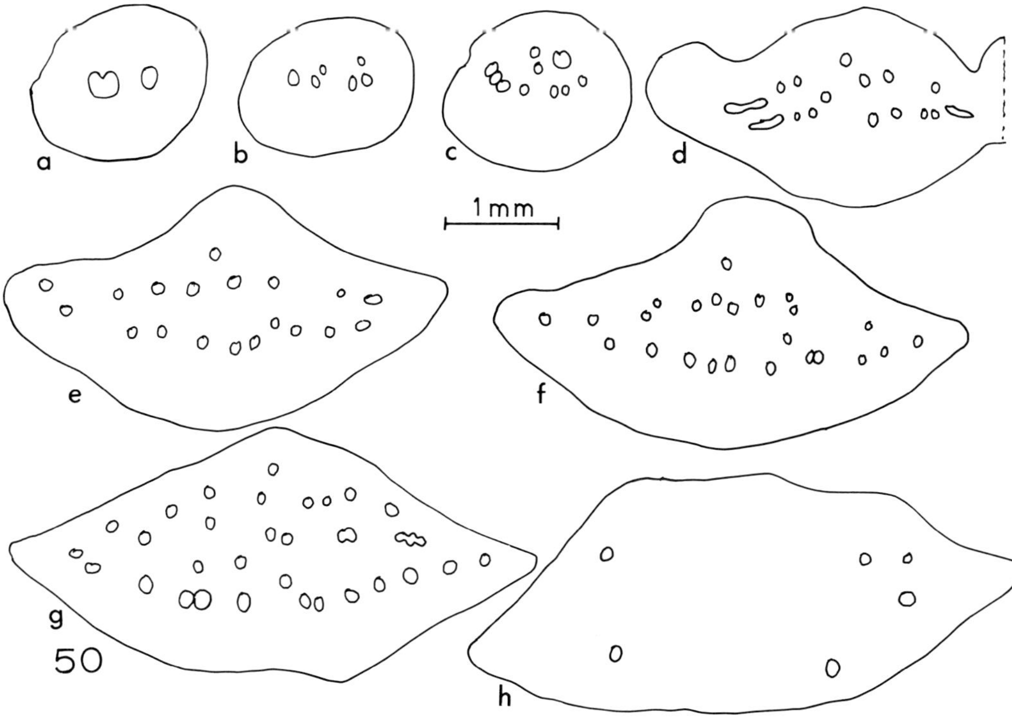

50

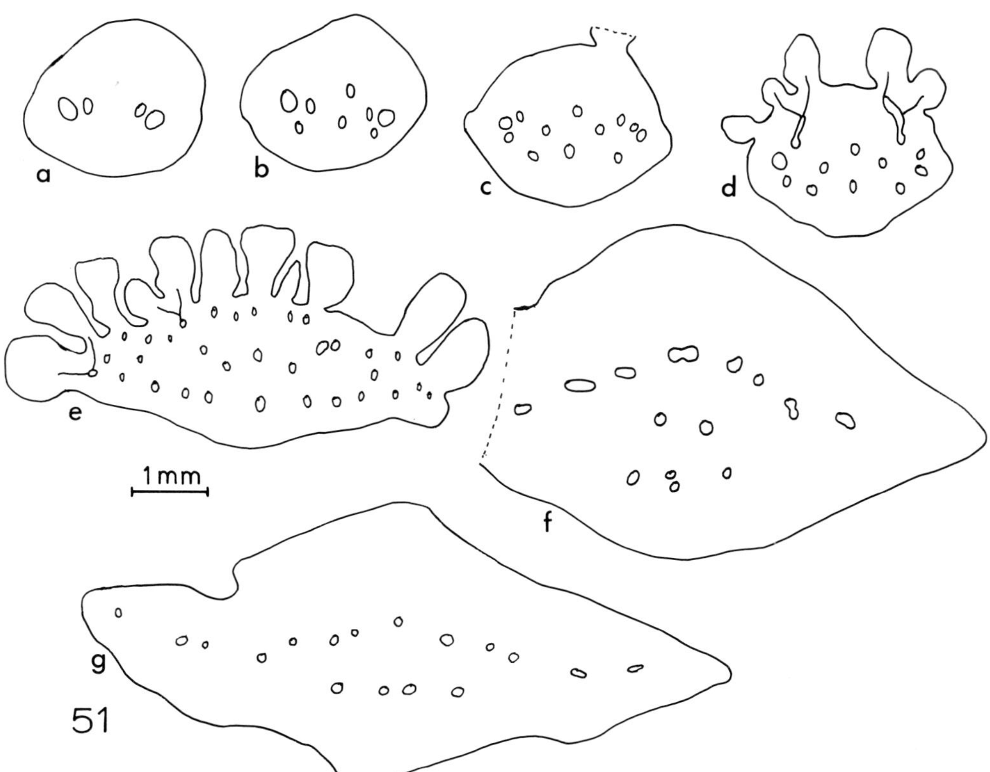

51

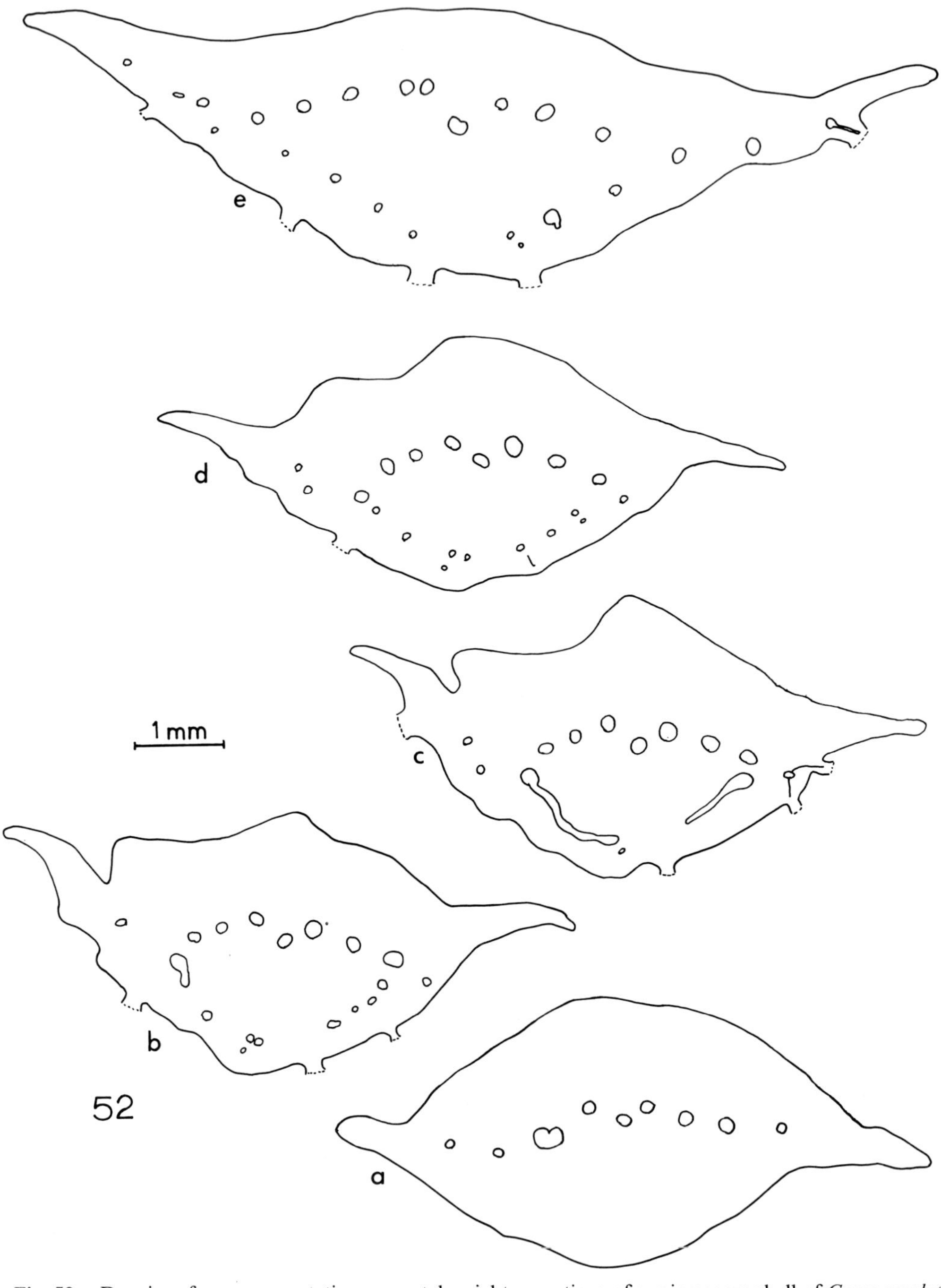

Fig. 52. Drawings from representative acropetal serial transections of a microsporophyll of *Cycas revoluta*. Sporangia have been removed. **a.** Middle region of basal sterile portion. **b–e.** Acropetal series of fertile portion. Sporangial stalks appear widely spaced because the sporangial clusters are loosely aggregated. **b.** Bundles near stalks on adaxial surface originate from the adaxial recurved bundles shown in c. **c.** Adaxial bundles, seen here in longisection, give rise to the sporangial vascular supply by branches that develop basipetally (a) and by branches that develop acropetally (d and e). Note that the vasculature is planar, except for sporangial supplies, in conformity with sporophyll shape.

of a sporangium (Figs. 52, 51, 53, respectively). All other genera show some basic degree of sporangial fusion, ranging from only a common stalk in the basal region in *Dioon, Encephalartos, Lepidozamia, Microcycas,* and *Ceratozamia,* to stalks more or less completely fused along their entire length in *Zamia,* to completely fused stalks with some or all of the sporangia fused laterally, in *Macrozamia.* The situation in *Chigua* is not yet known. When only the bases of the sporangial stalks are fused, then the common bundle approaches the fused bases and produces short branches which enter the stalks as illustrated for *Dioon* (Fig. 54). In *Zamia,* where the stalks are completely fused (or nearly so), the common bundle enters the base of the common stalk and then produces very short branches, one branch per sporangium (Fig. 55d). In *Macrozamia,* where the stalks are completely fused and some or all of the sporangia are fused, the common bundle enters the short common stalk and terminates without branching (Fig. 56c). Thus, fusion of the sporangia and their stalks is correlated with simplification of vasculature. Because the degree of fusion of the sporangia and stalks shows some variability in a given microsporophyll, the degree of vascular simplification varies accordingly. *Cycas* is unique because the major lateral bundles that supply the microsporangia traverse the microsporophyll longitudinally and then dichotomize with one shank recurving basipetally and centripetally while giving rise to the common bundles which in turn branch to ultimately enter the sporangial stalks (Fig. 52). Thus, in acropetal serial sections one observes common bundles that appear to have no other vascular connections (Fig. 52b), but when traced acropetally can be seen to be branches of the two lateral major bundles in Figure 52c.

Several features of the aggregation of the microsporangia need further study. Certainly, the comparative ontogenetic development of synangia with respect to postgenital or congenital fusion would extremely interesting. Also, the eusporangiate development of the microsporangium has only been demonstrated in one species of *Zamia* (Smith, 1907) and in the monotypic *Stangeria* (Lang, 1897).

Pollen

There have been some 21 papers published by Audran and Audran and co-workers on several aspects of pollen development and structure in the Cycadales (see Read & Stolt, 1986 for the complete list). The majority of the publications by Audran concern a series of papers on the ontogeny, histochemistry, and ultrastructural details of pollen formation and wall deposition in *Ceratozamia mexicana.* Only the two papers by Audran and Masure (1976, 1977) that deal with the comparative aspects of the pollen of the cycads will be discussed here.

Audran and Masure (1976) published a survey of the ultrastructure using transmission electron microscopy (TEM) of the pollen of five cycad genera; *Cycas, Dioon, Encephalartos, Stangeria,* and *Zamia.* Later, they (1977) extended their observations to include two more genera, *Ceratozamia* and *Macrozamia* plus observations on all taxa with light microscopy and scanning electron microscopy (SEM). From their extensive work on these seven of the ten known genera, Audran and Masure (1977) concluded that the most significant differences at the generic level were between *Cycas* on the one hand and all of the other genera on the other. More specifically, *Cycas* exhibits an exine structure that appears less specialized, particularly when compared against the fossil record. The exine of *Cycas* pollen appears to lack a well organized alveolate layer when compared with the other genera (see character 1.11 in Table I). They also recognized similarities in exine structure in groups of genera. That is, *Dioon* and *Stangeria* share similar pollen wall sculpturing, as do *Ceratozamia* and *Zamia,* and *Macrozamia* and *Encephalartos.* In addition, Audran and Masure (1977) considered the latter two groups to be more similar to each other than either to the former.

More recently, Dehgan and Dehgan (1988) independently repeated this work, claiming that no SEM survey had yet been done on cycad pollen. They do not cite the 1977 paper by Audran and Masure (although they do cite the 1986 bibliography by Read & Stolt that contains the Audran & Masure reference). While repeating the earlier work (Audran & Masure, 1977), Dehgan and Dehgan also extended their work to include *Bowenia, Lepidozamia,* and *Microcycas.* Their conclusions are basically the same as those of Audran and Masure (1977). They recognize the distinctness of *Cycas* from all other genera and similiarities between *Encephalartos, Macrozamia,* and *Lepidozamia* (not previously studied) and these three genera to *Bowenia.* They also

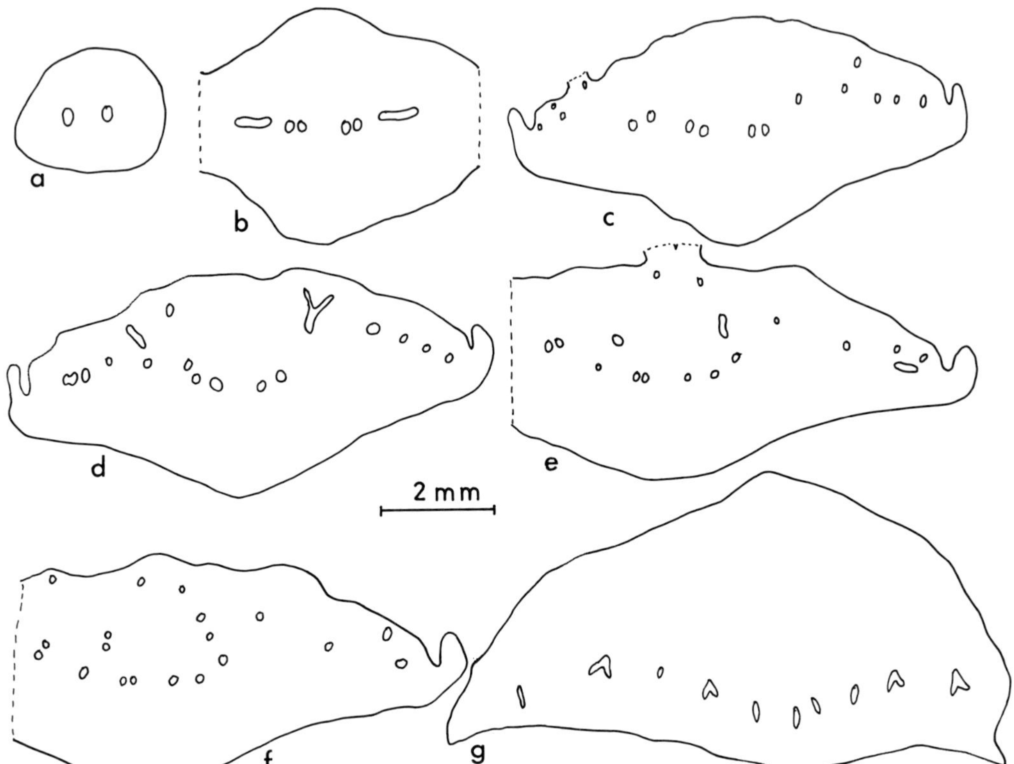

53

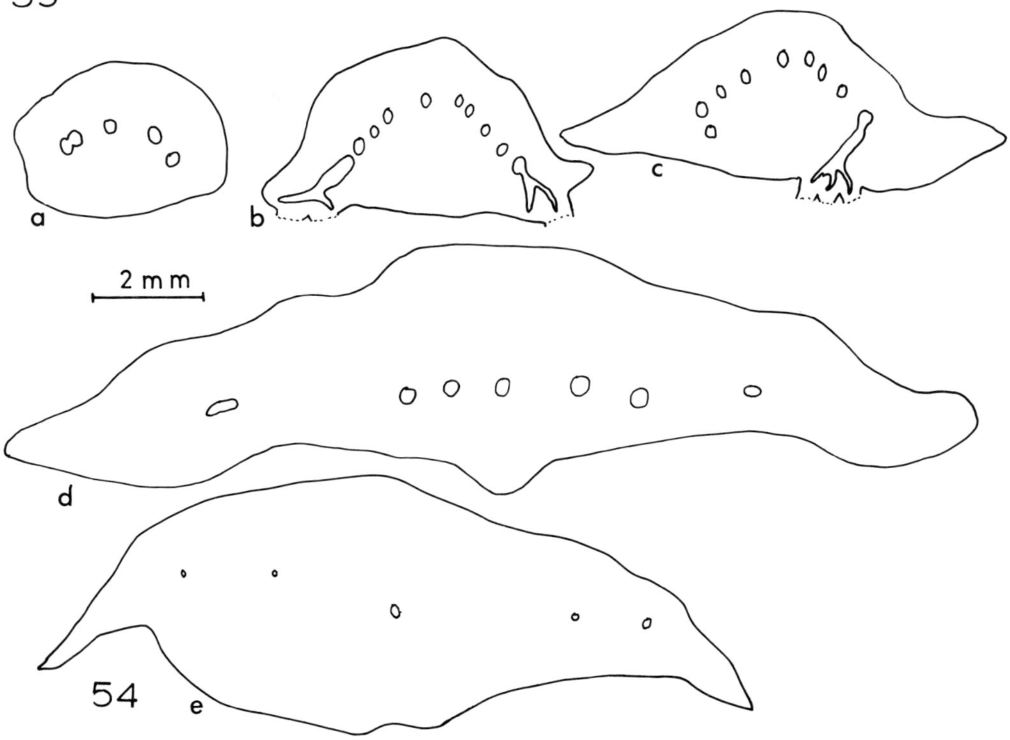

54

recognize similarity in pollen structure between *Dioon* and *Stangeria* and both to *Cycas*. Their work also supported the similiarity of *Ceratozamia* and *Zamia* to which they would add *Microcycas* (not previously studied). Dehgan and Dehgan (1988) go even further with their analysis of pollen characters and suggest that they help support the elevation of *Dioon* to the family level as well as the group of genera containing *Encephalartos, Lepidozamia,* and *Macrozamia.* This would obviously result in taxonomic inflation with six families in the Cycadales and would be similar to the system of Schuster (1932) that does little or nothing toward establishing a taxonomic system that reflects relationships between genera within the order.

Systematics

Historical

The history of classification systems for the Cycadales has been summarized by Johnson (1959) and Pant (1973). I shall not discuss the details of the earlier systems that appeared before Schuster's (1932) monograph, but rather will focus on common general aspects and the placement of problematic genera. Also, I will consider cycad systematics at the generic and higher levels. This is, in part, because reliable monographs of the genera at the specific level are noticeably absent except for the endemic Australian genera, *Bowenia, Lepidozamia,* and *Macrozamia* (Johnson, 1959). The monograph by Schuster (1932) of the order to the specific level has been severely and justifiably criticized by Johnson (1959). As far as I can determine, the only redeeming value of Schuster's monograph is the thorough bibliography, particularly with respect to old obscure references.

Before 1853, when *Stangeria* was recognized as a cycad, two groups were commonly recognized. One group consisted of *Cycas* and the other group consisted of *Zamia* and the other genera that had been described to that time. These two groups were attributed tribal, subfamilial, or even familial rank depending on the author. After, the discovery of *Dioon, Stangeria,* and *Bowenia,* classifications underwent change because these genera did not fit well into the traditional scheme. Subsequent nineteenth century authors recognized 3–4 major groups. Miquel (1861) was the first to do so when he established four tribes within a single family, the Cycadaceae. The Cycadeae and Stangerieae were monogeneric; the Encephalarteae contained *Encephalartos, Macrozamia,* and *Lepidozamia* (as *genus dubium*); and the Zamieae contained *Zamia, Ceratozamia,* and *Dioon.* Later, Miquel (1869) slightly revised this system, placing *Lepidozamia* in *Macrozamia* and the newly described *Bowenia* in Encephalarteae. The previous year, de Candolle (1868) published a system with a single family, Cycadaceae, with three tribes: Cycadeae, Encephalarteae, and Zamieae. Like Miquel (1861, 1869), the Cycadeae of de Candolle was monogeneric and *Lepidozamia* was considered a *genus dubium.* However, unlike Miquel, de Candolle divided the Encephalarteae into two subtribes, one containing only *Stangeria* and the other containing *Encephalartos, Macrozamia, Bowenia,* and *Dioon.* All New World genera were placed in the same tribe, the Zamieae, by Miquel (1861, 1869) but de Candolle transferred one of these, *Dioon,* to a tribe containing only African and Australian genera. Hutchinson (1924) modified de Candolle's (1868) classification in two ways. First, he recognized only two tribes, a monogeneric Cycadeae and an Encephalarteae with three subtribes—the Encephalartinae, the monogeneric Stangerinae, and the Zamiinae. The latter reduction represents the demotion of what was a coordinate tribe in de Candolle's system to the level of a subordinate subtribe. Secondly, Hutchinson transferred *Bowenia* from the subtribe Encephalartinae to the Zamiinae. The salient as-

←

Figs. 53, 54. Drawings from representative acropetal serial transections of microsporophylls. Microsporangia removed for sectioning. **53.** *Stangeria eriopus.* **a–g.** Series from base (a) through slighly inflexed sterile tip (g). **a, b.** Base of microsporophyll with two entering traces that undergo dichotomies in b in one plane. **c–f.** Fertile region with fused microsporangial stalks (f). Note common bundle dichotomizing with resultant branches entering common sporangial stalks in e. Elaborated vasculature is the result of sporangial supplies. The basic pattern is planar following the sporophyll shape. **g.** Planar sterile tip with planar vasculature. **54.** *Dioon edule.* **a–e.** Series from base (a) through planar sterile tip (e). **b, c.** Fused sporangial stalks evident. Note common vascular bundle extends to the base of the fused stalks (compare with Fig. 51). **d, e.** Planar sterile tip with planar vasculature.

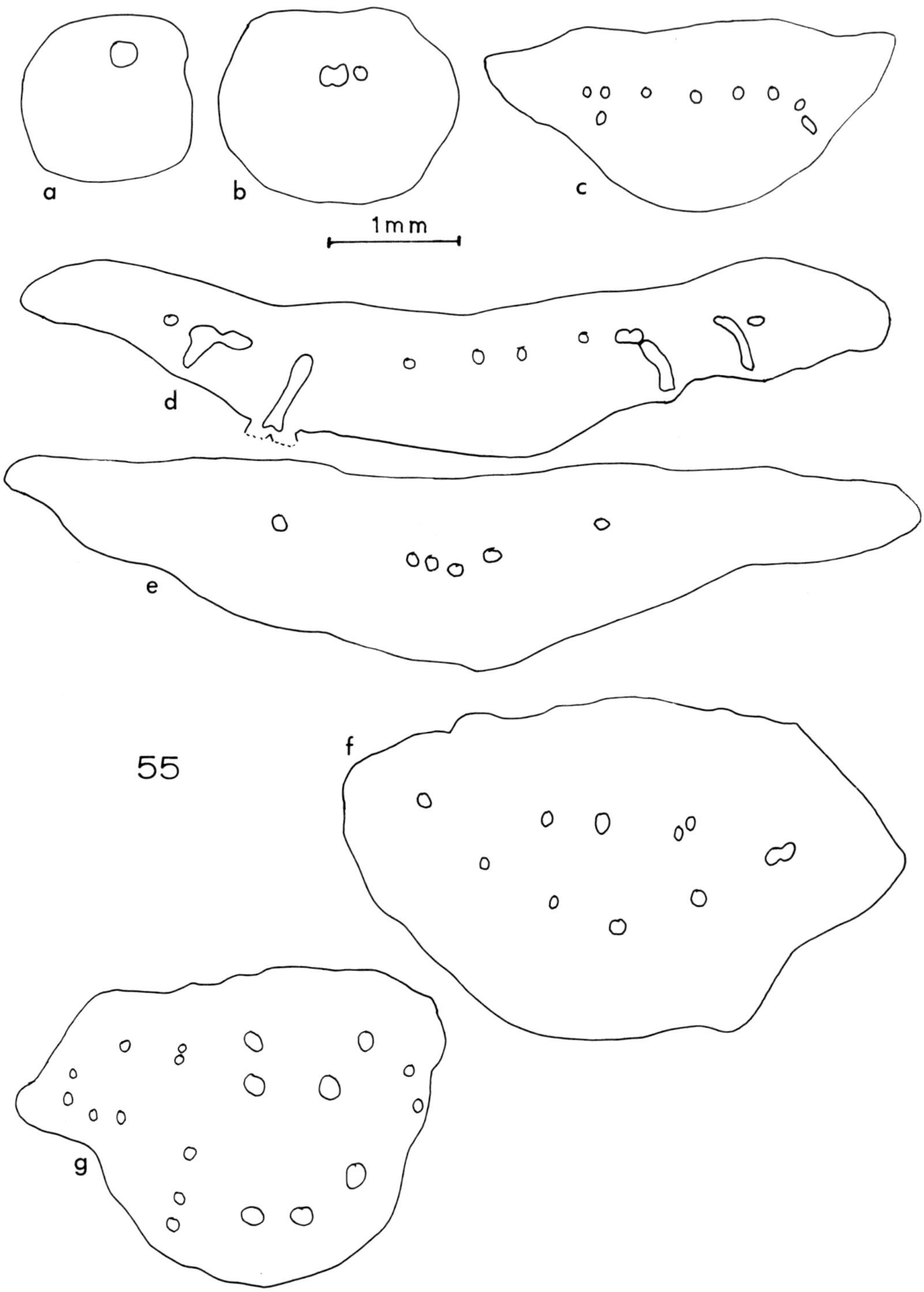

Fig. 55. Drawings from representative acropetal serial transections of a microsporophyll of *Zamia furfuracea*. **a.** Base of sporophyll with single trace. **b, c.** Basal sterile region with dichotomizing vasculature. **d.** Planar fertile region with fused sporangial stalks. Note that the common bundle enters the base of the common stalk and produces very short branches (compare with Fig. 54). **f, g.** Expanded sterile tip with elaborated vasculature.

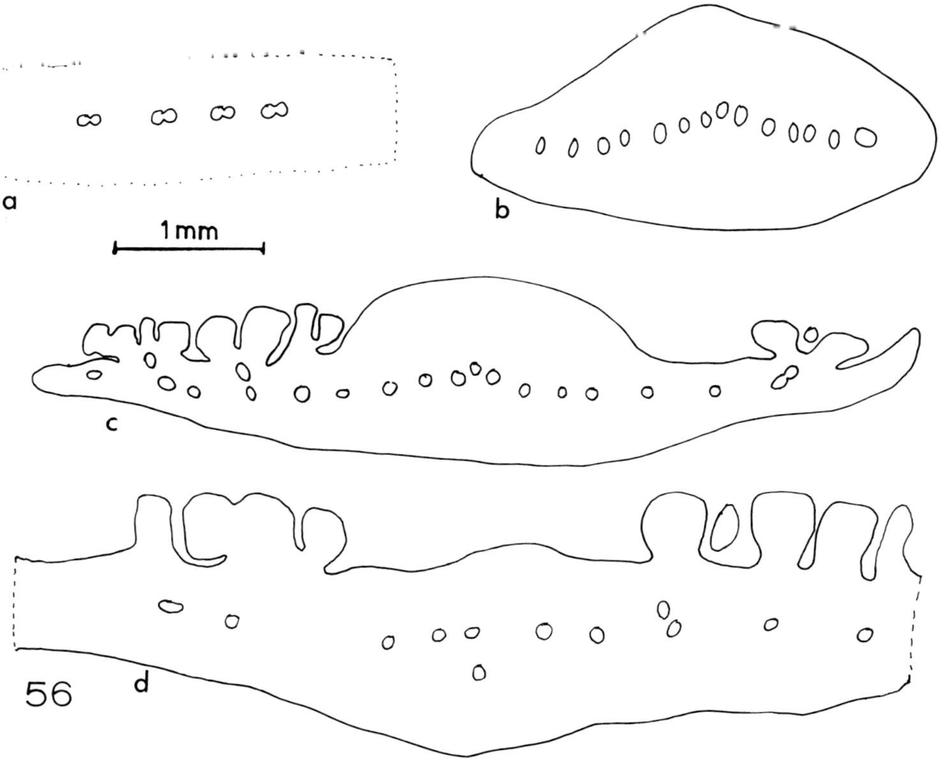

Fig. 56. Drawings from representative acropetal serial transections of a microsporophyll of *Macrozamia communis.* **a.** Region of sporophyll attachment to strobilar axis. All of the veins enter the sporophyll base. **b.** Planar basal sterile region with dichotomizing vasculature. **c, d.** Slightly expanded fertile region with fused sporangial stalks and some fused sporangia. **c.** Fused sporangia at left were plurilocular. Note that common bundle enters the fused sporangial stalks without branching (compare to Figs. 54 and 55). **d.** Slightly elaborated abaxial vasculature of slightly expanded region in center that is unrelated to sporangial supply.

pects of the preceding discussion are that *Stangeria* and *Cycas* have always been considered isolated, whilst *Dioon* and *Bowenia* were enigmatic enough to receive different placements by different authors. On the other hand, certain genera have always been grouped together, that is, *Zamia, Ceratozamia,* and *Microcycas* (Miquel considered this genus to be part of *Zamia*) in one group and *Encephalartos, Macrozamia,* and *Lepidozamia* (considered to be part of *Macrozamia,* but see Johnson, 1959) in another. Although Hutchinson's classification would seem to return to the pre-Miquel classifications with only two subdivisions in the cycads, it differs significantly in that he had a hierarchy of subtribes in his Encephalarteae.

The already recognized isolation of *Cycas* and *Stangeria* coupled with the problem of suitably placing *Dioon* and *Bowenia* with other genera led Pilger (1926) to place all four of these genera in their own monogeneric subfamilies. Schuster (1932) returned to the basic two taxon subdivision of the cycads. He recognized one family with two subfamilies, the monogeneric Cycadoideae and the Zamioideae with the rest of the genera. Although Schuster (1932) did recognize tribes in Zamioideae, it had no meaning, because each genus was put into its own tribe without any tribal hierarchy. Thus, more than one hundred years after Reichenbach's (1828) treatment, cycad systematics had returned full circle, with Schuster's (1932) treatment, except for taxonomic inflation.

As a result of Schuster's (1932) incomprehensible taxonomy, the systematics of the cycads received no significant attention until 1959, when Johnson published his monograph of the Australian genera and a new classification of the Cycadales. In this treatment, Johnson (1959) recognized three families, the Cycadaceae,

Stangeriaceae, and Zamiaceae. The first two are monogeneric, whereas the Zamiaceae are divided into three tribes: Encephalarteae with *Encephalartos, Macrozamia,* and *Lepidozamia*; Dioeae with *Dioon*; and Zamieae with *Zamia, Ceratozamia, Microcycas,* and *Bowenia.* This system resolved certain issues in that *Stangeria* is recognized as truly distinctive and Encephalarteae are seen as a discrete group. On the other hand, as pointed out by Crane (1988), certain other relationships remain unresolved at the familial, tribal, and generic levels. For example, the relationships between the three families are not clear, nor are the relationship(s) between the three tribes in the Zamiaceae or their constituent genera. In particular, the placement of *Bowenia* in the Zamieae is still problematic. Johnson (1959) noted that *Bowenia* is incongruous in this tribe, but until sufficient comparative data are obtained on *Bowenia* and other Zamieae its status would remain so. In 1981, I realized, from a comparative study of cycad genera, that *Bowenia* shared few characters with the Zamieae and so recognized the monogeneric Boweniaceae. However, this work (Stevenson, 1981) only recognized the distinct nature of *Bowenia* and did not resolve familial relationships among the four families. More recently, I (Stevenson, 1985) presented a tentative revised classification of the cycads in which I tried to resolve relationships at the family, subfamily, and tribal levels. Thus, I recognized two suborders in the Cycadales, the Stangerineae containing the Stangeriaceae and Boweniaceae and the Cycadineae containing the Cycadaceae and Zamiaceae. The Cycadaceae was monogeneric and the Zamiaceae consisted of two subfamilies, the monogeneric Diooiodeae and the Zamioideae. The Zamioideae was in turn divided into two tribes: the Zamieae, with *Zamia, Ceratozamia,* and *Microcycas*; and the Encephalarteae, with *Encephalartos, Lepidozamia,* and *Microzamia.* No attempt was made at that time to perform a cladistic analysis or to resolve relationships within the tribes. In fact, the classification was flawed from the start for both these reasons and the fact that I was too impressed by the similarities of *Stangeria* and *Bowenia* to each other to recognize their similarities to the Zamiaceae and the fact that the Cycadaceae was the most probable sister group to all other cycads, as demonstrated by Crane (1988). However, the classification was successful to the extent that I recognized that the Stangeriaceae and Boweniaceae are a monophyletic unit as well as the monophyletic nature of the Zamieae and Encephalarteae. Also, I considered this a tentative classification for comparison to future cladistic analyses. That is not to say that a cladistic analysis had not yet been done, because Petriella and Crisci (1977) had published just such an analysis. Although their analysis also identified *Stangeria* and *Bowenia* as a monophyletic group, it failed to identify the Zamieae and included *Zamia* in a larger group with *Stangeria* and *Bowenia.* This latter aspect is difficult to understand until one looks at the characters and character states used by Petriella and Crisci (1977). For example, there are no cycads with reticulately veined pinnae (their character 2). They also use the presence of subterranean stems as derived (their character 1), but as discussed in the morphology section of the present paper, both arborescent and subterranean forms occur in most genera. Thus this character simply cannot be used and because Petriella and Crisci (1975, for the data matrix in their 1977 paper) score *Zamia* as having only subterranean stems, it ends up in their analysis as the sister taxon to *Stangeria* and *Bowenia.*

Crane (1988) recently published another cladistic analysis of all cycad genera except *Chigua* (Stevenson, 1989), which was not known at that time. Crane (1988) presented this as a tentative hypothesis of phylogenetic relationships within the extant Cycadales that can be tested with future expanded data bases. The result of Crane's analysis (Fig. 57) is the placement of the monogeneric Cycadaceae as the sister taxon to all other extant cycads. *Stangeria* and *Bowenia* form a clade that is a sister taxon to Zamiaceae. There are three clades in the Zamiaceae; a clade composed solely of *Dioon*; a clade composed of *Encephalartos, Lepidozamia,* and *Macrozamia*; and a clade composed of *Ceratozamia, Microcycas,* and *Zamia. Dioon* is the sister taxon to these other two clades in the Zamiaceae. Thus, the results for the Zamiaceae and the relationship of *Stangeria* with *Bowenia* are completely compatible with the classification proposed by Stevenson (1985).

Cladistic Analysis

I present a new cladistic analysis based on the characters used by Crane (1988) and additional characters, of which many were discussed in the preceding morphological section of this paper. I have retained the same characters and numbering system of the characters used by Crane (1988) with the following exceptions. I have changed characters 1.3 (sympodial growth) and 1.19 (lateral cones) by omitting them and substituting new characters. The reason for this is that these two characters, sympodial growth and lateral cones, are no longer useable as a result of a recent study on reproductive branching (Stevenson, 1988), which appeared after Crane's (1988) paper. Neither sympodial growth nor lateral cones in fact occur in the cycads as discussed in the morphological section of the current paper. Moreover, this character usually refers to cones that occur on trunks and as such does not occur at all in *Zamia* and may or may not occur on the same individual in species of all the genera of the Encephalarteae. In the sense that it is used by Crane (1988), it refers to multiple cone production and this occurs in all cycads except ovulate plants of *Cycas*. It should be noted that character 1.3 (sympodial growth) is a synapomorphy present in all of the Cycadales in Crane's analysis. I have substituted the presence of coralloid roots, which is also a synapomorphy in all of the Cycadales, for sympodial growth. Thus, this character substitution causes no changes in Crane's cladogram. For character 1.19 (lateral cones), I substituted the "presence of a midrib in the pinnae." The removal of lateral cones also has no effect on Crane's cladogram (Fig. 57) because the character is only one of two synapomorphies that defines a clade with the Encephalarteae and Zamieae and simply removes it as a reversal for *Microcycas*. I also reversed the polarity of character 1.9 (radiospermic seeds) because I think that the platyspermic seed symmetry is derived, for the reasons discussed previously in the section on ovules. However, this reversal of the polarity in character 1.9 causes no changes in the branching pattern of the cladogram, but rather removes radiospermic seeds as one of seven synapomorphies (thus leaving six synapomorphies) for the other cycads as a sister group to *Cycas*, and places platyspermic seeds as an autapomorphy for *Cycas*.

Below is the list of characters used with a brief discussion of the rationale for each character. The data matrix is presented in Table I. The outgroup for cycads is not clear, particularly among extant plants and there are obvious problems with fossil groups when one tries to use cytological or phytochemical characters. I have followed Crane (1988) in using "other seed plants" as the outgroup. Crane was not clear as to what he meant by "other seed plants." I presume this group did not include the flowering seed plants because otherwise it would have been impossible to polarize many of the characters that occur in both states in the flowering plants. I have excluded the flowering plants in my analysis. In some instances, for example character 1.29, where the distribution of a character is unknown in the "other seed plants," I have considered only the state of that character in the medullosans because the cycads appear to be the sister group of the medullosans based on the cladistic analyses of Crane (1985) and Doyle and Donoghue (1986). Characters 1.1–1.23 follow Crane (1988) except for 1.3 and 1.19 as previously discussed. Thus, characters 1.3, 1.19, and 1.24–1.30 are newly added. The autapomorphic characters of each genus are not included because as autapomorphies they are uninformative in determining intergeneric relationships. The new genus, *Chigua* Stevenson (1989), has been omitted from the analysis because the states of only 15 of the 30 characters used here are known at this time (Table I). However, I included *Chigua* in the data matrix so that comparisons could be easily made. Only four plants of *Chigua* are in cultivation and thus it may be some time before certain features can be ascertained. However, comments concerning *Chigua* and the characters are included in the following list of characters.

Characters

1.1 Presence of Cycasin.—Cycasin is ubiquitous in the extent cycads but unknown in any other group of plants (Siniscalco Gigliano, 1989). Its presence has yet to be determined for *Chigua.*

1.2 Girdling Leaf Traces.—The girdling leaf traces of cycads are not found in any other group of plants. Girdling leaf traces are found in the adult stems of all cycads but are absent from the seedlings of some taxa (Chamberlain, 1935; Worsdell, 1906) and the strobilar axes and peduncles of all cycads (Matte, 1904; Worsdell, 1898), including the trunk portion bearing the megasporophylls in *Cycas*. The nature of this character has yet to be determined for *Chigua*.

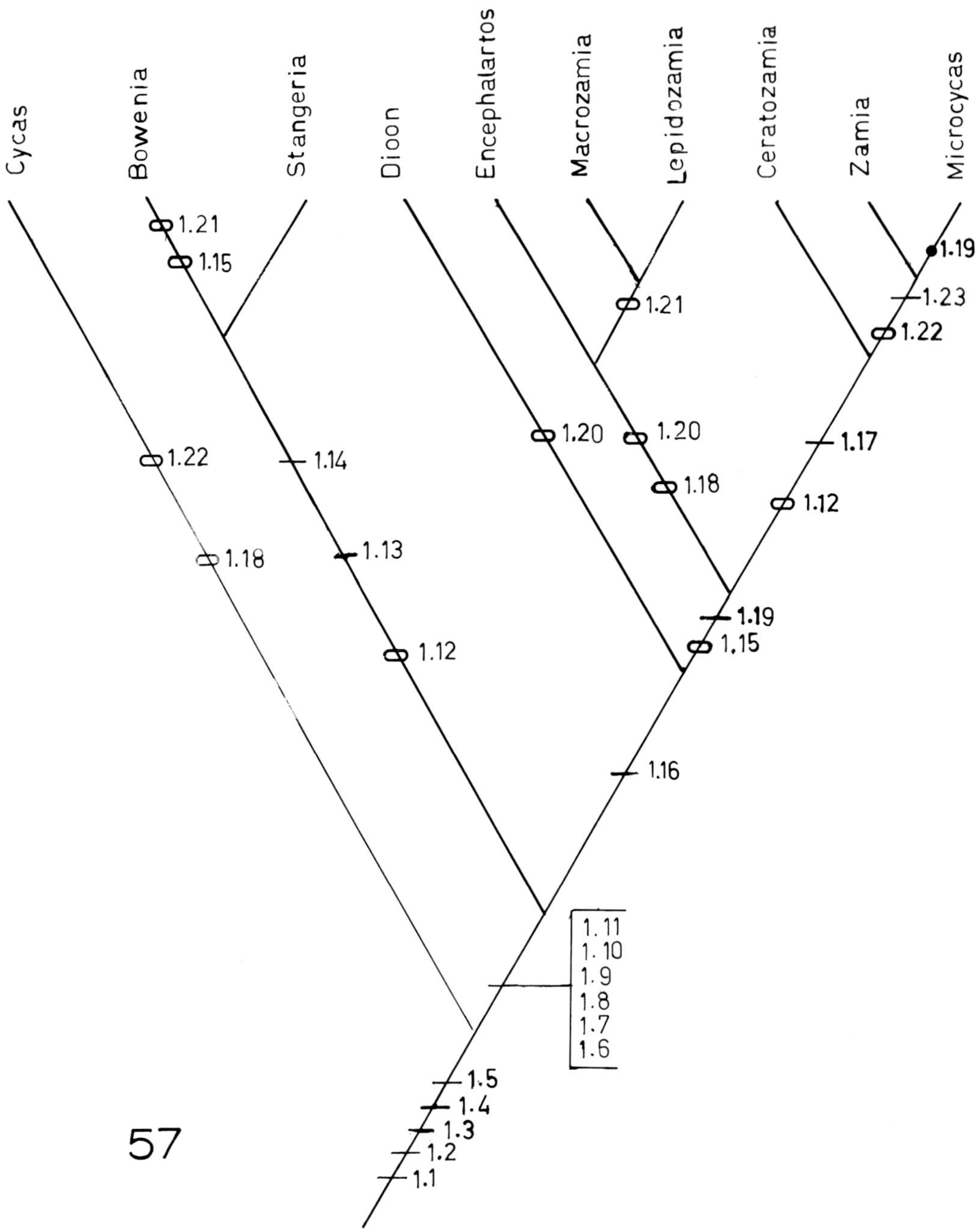

Fig. 57. Cladogram of the Cycadales based on the data matrix of Crane (1988). Parallelisms are indicated by an oval. Reversals are indicated by a solid circle. Character 1.20 can be interpreted as a parallelism as indicated or as a reversal in the clade composed of *Ceratozamia, Zamia,* and *Microcycas* without changing the cladogram. See text for details.

Table I

Data matrix for characters of extant cycad genera (except *Chigua*). Apomorphic characters = +, plesiomorphic characters = −, and unknown = ?. BOW = *Bowenia*, CER = *Ceratozamia*, CHI = *Chigua* (presented here for comparative purposes but omitted from the cladograms because the genus is insufficiently known), CYC = *Cycas*, DIO = *Dioon*, ENC = *Encephalartos*, LEP = *Lepidozamia*, MAC = *Macrozamia*, MIC = *Microcycas*, OSP = Other Seed Plants, STA = *Stangeria*, and ZAM = *Zamia*

		OSP	CYC	BOW	STA	CER	ZAM	MIC	DIO	ENC	MAC	LEP	{CHI}
1.1	Presence of cyacasin	−	+	+	+	+	+	+	+	+	+	+	?
1.2	Girdling leaf traces	−	+	+	+	+	+	+	+	+	+	+	?
1.3	Coralloid roots	−	+	+	+	+	+	+	+	+	+	+	+
1.4	Simple ovulate strobilus	−	+	+	+	+	+	+	+	+	+	+	+
1.5	Primary thickening meristem	−	+	+	+	+	+	+	+	+	+	+	?
1.6	Simple megasporophylls	−	−	+	+	+	+	+	+	+	+	+	+
1.7	Biovulate megasporophylls	−	−	+	+	+	+	+	+	+	+	+	+
1.8	Ovules with micropyles oriented proximally	−	−	+	+	+	+	+	+	+	+	+	+
1.9	Platyspermic seeds	−	+	−	−	−	−	−	−	−	−	−	−
1.10	Colored trichomes	−	−	+	+	+	+	+	+	+	+	+	?
1.11	Specialized alveolate exine structure	−	−	+	+	+	+	+	+	+	+	+	?
1.12	Stipules on leaves	−	−	+	+	+	+	+	−	−	−	−	+
1.13	Vascularized stipules	−	−	+	+	−	−	−	−	−	−	−	?
1.14	Petiole with circular arrangement of vascular bundles	−	−	+	+	−	−	−	−	−	−	−	?
1.15	Sporophyll vasculature in more than one plane	−	−	+	−	+	+	+	−	+	+	+	?
1.16	Developing pinnae flat	−	−	−	−	+	+	+	+	+	+	+	+
1.17	Articulate pinnae	−	−	−	−	+	+	+	−	−	−	−	+
1.18	Polyxylic stele	−	+	−	−	−	−	−	−	+	+	+	?
1.19	Pinnae with a midrib	−	+	+	+	−	−	−	−	−	−	−	+
1.20	Strobilar peduncles with cataphylls	−	−	−	−	−	−	−	+	+	+	+	−
1.21	Short, curved, colored trichomes	−	−	+	−	−	−	−	−	−	+	+	?
1.22	Branched trichomes	−	+	−	−	−	+	+	−	−	−	−	?
1.23	Equally branched trichomes	−	−	−	−	−	+	+	−	−	−	−	?
1.24	Reduced spinose pinnae on rachis	−	+	−	−	−	−	−	+	+	+	+	−
1.25	Prickles on rachis and petiole	−	−	−	−	+	+	−	−	−	−	−	+
1.26	Megasporophylls with lateral lobes	−	−	−	−	−	−	−	−	+	+	+	−
1.27	Medullary vascular bundles	−	−	−	−	−	−	−	−	+	+	+	?
1.28	Concentric vascular bundles in cotyledons	−	−	+	+	−	−	−	−	−	−	−	?
1.29	Extrafascicular bundles not present	−	−	−	+	+	+	+	+	−	−	−	?
1.30	Cataphylls absent on vegetative shoots	−	−	+	+	−	−	−	−	−	−	−	−

1.3 Coralloid Roots.—Coralloid roots are the ageotropic roots that contain nitrogen-fixing Cyanobacteria. These coralloid roots are found in all cycads, including *Chigua,* and are unknown for any other group of plants.

1.4 Simple Ovulate Strobilus.—All extant cycads, including *Chigua,* produce simple well defined clusters of megasporophylls. Except for *Cycas,* the megasporophylls are organized into discrete strobili. As discussed by Crane (1988) and earlier in this paper, the organization of the megasporangiate axis in *Cycas* may very well be derived in the extant cycads.

1.5 Primary Thickening Meristem.—A unique primary thickening meristem is present in all cycad genera (see previous discussion), although its presence has not been determined for *Chigua.* In Crane's analysis, the presence of a primary thickening had not yet been determined for *Lepidozamia.* However, I recently sectioned shoot apices of this genus and observed its presence. I suspect from the growth form of *Chigua,* that a primary thickening meristem is also present in this genus.

1.6 Simple Megasporophylls.—Except for *Cycas,* all extant cycad genera, including *Chigua,* have simple megasporophylls. I follow Crane (1988) in interpreting the pinnatifid to serrate megasporophyll of *Cycas* as plesiomorphic. *Cycas* does exhibit a transition from obviously pinnatifid megasporophylls in *C. revoluta* to simply serrate megasporophylls in *C. circinalis.* Also, in *Encephalartos,* there are some species with minute serrations on the lateral margins of the sterile tip of the megasporophyll.

1.7 Biovulate Megasporophylls.—Except for *Cycas,* all genera, including *Chigua,* have biovulate megasporophylls. I consider the occasional occurrence of only one or two ovules on individual megasporophylls of *Cycas* and one or three ovules on the megasporophylls of biovulate genera as teratisms.

1.8 Ovules with Micropyles Oriented Proximally.—Except for *Cycas,* all cycad genera, including *Chigua,* have their ovules parallel to the megasporophyll stalk with the micropyle pointing towards the strobilus axis. In *Cycas,* the ovules are oriented obliquely towards the megasporophyll apex, at least at the time of pollination. As discussed by Crane (1988), this character may well be derived in *Cycas,* but in a manner different from the other genera. If ovule orientation is derived in *Cycas,* it may be in response to pollination (Niklas & Norstog, 1984).

1.9 Platyspermic Seeds.—Crane (1988) listed this character as radiospermic seeds because he considers the radiospermic condition to be derived from the platyspermic condition. My conclusion runs counter to this and accordingly I consider platyspermic seeds to be derived in the extant cycads. For a rationale of my conclusion see the previous discussion on ovules. *Chigua* has radiospermic seeds.

1.10 Colored Trichomes.—All cycads have colored trichomes except *Cycas.* Transparent trichomes occur in all cycads, including *Cycas,* and are widely found throughout the outgroup, whereas colored trichomes are unknown in the outgroup. The colored condition is considered apomorphic. Although the trichomes of *Chigua* are unknown, I suspect colored ones are present.

1.11 Specialized Alveolate Exine Layer of the Pollen.—This character is derived in the cycads with respect to other seed plants (see Crane, 1988). *Chigua* has not been examined.

1.12 Stipules on Leaves.—Stipules have been discussed in the section on leaves. The presence of stipules is considered to be derived. Stipules are present in *Chigua.*

1.13 Vascularized Stipules.—Vascularized stipules occur in *Stangeria* and *Bowenia.* They appear to be morphologically distinct from the unvascularized stipules of *Zamia, Ceratozamia,* and *Microcycas.* See the section on leaves for more discussion. It is not known whether the stipules of *Chigua* are vascularized, but from their morphology I predict that they are not.

1.14 Petiole with Circular Arrangement of Vascular Bundles.—Most cycad genera have an omega arrangement of vascular bundles in their petioles and only *Stangeria* and *Bowenia* have a circular arrangement (Matte, 1904; see also the discussion in the section on leaves). This character is not yet known for *Chigua.*

1.15 Sporophyll Vasculature in More than One Plane.—Because the degree of elaboration of sporophyll vasculature is possibly related to sporophyll size and shape, this character is most likely homoplasic (see previous discussion). This character is not yet known for *Chigua.*

1.16 Developing Pinnae Flat.—The succubous arrangement of pinnae is common to all genera except *Cycas, Stangeria,* and *Bowenia* which have, respectively, circinnate, conduplicate, and circinnate pinnae pytxis (Stevenson, 1981). Flat pinnae are here considered derived with respect to what little is known in fossil and extant seed plants. *Chigua* has flat pinnae.

1.17 Articulate Pinnae.—The pinnae of *Ceratozamia, Microcycas,* and *Zamia* have a distinct articulation at their region of attachment to the rachis. Articulated pinnae also occur in *Chigua.* For a more extensive discussion see the section on leaves.

1.18 Polyxylic Stele.—The polyxylic steles that occur in the cycads are unique in the seed plants and are considered derived. This character occurs in *Cycas, Encephalartos, Lepidozamia,* and *Macrozamia* (Greguss, 1968). For a description and comments on this character see the section on stems. The state of this character is unknown for *Chigua.*

1.19 Pinnae with a Midrib.—Midribs occur in the pinnae of *Cycas, Stangeria,* and *Bowenia.* I have considered the pinnae of the bipinnately compound leaves of *Bowenia* to have a midrib because the rachilla of the pinnules would be a midrib if these leaves were to become pinnate by reduction. Conversely, based on the vascular pattern of the leaves of *Bowenia,* if they are derived from a pinnate leaf then this pinnate leaf would have a midrib. Based on comparison with other gymnosperms, I consider the presence of a midrib in the pinnae to be derived within the cycads more than once. *Chigua* also has a midrib in its pinnae.

1.20 Strobilar Peduncles with Cataphylls.—While all cycads have a series of cataphylls surrounding the base of the peduncle of the strobilus, only *Dioon, Encephalartos, Lepidozamia,* and *Macrozamia* have cataphylls borne directly along the peduncle. As discussed in the section on reproductive morphology, atrophied cataphylls do rarely appear on the peduncles of some species of *Zamia,* but these are considered teratisms. *Chigua* does not have cataphylls on its strobilar peduncles.

1.21 Short, Curved, Colored Trichomes.—Only the endemic Australian genera, *Bowenia, Lepidozamia,* and *Macrozamia,* have short, curved, colored trichomes (Stevenson, 1981). The short curved morphology of these colored trichomes is quite distinct fron the long slender colored trichomes of other genera and in character 1.10. The trichomes of *Chigua* are unknown.

1.22 Branched Trichomes.—Branched trichomes are found only in *Cycas, Microcycas,* and *Zamia* (Stevenson, 1981). In *Cycas,* these branched trichomes are uncolored in contrast to the colored, branched trichomes of the other two genera (Stevenson, 1981). The trichomes of *Chigua* are unknown.

1.23 Equally Branched Trichomes.—In *Cycas,* all branched trichomes are unequally branched. In contrast, in *Zamia* and *Microcycas,* there are both unequally and equally branched trichomes. The equally branched trichomes are considered derived. The condition in *Chigua* is unknown.

1.24 Reduced Spinose Pinnae on Rachis.—The gradual transition from foliar to spinose pinnae in a distal series is unique to *Cycas, Dioon, Encephalartos, Lepidozamia,* and *Macrozamia.* Reduced spinose pinnae do not occur in *Chigua.*

1.25 Prickles on Rachis and Petiole.—Prickles are morphologically defined as sharp epidermal emergences such as occur in the common garden rose. In the cycads, this character is unique to *Ceratozamia* and *Zamia.* However, not all species of *Zamia* have prickles but the majority do. All species of *Ceratozamia* have prickles, but in some, such as *C. keusteriana,* they may be uncommon and diminutive (Stevenson et al., 1986). Stout prickles occur in both known species of *Chigua.*

1.26 Megasporophylls with Lateral Lobes.—Megasporophylls with lateral lobes projecting distally along the ovules are found in *Encephalartos, Lepidozamia,* and *Macrozamia.* They do not occur in the other previously known seven genera or in *Chigua.* A more complete description of this character is given in the section on megasporophylls.

1.27 Medullary Vascular Bundles.—Vascular bundles occur in the pith of *Encephalartos, Lepidozamia,* and *Macrozamia.* These medullary bundles are not related to the presence of the cone domes found in stems of most cycad genera (see the section on stems). The state of this character is unknown for *Chigua.*

1.28 Concentric Vascular Bundles in Cotyledons.—Concentric cotyledonary bundles, with endarch protoxylem throughout, occur in *Stangeria* and *Bowenia.* Other cycads have collateral cotyledonary bundles with exarch protoxylem in the upper half of the cotyledons. The condition of the cotyledonary bundles in *Chigua* is unknown.

1.29 Extrafascicular Vasculature not Present.—Extrafascicular vasculature in the form of independent steles and/or bundles in the cortex occurs in *Bowenia, Cycas, Encephalartos, Lepidozamia,* and *Macrozamia.* Because of the similarity of the extrafascicular vasculature in these cycad genera to that in the medulloseans, the absence of extrafascicular vasculature is considered derived. There is some disagreement concerning the presence of extrafascicular vasculature in *Ceratozamia* (see the section on stems). My conclusion for *Ceratozamia* is that extrafascicular vasculature is not present in this genus. There is no information available concerning this character in *Chigua.*

1.30 Absence of Cataphylls on Vegetative Shoots.—*Bowenia* and *Stangeria* lack cataphylls on their vegetative axes and their large fleshy stipules function as cataphylls. Both genera have cataphylls enclosing their reproductive axes. I consider the absence of cataphylls in the cycads to be derived. Cataphylls are present on the vegetative shoots of *Chigua.*

Results

Cladistic analyses, using Hennig 86, were performed using Crane's data matrix and the data matrix in Table I. Only one cladogram (Fig. 57) was produced using Crane's data matrix. This cladogram was, of course, identical to that published by Crane (1988) and was completely resolved. The analysis using the expanded data matrix in Table I produced three equally parsimonious cladograms of 41 steps, all of which are also completely resolved (Figs. 58–60).

One of these cladograms (Tree 1, Fig. 58) is nearly congruent with that of Crane (1988) and differs only in the placement of *Dioon.* Whereas, in the analysis by Crane (1988) *Dioon* is the sister taxon to the rest of the Zamiaceae, in Tree 1 of my analysis *Dioon* is the sister taxon within the Zamiaceae to a clade containing only *Encephalartos, Lepidozamia,* and *Macrozamia.* The other two cladograms (Tree 2, Fig. 59 and Tree 3, Fig. 60) differ from that produced by Crane in two ways. Both Tree 2 and Tree 3 have the clade containing *Stangeria* and *Bowenia* as the sister taxon to the clade composed only of *Ceratozamia, Microcycas,* and *Zamia.* In contrast, in Tree 1 and Crane's cladogram, the clade with *Stangeria* and *Bowenia* is the sister taxon to all other genera except *Cycas.* The placement of *Dioon* in Tree 2 and Tree 3 also differs from its placement in Crane's cladogram where *Dioon* was the sister taxon to the rest of the Zamiaceae. In Tree 2, *Dioon* is the sister taxon to all other genera except *Cycas.* In Tree 3, *Dioon* is the sister taxon to a clade composed of only *Encephalartos, Lepidozamia,* and *Macrozamia.* Thus, the placement of *Dioon* is the same as in Tree 1, but in none of these three cladograms is the placement of *Dioon* the same as in Crane's cladogram.

Having discussed the differences between the three cladograms presented here as well as their differences from that of Crane (1988), the similarities need consideration. The same five clades occur in all four cladograms (Figs. 57–60) and each one is defined by at least one synapomorphy except where noted in the cladogram by Crane (1988). These five clades are: 1. all of the cycads defined by characters 1.1–1.5; 2. *Cycas* defined by character 1.9 in my cladograms but not in Crane's; 3. all genera except *Cycas,* defined by characters 1.6–1.8 and 1.10–1.11; 3. *Stangeria* and *Bowenia* defined by characters 1.13, 1.14, 1.28, and 1.30; 4. *Encephalartos, Lepidozamia,* and *Macrozamia* defined by characters 1.26 and 1.27 (these characters were not used by Crane and the clade is defined only by homoplasies in his cladogram); and 5. *Ceratozamia, Microcycas,* and *Zamia* defined by character 1.17. Thus, these clades as defined by Crane (1988) are congruent with the consensus tree derived from the three cladograms presented here. Moreover, these clades, with the exception of item three above, are the same as found in my previous classification (Stevenson, 1985). For the purposes of the following discussion on Trees 1, 2, and 3, I shall refer to these clades using the following notation: the clade containing only *Cycas* as the *Cycas* Clade; the clade containing *Stangeria* and *Bowenia* as the *Stangeria* Clade; the clade containing *Ceratozamia, Microcycas,* and *Zamia* as the *Zamia* Clade; and the clade containing *Encephalartos, Lepidozamia,* and *Macrozamia* as the *Encephalartos* Clade.

Homoplasies occur in all the cladograms as parallelisms and reversals. The homoplasies common to all three trees are characters 1.15, 1.18, 1.19, 1.21, 1.22, 1.24, 1.25, and 1.29. Each tree has four homoplasious characters that are parallelisms. Characters 1.18, 1.21, and 1.22 are congruent parallelisms in all of the cladograms,

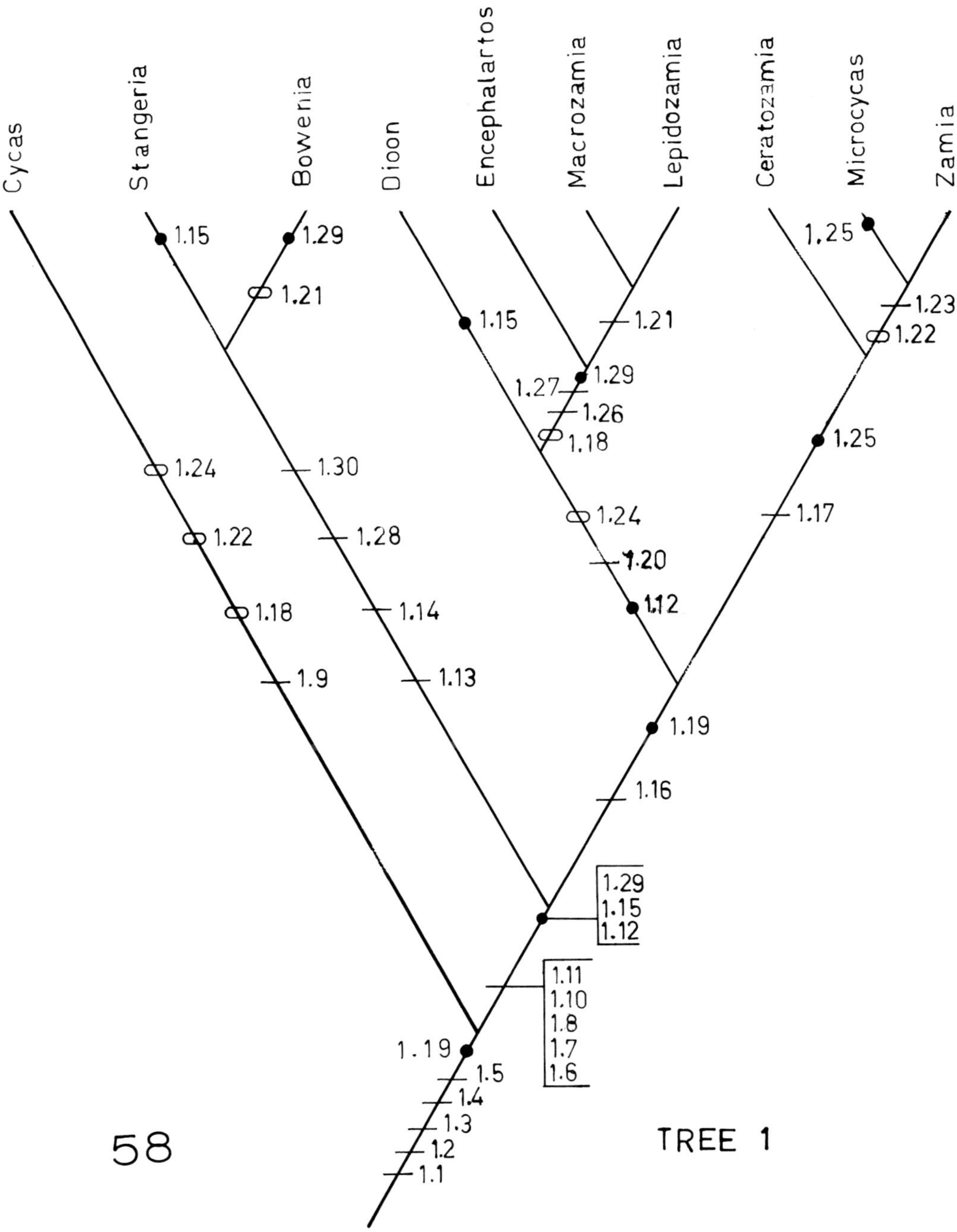

Fig. 58. One of three equally parsimonious cladograms of the Cycadales based on the data matrix in Table I. Characters exhibiting parallelism are indicated by an oval. Characters undergoing reversal in the tree are indicated by a solid circle. The provisional classification of the Cycadales is derived from this cladogram. See text for details.

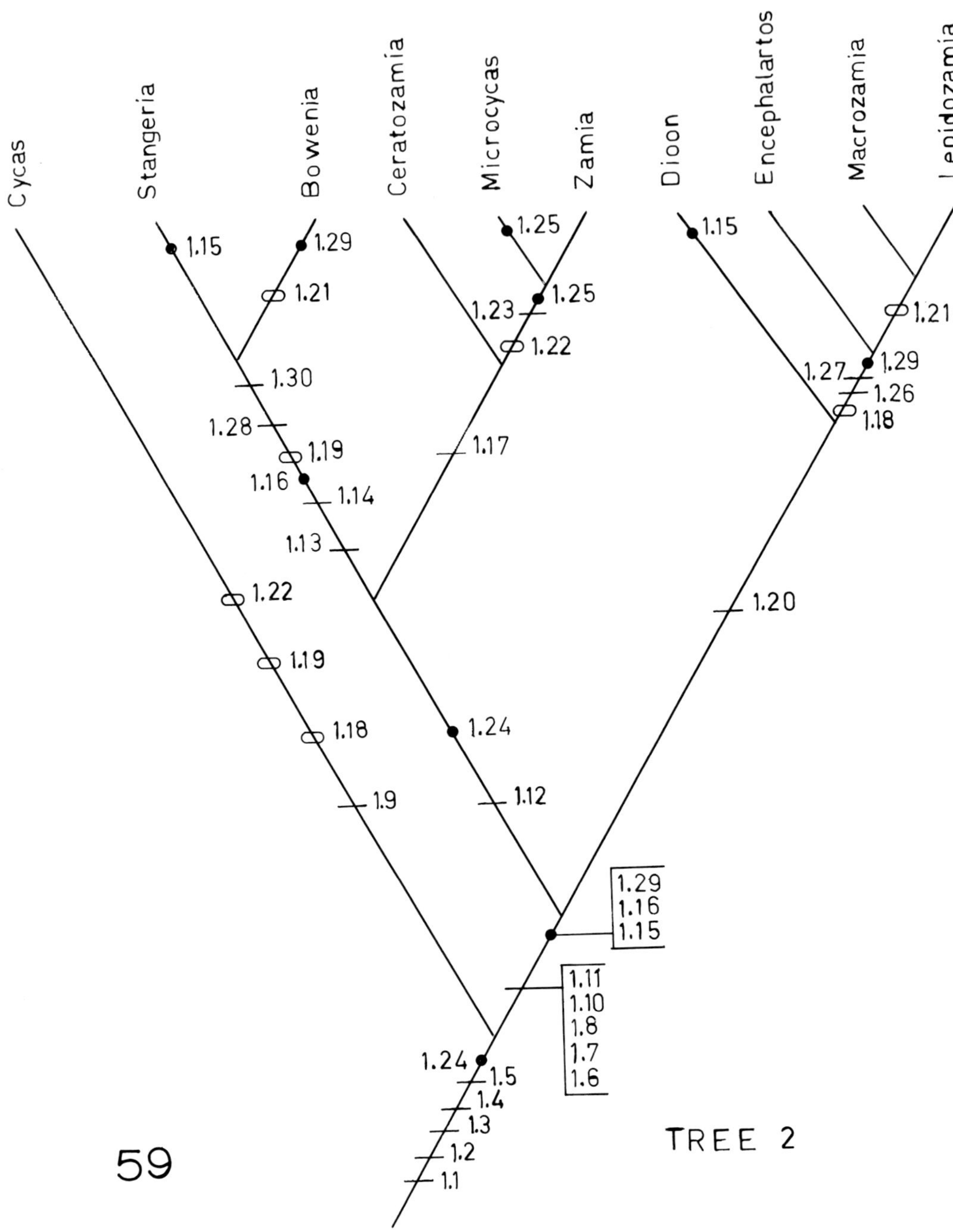

59

Fig. 59. One of three equally parsimonious cladograms of the Cycadales based on the data matrix in Table I. Characters exhibiting parallelism are indicated by an oval. Characters undergoing reversal in the tree are indicated by a solid circle. See text for details.

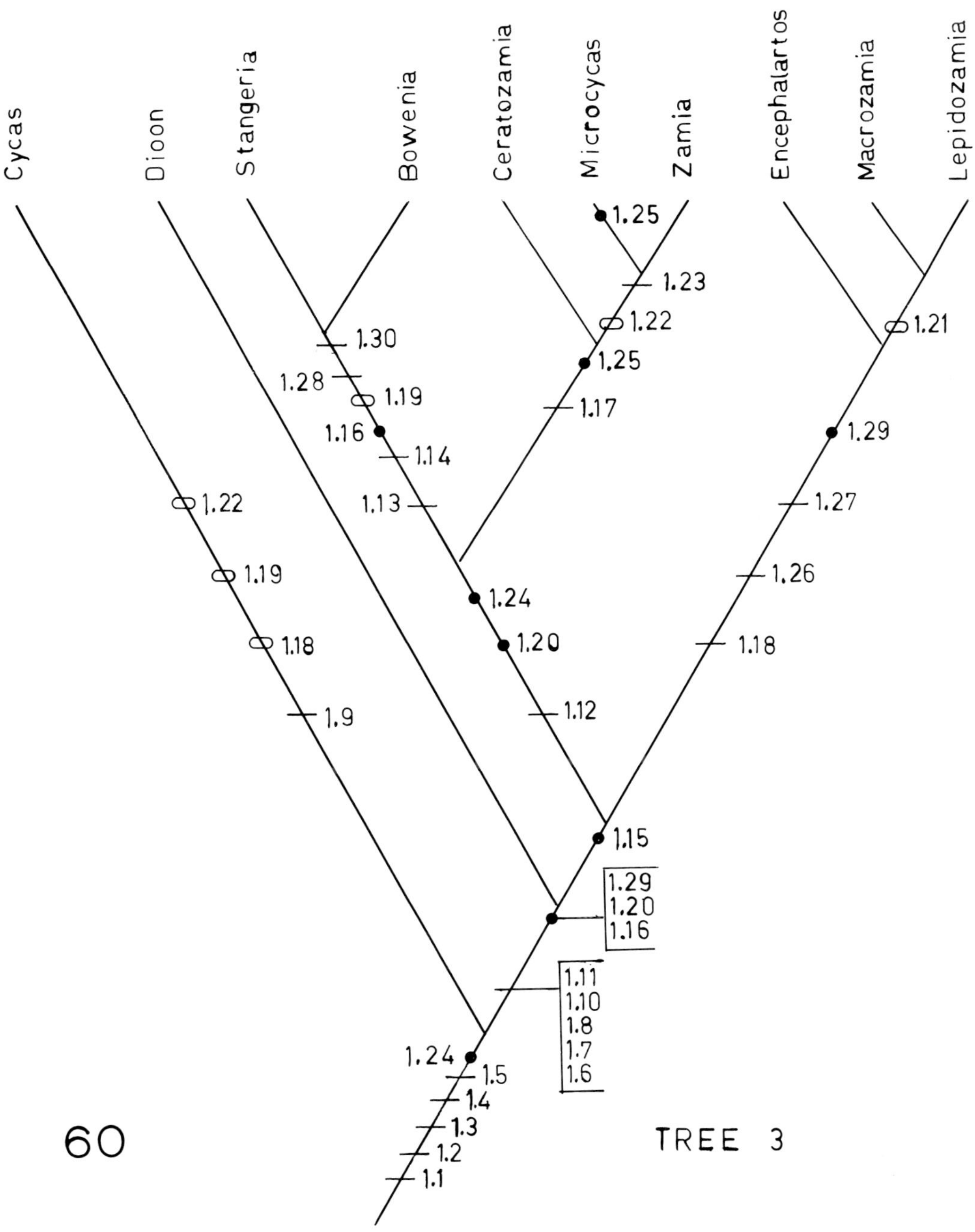

Fig. 60. One of three equally parsimonious cladograms of the Cycadales based on the data matrix in Table I. Characters exhibiting parallelism are indicated by an oval. Characters undergoing reversal in the tree are indicated by a solid circle. See text for details.

including the one by Crane (1988). Each tree has an additional parallelism, character 1.19 in Tree 1 and Tree 2 and character 1.24 in Tree 3. In Crane's cladogram, there are three additional parallelisms; characters 1.12, 1.15, and 1.20. Tree 1 and Tree 2 have five additional homoplasies that are reversals and Tree 3 has six homplasious characters that are reversals. Characters 1.15, 1.25, and 1.29 are reversals in all three trees. The latter two are congruent in all three trees and the former (1.15) is congruent in Tree 1 and Tree 2 but not with Tree 3. This concerns the placement of *Dioon* in the trees and will be discussed later. Other reversals occur in characters 1.16 and 1.24, which occur in and are congruent in Tree 2 and Tree 3; character 1.20, which occurs only in Tree 3; and characters 1.12 and 1.19, which occur only in Tree 1.

The important issue in interpreting these homoplasious characters is their occurrence regarding the placement of *Dioon* and the relationship of the *Stangeria* Clade. Thus, those homoplasious character states that are congruent in all of the trees need not be discussed because they are not germane to the placement of *Dioon* or the *Stangeria* Clade. In Tree 2 and Tree 3, where the *Stangeria* Clade is placed in a clade as the sister taxon to the *Zamia* Clade, the synapomorphic characters are characters 1.12 and 1.24 in Tree 2 and characters 1.12, 1.20, and 1.24 in Tree 3. In comparing these two trees with Tree 1, it can be seen that all of these synapomorphies are homoplasious in at least one of the three trees. I choose to interpret characters 1.12., 1.20, and 1.24 as parallelisms as they appear in Tree 1 for the following reasons.

Character 1.24, reduced spinose pinnae on rachis, is a synapomorphy for all extant cycads and a reversal in Tree 2 and Tree 3 for a clade where the *Stangeria* Clade is the sister taxon to the *Zamia* Clade. In contrast, in Tree 1, character 1.24 is a parallelism occurring in the *Cycas* Clade and in the clade containing *Dioon* as the sister taxon to the *Encephalartos* Clade. The gradual reduction of pinnae to spines in the basal portion of the rachis is a character of leaves on adult plants of *Cycas, Dioon,* and the genera of the *Encephalartos* Clade. It is not found in seedlings or juvenile plants of these genera. Thus, the character is added in the ontogeny of the individual. Those taxa that lack spinose pinnae in the basal region of the rachis as adult plants also lack them in seedling and juvenile stages. Moreover, this character never occurs teratologically. Therefore, given a choice, I do not interpret this as a character that has undergone reversal, but rather as an independently derived character in *Cycas* and a clade with *Dioon* as the sister taxon to the *Encephalartos* Clade.

Character 1.20, strobilar peduncles with cataphylls, is a synapomorphy for the clade composed of *Dioon* with the *Encephalartos* Clade as a sister group in Tree 1 and Tree 2. In contrast, this character is a reversal for the clade composed of the *Stangeria* Clade and *Zamia* Clade in Tree 3 because in this tree *Dioon* is the sister taxon to both of these clades plus the *Encephalartos* Clade. I prefer to interpret this character as a synapomorphy as in Tree 1.

Character 1.12, stipules on leaves, is a synapomorphy for the clade containing the *Stangeria* Clade and *Zamia* Clade as sister groups in Tree 2 and Tree 3. In Tree 1, this character is a synapomorphy for all genera except *Cycas* with reversal as a synapomorphy for the clade composed of *Dioon* and the *Encephalartos* Clade. As with character 1.20, character 1.12 as a reversal is contingent upon the placement of *Dioon.* In Crane's (1988) cladogram, character 1.12 is a parallelism occurring independently in the *Stangeria* Clade and the *Zamia* Clade. This again is the result of the placement of *Dioon.* It is interesting that the placement of *Dioon* in Crane's cladogram does not occur in the three trees presented here with the expanded data set. From the results, I suspect that the stipules found in the *Stangeria* Clade and the stipules found in the *Zamia* Clade may turn out not to be homologous and this is somewhat reflected in my discussion of stipules in the section on leaves. If these stipules are scored as two separate characters, one is a synapomorphy for the *Stangeria* Clade and the other a synapomorphy for the *Zamia* Clade and the results do not change.

Interpretations of characters 1.16 and 1.19 provide additional reasons for my preference of Tree 1 over the other two trees. Character 1.16, developing pinnae flat, is a reversal for the *Stangeria* Clade in Tree 2 and Tree 3, whereas in Tree 1 it is a synapomorphy for the clade containing *Dioon,* the *Zamia* Clade and the *Encephalartos* Clade, i.e., the Zamiaceae. As this character involves leaf ontogeny and histogenesis, the interpretation of it as a synapomorphy without reversal seems more tenable to me.

In Tree 1, character 1.19, pinnae with a midrib, shows reversal for the Zamiaceae. In Tree 2 and Tree 3, character 1.19 is a parallelism for the *Cycas* Clade and for the *Stangeria* Clade. An interesting feature of the presence of a midrib in the pinnae concerns the vascular pattern of cycad leaves. In all cycads, the vascular supply to a pinna is derived from a single anisotomous branch from the marginal rachis bundle (Matte, 1904). In *Cycas*, which has only a midrib, the bundle entering the pinna remains unbranched. In those genera that have an open dichotomous venation pattern in the pinnae, the single bundle undergoes repeated dichotomous branching at or near the point of pinna attachment. In *Stangeria* and *Bowenia*, the single bundle begins to undergo anisotomous branching upon entering the pinnae and thus a midrib remains. Thus, the reversal in Zamiaceae is really a derivation of the *Cycas* ground plan. A midrib also occurs in *Chigua* which is probably a member of the *Zamia* Clade as discussed below. Thus, in this sense, the dichotomous venation of pinnae is a reversal with respect to the outgroup. No doubt a change in the polarity of this character would make the issue simpler but would not change the cladogram. The inclusion of *Chigua* when all of its character states are known should clarify this.

As previously mentioned, the new genus *Chigua* was not included in the analysis because the states of 15 of the 30 characters were not known. However, a separate analysis was conducted with *Chigua* added to the data matrix as in Table I and its unknown characters entered as such. The trees obtained are not presented here, but some comments on the analysis are appropriate, in particular, for prediction. Fifteen equally parsimonious trees were obtained. In all of these trees and the consensus tree, *Chigua* was placed in the *Zamia* Clade. The three trees of the analysis performed without *Chigua* were obtained with five different arrangements for *Chigua* within the *Zamia* Clade; thus, fifteen trees. The important point is that *Chigua* is most likely a member of the *Zamia* Clade.

The provisional classification given below is based on Tree 1 (Fig. 58) of the cladistic analysis. *Chigua* is placed in the Zamieae based on the results of including it in a separate analysis based on the limited knowledge of its characters (see Table I). Because of the disparity between the three trees obtained, I have refrained from formalizing this classification at this time pending further analysis with a more expanded data set and the resolution of the polarity of some of the characters used for reasons previously discussed. I have tried throughout this paper to point to where new data are needed in order to further enhance our understanding of the morphology and systematics of the Cycadales. I am confident that certain aspects of the following classification will remain unchanged. These aspects are the recognition of the following groups: the monogeneric Cycadaceae as a monophyletic group, the rest of the Cycadales as a monophyletic group, the *Stangeria* Clade, the clade containing *Dioon* and the *Encephalartos* Clade as a monophyletic group, and the *Zamia* Clade.

Provisional Classification

Order Cycadales
- Suborder Cycadineae
  - Family Cycadaceae
    - *Cycas*
- Suborder Zamineae
  - Family Stangeriaceae
    - Subfamily Stangerioideae
      - *Stangeria*
    - Subfamily Bowenioideae
      - *Bowenia*
  - Family Zamiaceae
    - Subfamily Encephalartoideae
      - Tribe Diooeae
        - *Dioon*
      - Tribe Encephalarteae
        - Subtribe Encephalartinae
          - *Encephalartos*
        - Subtribe Macrozamiinae
          - *Lepidozamia*
          - *Macrozamia*
    - Subfamily Zamioideae
      - *Ceratozamia*
      - *Chigua*
      - *Microcycas*
      - *Zamia*

Acknowledgments

I greatly appreciate the many helpful discussions on cycads that I have had with Peter Crane, Andrew Drinnan, Knut Norstog, and the authors of the other papers in this volume. This paper has benefitted from the critical and constructive review of Rudolf Schmid, whom I thank. Without the help, encouragement, and generosity of John Popenoe, Director of the Fairchild Tropical

Gardens, I would not have been able to conduct this cycad work. Thanks for the plants! I thank Pauline Ladiges for help with the cladistics and Donald Black for the photographic plates. Special thanks to Michele Albury, who cut seemingly endless sets of serial sections of young leaves, strobili, and mega- and microsporophylls. The research presented here was supported in part by National Science Foundation Grants 7721495, 7921130, BSR-8607049, and BSR-8796279.

Literature Cited

Audran, J. & E. Masure. 1976. Precisions sur l'infrastructure de l'exine chez les Cycadales (Prespermaphytes). Pollen et Spores **18:** 5–26.

——— & ———. 1977. Contribution à la connaissance de la composition des sporodermes chez les Cycadales (Prespermaphytes). Etude en microscopie electronique à transmission (M.E.T.) et à balayage (M.E.B.). Palaeontographica **162:** 115–158.

Bierhorst, D. 1971. Morphology of vascular plants. Macmillan Co., New York.

Boureau, E. 1946. La convergence des structure et des formes dans les organes femelles des Cycadées et de Coniferes. Rev. Sci. **84:** 137–144.

Candolle, A. de. 1868. Cycadaceae. Prodromus systematis naturalis **16(2):** 522–548.

Chamberlain, C. 1919. The living cycads. University of Chicago Press, Chicago.

———. 1935. Gymnosperms: Structure and evolution. University of Chicago Press, Chicago.

Coulter, J. & M. Chrysler. 1904. Regeneration in *Zamia.* Bot. Gaz. (Crawfordsville) **38:** 452–458.

Crane, P. 1985. Phylogenetic relationships in seed plants. Cladistics **1:** 329–348.

———. 1988. Major clades and relationships in "higher" gymnosperms. Pages 218–272 *in* C. Beck (ed.), Origin and evolution of gymnosperms. Columbia University Press, New York.

Dehgan, B. & N. Dehgan. 1988. Comparative pollen morphology and taxonomic affinities in the Cycadales. Amer. J. Bot. **75:** 1501–1516.

Delevoryas, T. 1982. Perspectives on the origin of cycads and cycadeoids. Rev. Palaeobot. Palynol. **37:** 115–132.

——— **& R. Hope.** 1971. A new Triassic cycad and its phyletic implications. Postilla **150:** 1–21.

——— & ———. 1976. More evidence for a slender growth habit in Mesozoic cycadophytes. Rev. Palaeobot. Palynol. **21:** 93–100.

Dorety, H. 1908. The seedling of *Ceratozamia.* Bot.Gaz. (Crawfordsville) **46:** 203–220.

———. 1909. The extrafascicular cambium of *Ceratozamia.* Bot. Gaz. (Crawfordsville) **47:** 150–152.

Doyle, J. & M. Donoghue. 1986. Seed plant phylogeny and the origin of angiosperms: An experimental cladistic approach. Bot. Rev. **52:** 321–431.

Gifford, E. M., Jr. & A. Foster. 1989. Morphology and evolution of vascular plants. 3rd ed. W. H. Freeman and Co., New York.

Greguss, P. 1968. Xylotomy of the living cycads, with a description of their leaves and epidermis. Akademiai Kiado, Budapest.

Hooft, J. 1970. *Zamia* from seed. Carolina Tips **33:** 21–22.

Hutchinson, J. 1924. Contributions towards a phylogenetic classification of flowering plants. III. The genera of gymnosperms. Kew Bull. **1924:** 49–66.

Johnson, L. 1959. The families of cycads and the Zamiaceae of Australia. Proc. Linn. Soc. N.S.W. **84:** 64–117.

Lang, W. 1897. Studies in the development and morphology of cycadean sporangia. I. The microsporangia of *Stangeria paradoxa.* Ann. Bot. **11:** 421–438.

Matte, H. 1904. Recherches sur l'appareil libero-ligneux des Cycadacées. Mem. Soc. Linn. Normandie **22:** 1–233.

———. 1907. Note preliminaire sur germinations des Cycadacées. Assoc. Franc. Advancem. Sci. **1907:** 430–433.

Meeuse, A. D. J. 1963. The so-called "megasporophyll" of *Cycas*—A morphological misconception: Its bearing on the phylogeny and classification of the Cycadophyta. Acta Bot. Neerl. **12:** 119–128.

———. 1986. Anatomy of morphology. J. Brill, The Hague.

Miquel, F. 1861. Prodromus systematis cycadearum. C. v. D. Post, Jr. Amsterdam, C. G. v. D. Post, Utrecht.

———. 1869. Nieuwe bijdragen tot kennis der Cycadeen. Zesde gedeelte. Versl. Med. Kon. Akad. afd. Natuurk. ser. 2, **4:** 23–37.

Niklas, K. & K. Norstog. 1984. Aerodynamics and pollen grain deposition patterns on cycad megastrobili: Implications on the reproduction of three cycad genera (*Cycas, Dioon,* and *Zamia*). Bot. Gaz. (Crawfordsville) **145:** 92–104.

Norstog, K., D. Stevenson & K. Niklas. 1986. The role of beetles in the pollination of *Zamia furfuracea* L. fil. (Zamiaceae). Biotropica **18:** 300–306.

Pant, D. 1973. *Cycas* and the Cycadales. 2nd ed. Central Book Depot, Allahabad.

Petriella, B. & J. Crisci. 1975. Estudios numericos en Cycadales. 1. Cycadales actuales: Sistimica. Bol. Soc. Argent. Bot. **16:** 231–247.

——— & ———. 1977. Estudios numericos en Cycadales. 2. Cycadales actuales: Simulacion de arboles evolutivos. Obra Mus. La Plata **3:** 151–159.

Pilger, R. 1926. Cycadaceae. *In* A. Engler (ed.), Die naturlichen Pflanzenfamilien, 2. Aufl. **13:** 44–82. Engelmann, Leipzig.

Read, R. & M. Stolt. 1986. Bibliography of the living cycads (annotated). Lyonia **2:** 33–200.

Reichenbach, H. G. L. 1828. Conspectus regni vegetabilis. Arnold, Leipzig.

Rudall, P. 1989. Stem thickening growth in bulbous Iridaceae. Bot. Gaz. (Crawfordsville) **150:** 132–138.

Schuster, J. 1932. Cycadaceae. *In* A. Engler (ed.), Das Pflanzenreich **99(IV. 1):** 1–168.

Singh, H. 1978. Embryology of gymnosperms. Gebruder Borntraeger, Berlin.

Siniscalco Gigliano, G. 1990. Chemotaxonomic significance of MAM glycosides and mucilages in cycads. Mem. N.Y.B.G. **57:** 123–131.

Smith, F. 1907. Morphology of the trunk and development of the microsporangium of cycads. Bot. Gaz. (Crawfordsville) **43:** 187–204.

Stevenson, D. 1976. Shoot apex organization and ontogeny of the rhizome borne roots and their associated gaps in *Dennstaedtia cicutaria.* Amer. J. Bot. **63:** 673–678.

———. 1978. Observations on the shoot apices of eusporangiate ferns. Kew Bull. **33:** 279–282.

———. 1980a. Radial growth in the Cycadales. Amer. J. Bot. **67:** 465–475.

———. 1980b. Radial growth in *Beaucarnea recurvata.* Amer. J. Bot. **67:** 476–489.

———. 1980c. Observations on stem and root contraction in cycads (Cycadales) with special reference to *Zamia pumila* L. J. Linn. Soc., Bot. **81:** 275–281.

———. 1981. Observations on ptyxis, phenology, and trichomes in the Cycadales and their systematic implications. Amer. J. Bot. **68:** 1104–1114.

———. 1985. A proposed classification of the Cycadales. Amer. J. Bot. **72:** 971–972. (Abstract.)

———. 1988. Strobilar ontogeny in the Cycadales. Pages 205–244 *in* P. Leins, S. Tucker & P. Endress (eds.), Aspects of floral development. J. Cramer, Berlin.

———. 1990. *Chigua,* a new genus in the Zamiaceae with comments on its biogeographical significance. Mem. N.Y.B.G. **57:** 169–172.

———**, S. Sabato & M. Vazquez Torres.** 1986. A new species of *Ceratozamia* (Zamiaceae) from Mexico, with comments on vegetative morphology, habitats, and species distribution in *Ceratozamia.* Brittonia **38:** 17–26.

Tang, W. 1987. Insect pollination in the cycad *Zamia pumila* (Zamiaceae). Amer. J. Bot. **74:** 90–99.

Worsdell, W. 1896. Anatomy of the stem of *Macrozamia* compared with that of other genera of Cycadaceae. Ann. Bot. **10:** 601–620.

———. 1898. The vascular structure of the sporophylls of the Cycadaceae. Ann. Bot. **12:** 203–241.

———. 1900. The anatomical structure of *Bowenia spectabilis.* Ann. Bot. **14:** 159–160.

———. 1901. Contributions to the comparative anatomy of the Cycadaceae. Trans. Linn. Soc. London, ser. 2, **6:** 109–121.

———. 1906. The structure and origin of the Cycadaceae. Ann. Bot. **20:** 129–155.

Occurrence of Non-Coralloid Aerial Roots in *Cycas*

D. D. PANT AND KABITA DAS

Abstract

PANT, D. D. and K. DAS (Dept. of Botany, Allahabad University, Allahabad-211002, India). Occurrence of non-coralloid aerial roots in *Cycas*. Memoirs of the New York Botanical Garden **57**: 56–62. 1990.—Occurrence of non-coralloid aerial roots is reported for the first time in plants of *Cycas circinalis* L., *C. revoluta* Thunb. and *C. rumphii* Miq. Adventitious roots grow out from the lower side of young or old bulbils or leaf bases and some of them travel downwards around the columnar trunks. During their downward aerial course they may even produce one or more branches. The roots are 3- to 8-arch in *C. circinalis*, 2- to 5-arch in *C. revoluta*, and 2- to 7-arch in *C. rumphii*, and they often show a well developed parenchymatous pith with scattered xylem elements. Older roots show a good amount of secondary growth. Maceration of the aerial roots of *C. rumphii* shows unusually branched xylem elements, some of which have large pores like those of vessels.

Key words: *Cycas*; aerial roots; bulbils; xylem elements.

Introduction

In their account of the aerial roots of diverse plants, Gill and Tomlinson (1975) have mentioned the occurrence of coralloid roots in cycads which are partly aerial and partly subterranean. But they have not mentioned any non-coralloid aerial roots among the cycads although roots which could be termed aerial had already been mentioned as occurring in the bulbils of *Cycas* (Pant, 1962). These roots are adventitious and they are usually attached on the lower sides of bulbils while the latter are still attached to the aerial parts of the columnar trunks of *Cycas* trees. Since no account of the form and structure of such aerial roots is available, we decided to study them in some detail.

Material and Methods

Aerial roots of *Cycas circinalis* L., *C. revoluta* Thunb., and *C. rumphii* Miq. were observed in plants growing locally, particularly in the Roxburgh Botanical Garden of the University of Allahabad. Some of these roots were collected and fixed in F.A.A. Sections of thicker, and harder roots were microtomed at 8 to 11 μm after being embedded in wax. Sections were usually stained in a safranin and fast green combination and mounted in Canada balsam. Pieces of roots were also macerated in Jeffrey fluid for observing individual xylem elements and mounted in Canada balsam after staining in safranin.

Observations

The yellowish brown aerial roots are attached to leaf bases on the lower sides of young and old bulbils (Fig. 1B). These bulbils can become the branches of a tree if allowed to grow undisturbed. Occasionally a few of these roots may run almost parallel and vertically down the tree trunk between the leaf bases just outside the cortex (Figs. 1C, 3A, C). The roots are all adventitious. They

Fig. 1. **A.** *Cycas revoluta,* roots arising from a leaf base, ×0.2. **B.** *C. rumphii,* aerial roots (whitish) arising from old bulbils, ×0.3. **C.** *C. circinalis,* a number of downwardly travelling parallel aerial roots (whitish) exposed by removing the leaf base armour, ×0.4. **D.** *C. circinalis,* a downwardly travelling aerial root attached on lower side of a leaf base, ×0.5. **E.** *C. revoluta,* transverse section of an aerial root showing triarch condition with a wide pith and a well-defined endodermis, ×50.

may be up to 25 mm or more in thickness and may attain a length of 23 cm or more. Some of them are unbranched but others show a number of secondary or tertiary branches. Their apices are usually blunt and obtuse. Occasionally they may also have root hairs.

Transverse sections of the roots show that their cortex is wide (more than half of their radius), and their stele is di- to polyarch—up to 8-arch in *C. circinalis,* up to 5-arch in *C. revoluta,* and up to 7-arch in *C. rumphii* (see Figs. 1E, 2A, 3D–F, 4A, B). The center of the stele is usually occupied by a mixed, parenchymatous pith with scattered xylem elements among parenchyma cells (Fig. 2A) although some metaxylem strands may be joined to each other (Fig. 4A). The secondary xylem of older aerial roots shows radially seriated xylem elements, with numerous parenchymatous secondary rays between the rows of xylem cells. The secondary xylem cell is thus typically manoxylic (Fig. 4B).

Maceration of the roots of *C. rumphii* yielded xylem elements which are unevenly thick, being narrower in some places and wider at other points or they may be equally or unequally forked (Figs. 2B, 4E). Some of these xylem elements are obviously tracheids, but there are others which show a large open pore in the wall which may have been continuous with an adjacent xylem element (Fig. 4A). It is likely, that when such elements were joined to each other in a series, they formed branched vessels (Figs. 2B–F, 4C–E), and some of the vessels may have been unevenly wide. Consequently, after observing such vessel elements in maceration residues, we made a careful search for them *in situ.* However, we failed to find any definite intercommunicating vessel elements in sections of roots. Thus, the issue remains unresolved.

The phloem consists of thin-walled cells. Cells in its outer periphery generally contain tannin.

A multilayered parenchymatous pericycle, about six cells thick, surrounds the conducting tissue, and is in turn encircled by a single layer of clearly differentiated endodermis with Casparian thickenings in the radial walls of its cells (Fig. 1E). Sections passing through points where a root is branching show that the branches arise by proliferation of pericycle cells opposite the protoxylem poles, as is usual with branch roots.

The parenchymatous cortex appears to be bounded by a multilayered peripheral zone of radially seriated, suberized dead cells forming the phellem. The phellem cells may occasionally be disorganized and its periphery commonly shows lenticels (Fig. 3G). Inside the phellem lies a phellogen which produces a barely distinguishable phelloderm that merges with the cortex. The phelloderm and cortex show similar rounded parenchymatous cells. Together they may form a zone about 60 cells wide whose elements gradually increase in size centrifugally, so that the largest cells are adjacent to the phellogen. Frequently, the cortex also shows darkly colored tannin cells and crystalliferous cells.

Discussion

Occurrence of aerial roots is fairly common in angiosperms (Gill & Tomlinson, 1975). A few living and fossil tree ferns, e.g., *Cyathea, Psaronius, Tempskya,* as well as some fossil gymnosperms, e.g., *Lyginopteris* and *Medullosa* (Taylor, 1981), have aerial roots. However, the aerial roots in many of the above-mentioned living and fossil genera (except *Lyginopteris*) form a mantle of roots around the erect stem. The presently investigated aerial roots of *Cycas* similarly run almost parallel and vertically downwards between the leaf bases. Thus, they resemble similar roots in tree ferns and other forms, but they do not form anything like a mantle. However, the production of aerial roots by the trunks of *Cycas circinalis, C. revoluta,* and *C. rumphii* adds one more character to the fern-like features of *Cycas* which include a persistent armor of leaf bases, circinate vernation of pinnae and young leaf, multipinnate fronds, ramentum, grouping of microsporangia in sori on the lower surface of microsporophylls, and motile sperms.

The aerial roots of *Cycas* are comparable with coralloid roots of cycads, which are only partly

→

Fig. 2. **A.** *Cycas revoluta,* transverse section of a pentarch root having a mixed pith, ×50. **B.** *C. rumphii,* a branched xylem (vessel) element showing a pore, ×200. **C.** *C. rumphii,* unequally branched xylem element, ×200. **D.** *C. rumphii,* two joined xylem elements with a pore between them, ×200. **E, F.** *C. rumphii,* two unbranched xylem elements each with a large pore, ×200.

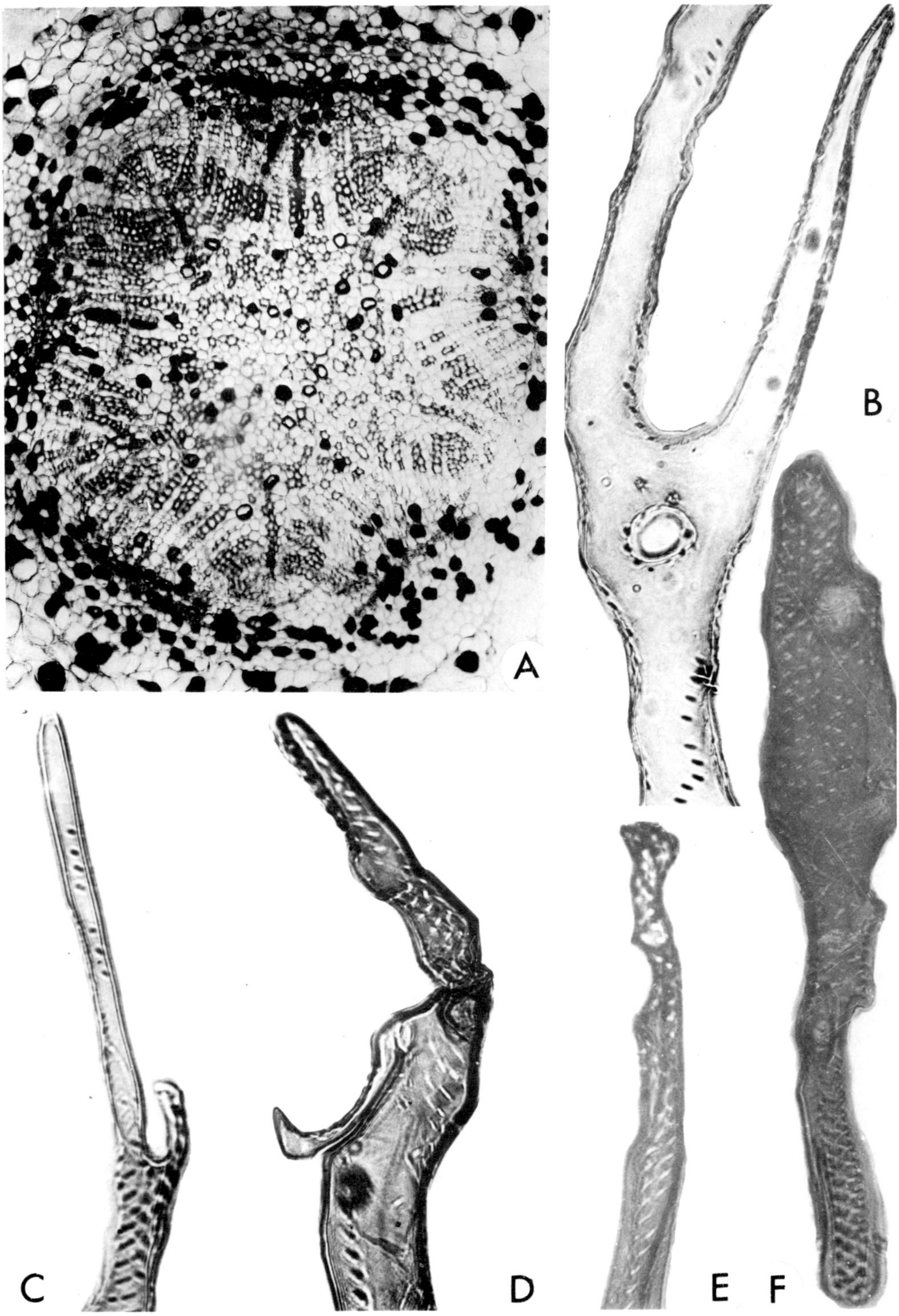
A
B
C
D
E
F

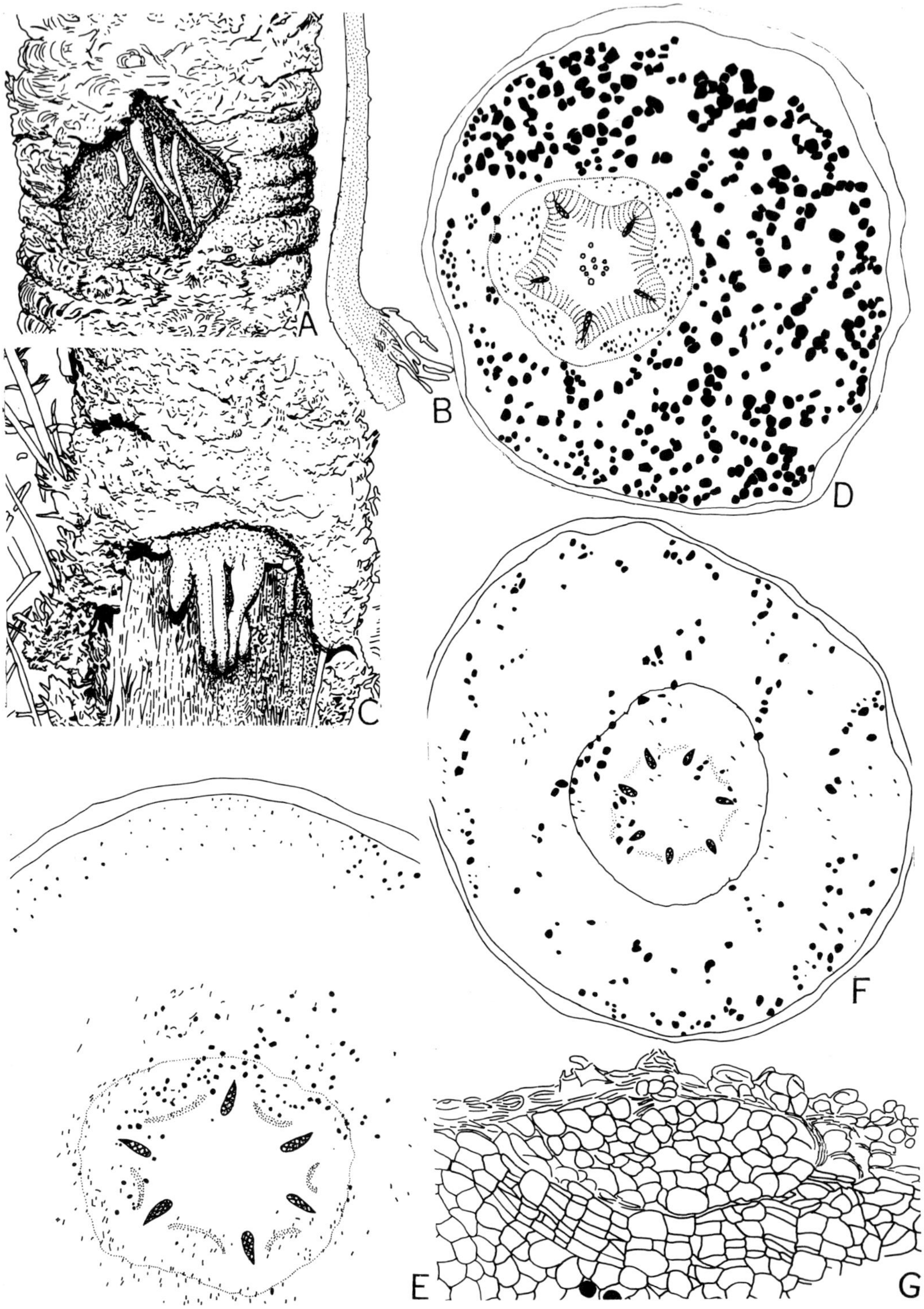

Fig. 3. **A–G. A–C.** A number of downwardly travelling parallel aerial roots exposed by removing the leaf base armours of *C. rumphii,* ×0.25, and *C. circinalis* (see also Fig. 1C), ×0.3, respectively. **B.** *C. revoluta,* ×0.45, roots coming out of a leaf base (see also Fig. 1A). **D–F.** Transverse sections showing 5- to 7-arch steles. **D.** *C. revoluta,* ×15. **E, F.** *C. rumphii,* ×10. **G.** *C. circinalis,* transverse section of root passing through a lenticel, ×100.

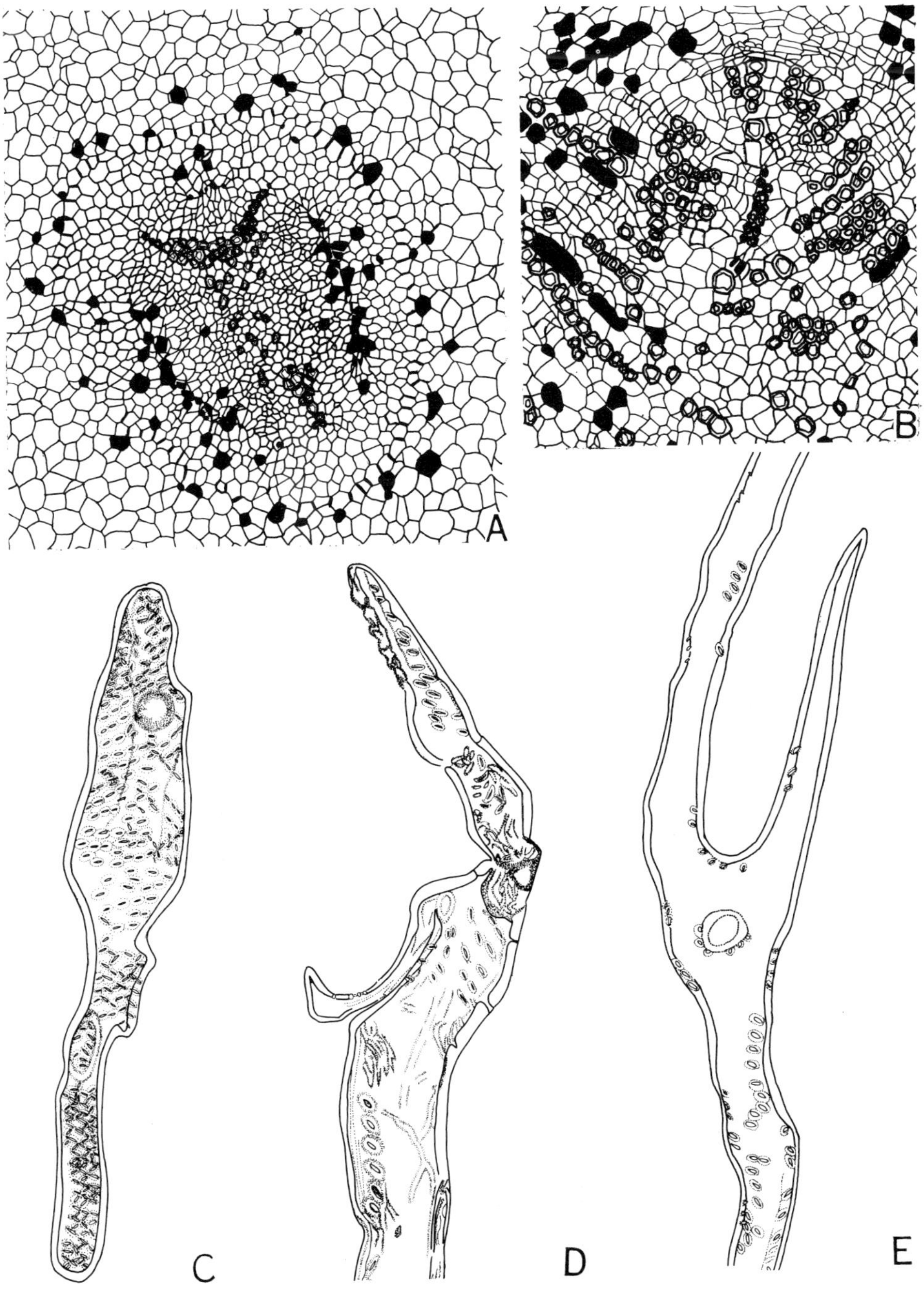

Fig. 4. **A–E.** **A.** *C. rumphii,* transverse section showing jointed metaxylem element, ×60. **B.** *C. revoluta,* transverse section of roots showing manoxylic condition of stele (see also Fig. 2A), ×75. **C.** *C. rumphii,* unbranched xylem element with simple pore (see also Fig. 2F), ×145. **D.** *C. rumphii,* two jointed xylem elements with a pore between them (see also Fig. 2D), ×200. **E.** *C. rumphii,* a branched xylem (vessel) element with a large pore (see also Fig. 2B), ×130.

Table I

Character states of the three types of roots occurring in *Cycas*

	Aerial roots	Coralloid roots	Normal roots
Geotropism	Positive	Negative	Positive
Cortex	Wide	Wide	Narrow
Algal zone	Absent	Present	Absent
Protoxylem poles	Di- to polyarch	Di- to triarch	Di- to tetrarch, rarely up to octarch (Atwood, 1936)
Pith	Well defined	Small or absent	Small or absent
Secondary tissues	Usually present	Absent or poorly developed	Well developed

aerial, in that both have a wide cortex. On the other hand, they differ in that they lack an algal zone, are often polyarch, and show a fair amount of secondary growth, whereas the coralloid roots are normally 2–3-arch and they show very little secondary growth or none at all. Table I summarizes the characters of these three kinds of roots.

The presence of vessels in the aerial roots of *Cycas rumphii* supports a previous report by Greguss (1958) concerning the occurrence of vessels in macerated xylem of the trunks of *Cycas revoluta* and *Zamia skinneri,* although the photographs given by Greguss (his fig. 1) do not show pores between vessel elements. In the present case the obvious occurrence of large rounded apertures in xylem elements is a clear indication of their being vessels, although we could not observe any *in situ* vessels in sections. Nevertheless, some of our putative vessel elements are remarkable and unique in being branched and in having an uneven width.

Another peculiarity of the primary xylem of the aerial roots is the presence of scattered xylem elements in their pith. Scott (1902) had suggested that the primitive metaxylem was all centripetal whereas centrifugal metaxylem was advanced and came later. Admittedly, a transitional stage in the evolution of medullated xylem is the occurrence of mixed pith with scattered centripetal metaxylem elements intermixed with parenchyma like that found in some fossil and living forms. It is generally considered that the roots of all plants, including many advanced angiosperms, are conservative and in most cases they have continued to retain the primitive state without a parenchymatous pith. The aerial roots of the presently investigated species of *Cycas* would seem to be transitional in having a mixed pith.

Acknowledgments

The authors are grateful to Drs. D. D. Nautiyal, D. R. Misra, N. Bhowmik, and Rekha Pant for their manifold help. Thanks are also due to the Department of Science and Technology, Government of India, for providing a Senior Research Fellowship to the junior author (K.D.).

Literature Cited

Atwood, S. 1936. The anomalous root structure of *Cycas revoluta.* Amer. J. Bot. **23:** 336–340.

Gill, A. M. & P. B. Tomlinson. 1975. Aerial roots: An array of forms and function. Pages 237–260 *in* Third Cabot Symposium, The development and function of roots.

Greguss, P. 1958. Some recent data on the xylotomy of *Cycas, Zamia* and *Ginkgo.* Acta. Biol., n.s. **4(3, 4):** 143–147.

Pant, D. D. 1962. Cycas and the Cycadales. Central Book Depot, Allahabad.

Scott, D. H. 1902. The old wood and the new. New Phytol. **1:** 25–30.

Taylor, T. N. 1981. An introduction to fossil plant biology. McGraw-Hill Book Company, London.

MARCH 1990 MEMOIRS OF THE NEW YORK BOTANICAL GARDEN 57: 63–81

Studies of Cycad Reproduction at Fairchild Tropical Garden

KNUT J. NORSTOG

Table of Contents

Abstract

NORSTOG, K. (Fairchild Tropical Garden, 11935 Old Cutler Road, Miami, Florida 33156, U.S.A.). Studies of cycad reproduction at Fairchild Tropical Garden. Memoirs of the New York Botanical Garden **57:** 63–81. 1990.—Fairchild Tropical Garden is located in a region of subtropical climate where occasional frosts inhibit growth of some tender tropical plants. However, all genera and species of cycads can be grown out-of-doors the year around.

A major problem in propagation of cycads at FTG is production of fertile seeds of a majority of exotic cycads. It is thought that this is due to absence of specific insect pollen vectors. With appropriate methods of artificial pollination, all species can be sexually propagated and certain hybrids can be produced. One such hybridization, *Zamia furfuracea* × *Z. spartea,* is described.

Studies on the role of wind and insects in pollination of cycads have been done at FTG and suggest a dual function for wind and insects—wind conveys pollen to metastrobili and insects further transport the pollen to the ovules. These studies suggest that cycad metastrobili are the functional equivalents of angiosperm carpels and further suggest an ancient origin for plant–insect coevolution.

Cycads produce multiflagellated male gametes. These differ from those of ferns, lycopods, and mosses mainly in redundancy of organelles and imply certain phylogenetic lines of evolution.

Resumé

Le jardin tropical de Fairchild (Fairchild Tropical Garden) est situé dans une région à climat subtropical où des gelées occasionnelles inhibent la croissance des plantes tropicales les plus fragiles. Cependant tous les genres ainsi que toutes les espèces de Cycadacées peuvent être cultivés a l'extérieur durant toute l'année.

La propagation des Cycadacées au jardin tropical de Fairchild est une problème essentiel qui ne peut être résolu qu'en assurant la production de graines fertiles par l'ensemble des Cycadacées exotiques. La difficulté pour obtenir des graines proviendrait de l'absence d'insecte spécifique qui serait le vecteur du pollen. Grâce à l'utilisation de certaines techniques particulières de pollinisation artificielle, il est possible de propager sexuellement l'ensemble des espèces et d'obtenir certains hybrides. C'est ainsi que l'on a pu obtenir l'hybride, *Zamia furfuracea* × *Z. spartea.*

Des études sur le rôle du vent et des insectes ont été entreprises au jardin tropical de Fairchild et elles suggèrent une fonction double du vent et des insectes. Le vent transporte le pollen aux mégastrobiles et le transport ultérieur vers les ovules est assuré par les insectes. Ces études indiquent que les mégastrobiles des Cycadacées ont fonction équivallente aux carpelles d'Angiospermes et suggèrent une origine commune ancienne pour une co-evolution plante-insecte.

Les Cycadacées produisent des gamètes mâles pluriciliés. Ceux-ci diffèrent de ceux des Fougères, des Lycopodes et des Mousses essentiellement par la répétition des mêmes organites et ceci implique la présence de certaines lignes d'évolution phylogénétique particulières.

Key words: Cycads; reproduction; pollination; fertilization; hybridization; spermatogenesis; spermatozoids.

Introduction

Fairchild Tropical Garden (FTG) maintains a large collection of living cycads. They include about 110 species representing all of the extant cycad genera, and many of these exist as small, breeding populations. The collection therefore is a valuable scientific resource in that it consists of a large assemblage of well-documented cycad germ plasm. The Garden's cycad collection is located at two arboreta: one being the public garden, which features tropical and subtropical vegetation and covers about 30 ha (Fig. 1A, B), and the Montgomery Foundation, a research garden of about 25 ha (Fig. 2A). The latter contains most of the research collection of cycads (Figs. 2B, 3–5), but non-destructive cycad studies are permitted at the public garden.

Fairchild Tropical Garden was founded 50 years ago by Robert Montgomery, a wealthy businessman and scholar, who was keenly interested in palms and cycads. Montgomery collected many unusual and rare specimens which he maintained on his estate in Miami (Fig. 2A). Eventually, fearing for the future of his plant collection, he founded a botanical garden, which he named for his friend David Fairchild, onetime Director of the United States Bureau of Plant Introduction. The new garden was stocked with Montgomery's own specimens, but these have since been greatly augmented through numerous worldwide collecting expeditions.

Colonel Montgomery's collections were for the most part single specimens and sometimes their origins were not well documented; however, they remain as both a useful and ornamental nucleus of the present cycad and palm collections (Figs. 1, 2A). Ultimately, the Montgomery Estate was given over to management by Fairchild Tropical Garden and as the Montgomery Foundation it

Fig. 1. Fairchild Tropical Garden. **A.** Entrance with view of palms and cycads in background. The specimens are among the contributions of Robert Montgomery to the garden's collections. **B.** The public garden's "rain-forest" area in which several genera of cycads are featured. *Cycas circinalis* at right and tree ferns at left.

Fig. 2. Montgomery Foundation of Fairchild Tropical Garden. **A.** Some of Robert Montgomery's original cycad collections are included in this view; a very large male specimen of *Microcycas* may be seen at the center of the photo, just to the left of a large *Phoenix*. **B.** William J. Robbins Research Center of the Montgomery Foundation. Note the colony of *Zamia furfuracea* at the base of the building. These plants, numbering some 40 individuals, are employed in pollination studies.

now functions as the Garden's plant research center (Fig. 2B). While much of the Montgomery Foundation grounds has been extensively landscaped, it is of interest that about 5 ha still are covered with the original pine-palmetto vegetation (Fig. 3A) and this area contains several hundred wild plants of the little Florida cycad, *Zamia pumila* (Fig. 4).

Fairchild Tropical Garden and its Cycad Collection

Because of its interest in the total biology of cycads, Fairchild Tropical Garden now places great emphasis on the establishment of breeding colonies of cycad species (Figs. 2B, 3B). Initially, for example, the *Microcycas* collection consisted of only two specimens, a female and a male (Fig. 2A). Unfortunately, the female plant died, but about 25 years ago three small plants of unknown sex were obtained in Cuba. One of these turned out to be a female (Fig. 5) and now has been pollinated several times (female cones of this species are produced in alternate years). The resulting seeds and seedlings, numbering more than 800, have been widely distributed and testify to the important role botanical gardens can play in the preservation of rare and endangered species of plants. In addition, the *Microcycas* collection is also serving as a source of material for an electron-microscopic study of spermatogenesis (Figs. 8, 9). Another colony of 40 individuals of *Zamia furfuracea* (Fig. 2B) has served in an experimental study of insect pollination of cycad cones (Norstog et al., 1986). Ideally, such colonies should consist of at least 20 individuals, evenly divided between males and females. In practice, however, this approach often is an impossibility due to the rarity of certain species. Nevertheless, we think that the concept that cycad specimens should first and foremost constitute an information source and secondarily function as a germ plasm pool is a valid one. The use of Fairchild Tropical Garden's collections of cycads in past and current research will now be described.

Reproductive Structures and Processes of Cycads

It may be helpful to the general reader if reproductive structures and processes of cycads are first briefly described.

All cycads are dioecious, having separate male and female individuals, and, with the exception of *Cycas,* these bear, respectively, male and female strobili or cones (Figs. 5, 6A). Males of all cycads bear pollen-producing strobili, commonly called pollen cones. The cones are composed of spirally arranged pollen organs called microsporophylls. These, in turn, bear pollen sacs or microsporangia. Most cycads produce copious quantities of pollen, which tends to be somewhat sticky. Females of all cycad genera except *Cycas* bear seed-producing strobili, or seed cones. The strobili are composed of ovule-bearing organs called megasporophylls (after pollination and fertilization ovules develop into seeds). In females of *Cycas,* however, the megasporophylls are more loosely arranged around the growing point of the stem (i.e., the apical meristem of the stem), and there is no true seed cone (Niklas & Norstog, 1984, refer to these as "pseudocones").

Upon transfer of pollen from the pollen sacs of the male cycad to the cone of the female and its ovules, pollen tubes slowly develop in the tissues of the ovule (Fig. 10). Then, about 5–6 months after pollination, large multiflagellated male sex cells (spermatozoids) are formed (Fig. 7); these swim to the egg cells in the ovule and fertilization is accomplished. Subsequently, the fertilized egg becomes an embryo and the ovule is transformed into a ripe seed.

State of Research into Cycad Biology

Although the reproductive processes of cycads have been understood in general for nearly a century, there are still enormous gaps in our knowledge. Most of the detailed study of cycad reproduction has been done just on *Cycas revoluta* and *Zamia pumila,* especially the latter. The reason for this emphasis is that only these two species occur in regions easily accessible to botanists who wish to observe complete reproductive cycles. One can understand the difficulty encountered in obtaining large quantities of living study material from cycads occurring in the wilds of Central and South America or remote regions of Asia. Thus we know almost nothing about pollination mechanisms, reproductive processes, or the genetics of some 98% of the species.

Because cycads are rare, slow-growing, and often inaccessible in the wild, it is obvious that botanical gardens must develop the capability to

A
B

Fig. 4. *Zamia pumila* in its natural state growing in the pine-palmetto woodland shown in Figure 3A.

provide research specimens if we are to more fully understand these uniquely interesting plants.

Pollination in Cycads

Most gymnospermous plants are monoecious—male and female organs are present on the same individual. So too are a majority of flowering plants. Since monoecy is so commonplace in seed plants, one wonders why cycads have retained or evolved a reproductive system based on dioecy. We do not know whether the seed ferns also were dioecious; if they were it may explain why cycads, the most direct descendents of seed ferns known to us, are dioecious, but it does not, of course, tell us why systems based on dioecy evolved in the first place.

The Significance of Dioecy

All plants, whether dioecious or monoecious, have evolved processes that tend to prevent self-fertilization and promote outcrossing. Dioecy is possibly the simplest solution to the problem of self-fertilization, but monoecious plants have evolved various strategems to reduce the incidence of self-fertilization and consequent inbreeding. These include the maturation of sex organs at different times, as occurs in a number of fern gametophytes, placement of female strobili at the top of a tree and the male strobili at the bottom, as in conifers, or morphological, anatomical, and sometimes complex biochemical incompatibility systems, which are found to suppress self-pollination and fertilization in many flowers (Faegri & van der Pijl, 1971).

←

Fig. 3. Research areas at the Montgomery Foundation. **A.** Longleaf pine (*Pinus elliotii*) and saw palmetto (*Serenoa repens*) are co-dominant species in a natural forest association in which *Zamia pumila* commonly occurs. **B.** A recently opened area in which small populations of exotic cycads are being planted. *Microcycas* in the foreground; *Dioon mejiae* in the background.

Fig. 5. A 30-year-old female specimen of *Microcycas calocoma.* This plant is the source of more than 800 seedling plants distributed to cycad collectors and botanical gardens the world over.

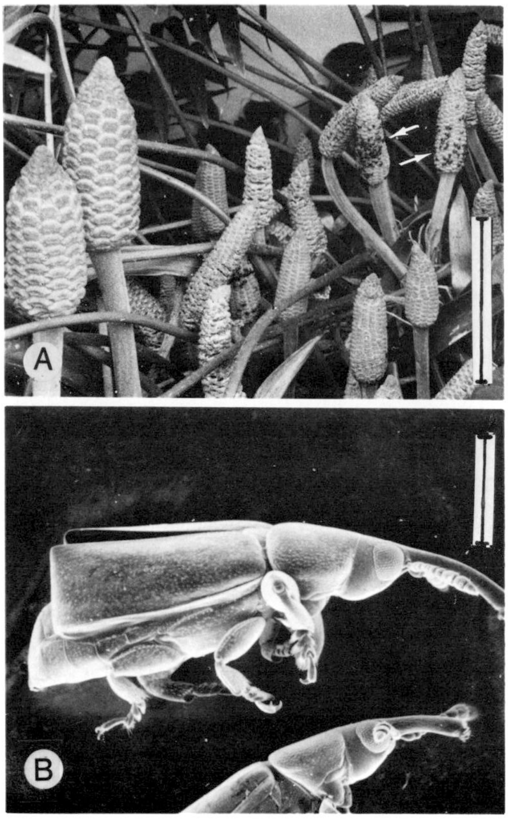

Fig. 6. Pollination of *Zamia furfuracea.* **A.** Female (left) and male (right) cones at time of pollination. Copulating weevils (*Rhopalotria mollis*) are indicated by arrows. Scale bar = 10 cm. **B.** The pollinator, *R. mollis* (photo courtesy of George Taylor). Scale bar = 0.5 mm.

Dioecy, though an effective mode of ensuring out-crossing, has certain disadvantages. For one thing, only the resources of female plants will be directly allocated to the next generation. Therefore, if the sex ratio is approximately 1:1, then half the population is non-productive of seeds and seedlings. This no doubt places a disproportionate energy load on female individuals of a cycad population and results in a net lowering of reproductive efficiency. In this context, Newell (1985) suggests that male and female individuals of *Zamia* in a wild population in Puerto Rico may differ in their within-plant allocation of resources, but notes that this hypothesis is untested. (In contrast, despite selfing barriers, all individuals of a monoecious seed plant population can be productive of seeds and seedlings.) Another problem with dioecy is that males and females may be so widely separated that the pollination of a single isolated female plant may not occur, especially if the pollination is anemophilous. That being the case, it is perhaps not surprising to find that some and perhaps most, cycads have developed coevolutionary insect-pollination systems, such as will be described.

Both problems outlined above may be related to sex ratios in wild cycad populations, of which, unfortunately, little is known. Net reproductive efficiency will be reduced if functional sex ratios are skewed in the direction of male individuals, as seems to be the case since, in their natural setting, female cycads produce cones less frequently than do male individuals. However, this reduction in numbers of coning females might be offset if pollination efficiency were to be increased as a result of relative abundance of male plants in the vicinity of female plants.

Sex Ratios

Although data on cycad sex ratios are sparse, it is possible that actual ratios may approach 1:1. Clark and Clark (N.D.) report low reproductive productivities for *Zamia skinneri* populations in primary and secondary rain forest habitats, with females producing cones less frequently than males. This differential in coning frequencies produces male-biased, observed sex ratios. Nevertheless, the authors suggest that the assumed sex ratio may approximate 1:1. Tang (1985) surveyed a population of *Zamia pumila* in a pine-palmetto habitat of approximately 3600 m^2 (a portion of this area is shown in Fig. 3). He identified 369 individuals having a spacing density of 9.8 individuals per 100 m^2. Of these, 142 were male, 82 were female, and 145 were non-coning. The coning sex ratio was 0.37 (female/male). Among the 82 females, 24 (0.29) had coned the previous year but not in the current year, 3 (0.04) had coned both years, and 55 (0.67) had coned in the current year only. Disregarding the very small percentage of females that had produced cones in successive years, it can be assumed from Tang's data that the maximum coning frequency per female plant in this population is probably on the order of one coning per 3 years. Thus, if one assumes that ⅔ of the reproductively mature females had produced cones during the sampling period, an additional 27 of the non-coning plants in the population are reproductive females. The total number of coning females in the population thus could be at least 109, giving a coning sex ratio of 0.43 (female/male). Because reproductively mature male individuals of *Z. pumila* appear to produce cones on an annual basis (Norstog, unpubl. obs.), it can be assumed that the remaining 118 individuals (145, less 27 mature but non-coning females) include approximately equal numbers of immature male and female plants which have not yet attained sexual maturity. If this is true, then the sex ratio for the entire population of 369 individuals approaches 1:1, based on the estimated proportion of 169 females and 201 males (0.46, female/male).

It may be, however, that because of male-coning precocity, immature females are in the majority in the *Zamia* population described by Tang (1985) and the actual sex ratio may thus be about 1:1. Experimental evidence suggests too, that male seedlings of *Zamia pumila* tend to exhibit coning precocity. (Among 88 unstressed seedlings, 2 male cones and no female cones were produced in year 2; 9 male cones and 2 female cones were produced in year 3. Among 47 water-stressed seedlings, no cones were produced in year 2, but 7 male cones and no female cones were produced in year 3 [Norstog, unpubl. data].) Further evidence for sex ratios approaching 1:1 is furnished by progeny of a *Zamia* hybrid cross described further on. The sex ratio for 44 F_1 plants of *Z. furfuracea* × *Z. spartea* was approximately 1:1 (17 females, 19 males, 8 undet.).

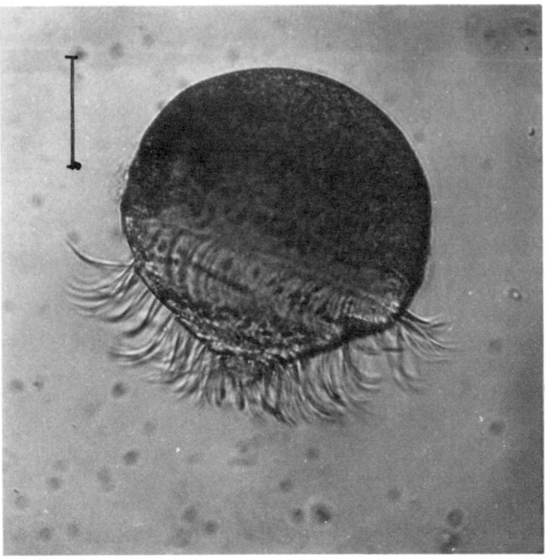

Fig. 7. A living spermatozoid of *Zamia pumila*. Clusters of flagella give an "eyelash" appearance to the cell's locomotory system. Scale bar = 0.1 mm.

Sex Determination and Sex Reversal

Before taking up the subject of pollination, it may be well to devote a few lines to the subject of sex determination and the possibility of sex reversal in cycads. In many dioecious organisms, including mosses, fruit flies and humans, certain chromosomes have been identified as being related to sex determination. The question of sex chromosomes in cycads is one that is often raised, and attempts to have been made to look for pairs of dissimilar chromosomes which conceivably could be sex determinants. Scattered references to this possibility are to be found in the literature (Abraham & Mathew, 1962; Marchant, 1968), but the putative sex chromosomes described in these reports appear to exhibit rather minor structural differences, which though possibly real could equally be artifactual. That this might indeed be the case is suggested by the puzzling coincidence of two articles in the same issue of "American Journal of Botany," each describing sex chromosomes of the likewise dioecious *Ginkgo biloba,* but each identifying a different pair as the sex chromosomes (Lee, 1954; Newcomer, 1954).

Related to the subject of sex determination is the question of sex reversal. Occurrences of sex reversal are sometimes encountered and, in recent times, van Wyk and Claasen (1981) have reported an instance of sex reversal in *Encephalartos umbeluziensis* which seems bona fide. Dennis Stevenson (pers. comm.) also has recently observed sex reversal in *Zamia pygmaea.* However, such reports are difficult to validate and interpret since they are isolated phenomena and often do not have a long-term history. Moreover, they appear to be stress related, resulting from necrosis and/or transplanting and as such may merely represent short-term, teratological phenomena. What is wanted is long-term monitoring of large number of plants that have been permanently labeled as male or female. In addition, location records should be kept, together with accompanying information as to sex, coning intervals, and occurrences of sex reversals. At Fairchild Tropical Garden, we have embarked on long-term studies of sex in two species of *Zamia,* beginning with statistically significant numbers of seedlings, noting their sex when cones appear, and undertaking long-term monitoring for sex reversal. In addition, some seedlings are being treated with gibberellic acid, which in some conifers (Longman, 1970) has been found to induce early formation of male cones. Among 40 seedlings of *Z. pumila,* only male cones have thus far been produced by plants in the GA-treated set, but both male and female cones have been produced by the control plants (i.e., untreated plants). This study has been underway since 1981, but the results thus far are inconclusive and have not been published. Nevertheless, this example brings up an important point: much cycad research pertaining to phenology, cytology, and reproductive behavior requires long periods of time for completion. Studies must be carefully planned and provisions made for long-term and precise monitoring. Since this may require the work of several generations of observers, it would be wise for investigators to establish their collections in botanical gardens where survival of experiments can be assured.

Wind Pollination

Charles J. Chamberlain, who was probably the most widely recognized authority on cycads, wrote "There have been many reports of insect pollination, but in a rather extensive field study in which all the genera have been examined, nothing has been observed which would indicate anything but wind pollination so characteristic of the whole group of gymnosperms" (Chamberlain, 1935). Others, including Lawson (1926), also considered wind to be the principal agent of pollination in cycads (e.g., *Bowenia*). Baird (1938), however, suggested that wind might carry pollen to female cones of *Macrozamia,* but insects probably transported the pollen within the cones. Rattray (1913) suggested that weevils might pollinate the ovules of *Encephalartos,* but thought *Stangeria* was wind-pollinated. Niklas and Norstog (1984) studied the dynamics of wind pollination in cycads by placing cones of *Dioon* and *Zamia,* as well as *Cycas* pseudocones, in a wind tunnel and blowing pollen past them at various velocities. When photographed by strobe-lighting, clear patterns of pollen flow to specific receptive areas of the cones of *Dioon* and *Zamia* were revealed, but pollen was rather uniformly distributed over the megasporophylls of *Cycas* and it was suggested that water might carry pollen to the ovules. In the cases of *Dioon* and *Zamia,* however, it was thought that insects might be required to convey pollen to ovules inside the cones.

Insect Pollination

Clear evidence of insect pollination in cycads has recently been published. Norstog et al. (1986) show that the snout weevil, *Rhopalotria mollis* (Fig. 6B), is an effective pollinator of *Zamia furfuracea,* and, further, that it utilizes male cones of that cycad for its full reproductive cycle including copulation, oviposition, larval feeding and pupation (Fig. 6A). This is a clear-cut case of symbiotic mutualism from which both organisms profit. Although the origin and antiquity of this particular mutualism is unknown, one can guess that it is a long-standing coevolutionary system. Snout weevils are an ancient group of beetles and there is evidence of their association with the reproduction of Mesozoic cycadeoids (Crowson, 1981). Tang (1987) also has provided solid evidence for weevil pollination of the native Florida cycad, *Zamia pumila*. In this case two weevils, *Rhopalotria slossoni* and *Pharaxonotha zamiae*, were shown to be effective pollinators. In neither of these recent studies (Norstog et al., 1986; Tang, 1987) was wind pollination ruled out completely, however, the morphology of cycad cones does not seem to be consistent with a wind-pollination syndrome because the ovules are protected by shield-like megasporophyll tips.

Evolutionary Significance of the Cycad Megastrobilus

Insect pollination often is associated with the evolution of flowering plants, and coevolution of flowers and pollinating insects is often discussed in the botanical literature (Faegri & van der Pijl, 1971). True, in many flowering plant species, highly complex flower–pollinator interactions are found, especially as in the case of certain bee-pollinated flowers. However, it is thought that the more primitive flowers of the Cretaceous may have been pollinated by beetles, and in fact, Crepet (1979) points out that one may assume that pollen was one of the first nutritional rewards for pollinators. Such early pollinators probably had mandibles suitable for chewing pollen, e.g., beetles (Coleoptera), which were already present in the Carboniferous. At first such predation may have simply been opportunistic, but undoubtedly specializations evolved both in plants and their insect pollinators that produced more intimate and mutually beneficial relationships. The development of coevolutionary pollination mechanisms would tend to make pollination less haphazard and would also have led to the evolution of protected ovules and seeds. Probably the "closed" seed cones of cycads resulted from such evolutionary tendencies. If so, it is quite likely that the plant–pollinator syndrome antedated flower–pollinator coevolution by a hundred million years. The early record of angiosperm evolution is unclear (Dilcher, 1979), but it seems likely that the earliest flowers were unisexual and that bisexual flowers evolved as an adaptation to insect pollination.

The primitive angiosperm flower may not have been specialized for insect pollination but rather depended partially on opportunistic foraging by pollen-eating coleopterans. Beetles are relatively clumsy pollinators. They are unable to land precisely on a target structure (e.g., cone or flower—see Faegri & van der Pijl, 1971). Beetle-pollinated flowers are unspecialized, open, bowl-shaped blossoms such as *Magnolia* and *Nymphaea*, and the transition from foraging upon cycad cones to such primitive flowers may have been one of simple opportunism. Subsequently, more complex pollination systems, which involved bisexual flowers with well-developed nectaries and "target" colors and patterns, coevolved together with insect specializations such as color vision, ability to perceive and remember target flowers, elongated tongues for lapping nectar, and pollen receptors and combs. These are most highly developed in bees and bee-flowers (Faegri & van der Pijl, 1971). It may be, however, that the earliest insect pollinators evolved primitive behavioral patterns of pollen foraging and life cycle dependencies upon cycad and cycadeoid reproductive organs prior to the origin of flowering plants. Thus a type of preadaptation of pollinators may have occurred, which in the Mesozoic produced an accelerated evolution of flower–pollinator systems. If so, the apparently rapid radiation of flowering plants in the Cretaceous is more readily explicable when viewed on the basis of adoption and refinement of preexisting patterns of insect pollination than on the *de novo* origin of insect–flower pollination.

Spermiogenesis, Fertilization, and Embryogeny

The discovery of multiflagellated male gametes (spermatozoids) in *Ginkgo* and *Cycas* has been hailed as one of the most significant botanical discoveries of the latter 19th Century (Ogura, 1967), for it testified to a close connection between primitive seed plants and cryptogams, such as mosses, ferns, and lycopods, which also have spermatozoids.

The Development of Multiflagellated Male Gametes in Cycads

Spermatozoids of cycads are spectacular cells (Fig. 7), ranging in size from about 80 μm in

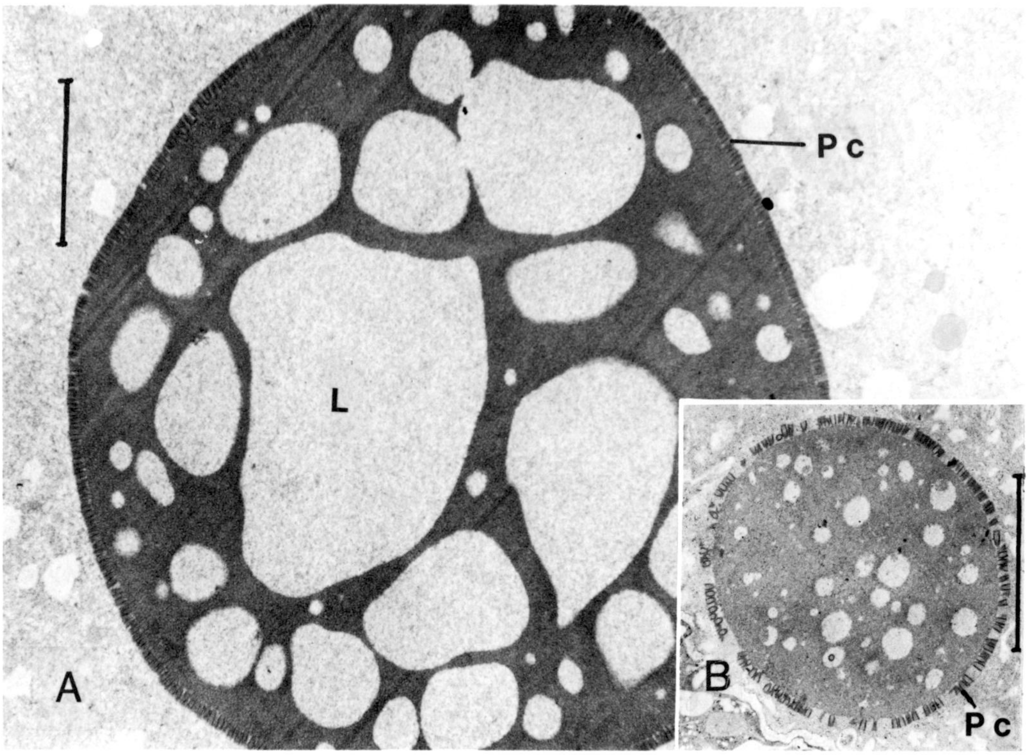

Fig. 8. Cycad blepharoplasts—multicentriolar bodies found in spermatocytes and spermatids prior to maturation of the spermatozoids. **A.** Blepharoplast of *Z. pumila.* Note dense-staining matrix matter and light-staining lacunae (L). About 40,000 procentrioles (Pc) compose the cortical layer. Scale bar = 5 μm. **B.** The much smaller blepharoplast of *Microcycas* has a similar construction but only about 5000 procentrioles (Pc). Scale bar = 5 μm.

Microcycas to around 400 μm in *Zamia roezlii,* and equipped with as many as 40,000 flagella (Norstog, 1977, 1986). They are produced toward the end of a period of slow development of the pollen tubes and are two in number in all cycads studied except *Microcycas.* In the latter, 16 or more spermatozoids are formed, leading to speculation that in this respect at least, *Microcycas* may be primitive because in ferns and other cryptogams, many spermatozoids are formed in each antheridium. Partly for this reason, and partly because the ultrastructure of only one other cycad has been described (*Z. pumila*), an ultrastructural study of spermiogenesis in *Microcycas* is now underway at Fairchild Tropical Garden (Fig. 9B).

Spermatozoids of cycads develop from a single spermatogenous cell (several in *Microcycas*) in which two conspicuous multicentriolar bodies, the blepharoblasts, are the most obvious cytoplasmic organelles (Fig. 8). In *Z. pumila,* the blepharoplast consists of a dense matrix and a cortex of about 40,000 procentrioles (Fig. 8A), each of which eventually gives rise to a basal body and its single flagellum. Recent electron-microscopic studies of spermiogenesis in *Microcycas* have shown that in this cycad the blepharoplast is considerably smaller than that of *Zamia* (cf. Fig. 8A, B), and has about 5000 procentrioles. This is consistent with the much smaller dimensions of the spermatozoids of this species. Conspicuous also in cycad spermatozoids is a spiral, flagellated band, in which the organizing element is the MLS (multilayered structure) underlying the flagella (Fig. 9A). It is the MLS and its attached flagella which gives the cycad spermatozoid its spiral structure. It has been shown that the cycad MLS and the organization of its basal bodies and flagella has phylogenetic importance for it is this system of

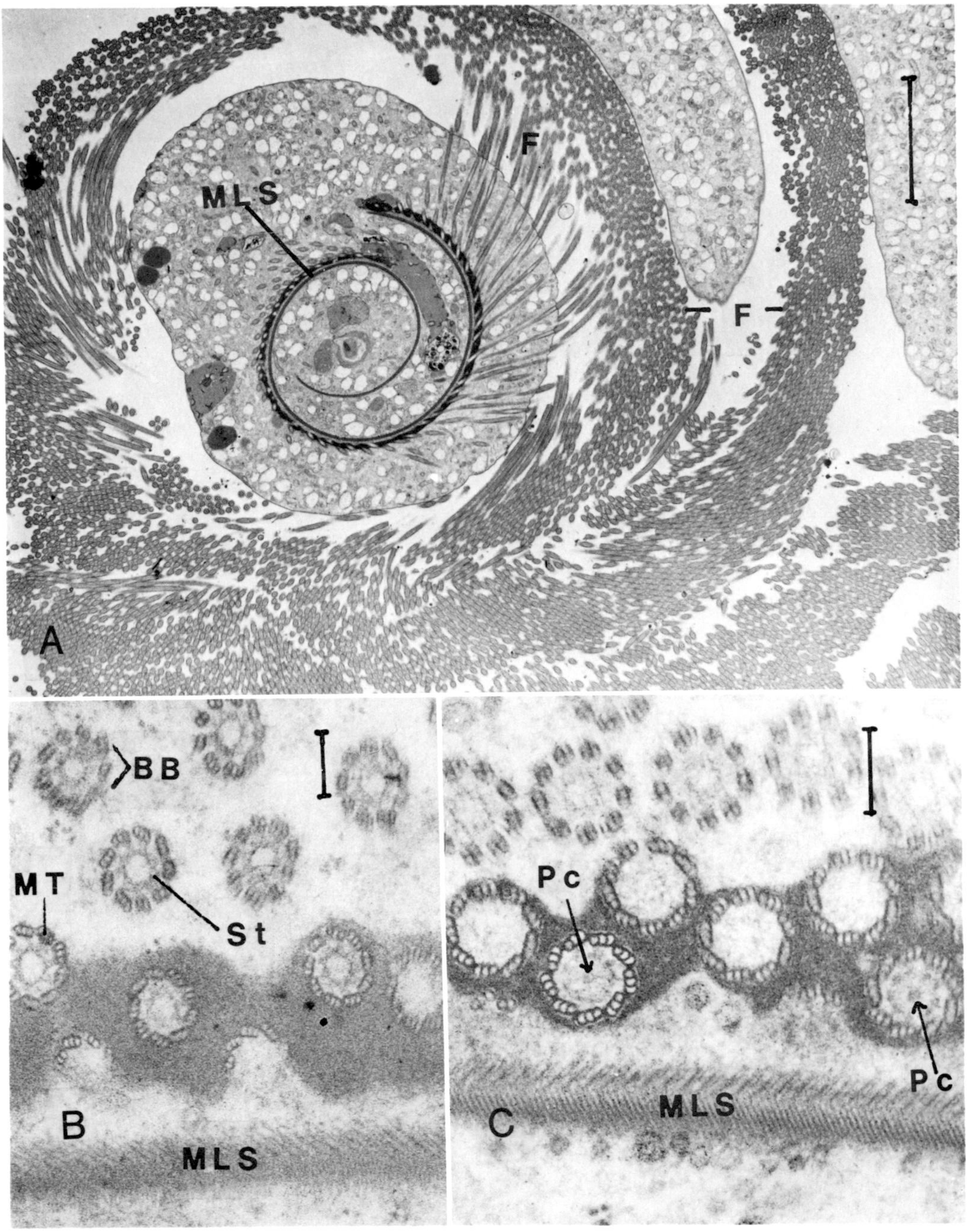

Fig. 9. Ultrastructural aspects of cycad spermatozoids. **A.** A cross section through the anteriormost gyres of a spermatozoid of *Zamia pumila*. There are about six flagellated gyres in the entire spermatozoid. F, flagella; MLS, multilayered structure. Scale bar = 10 μm. **B.** Proximal transverse sections of basal bodies (BB) of a *Microcycas* spermatozoid. There do not appear to be any basal procentriolar regions of the basal bodies (compare with Fig. 9C). MT, microtubular triplets of basal body; MLS, multilayered structure; St, stellate zone of basal body. Scale bar = 200 nm. **C.** Transverse section through proximal ends of basal bodies of a *Zamia pumila* spermatozoid. Note procentrioles (Pc) at basal ends of basal bodies. MLS, multilayered structure. Scale bar = 200 nm.

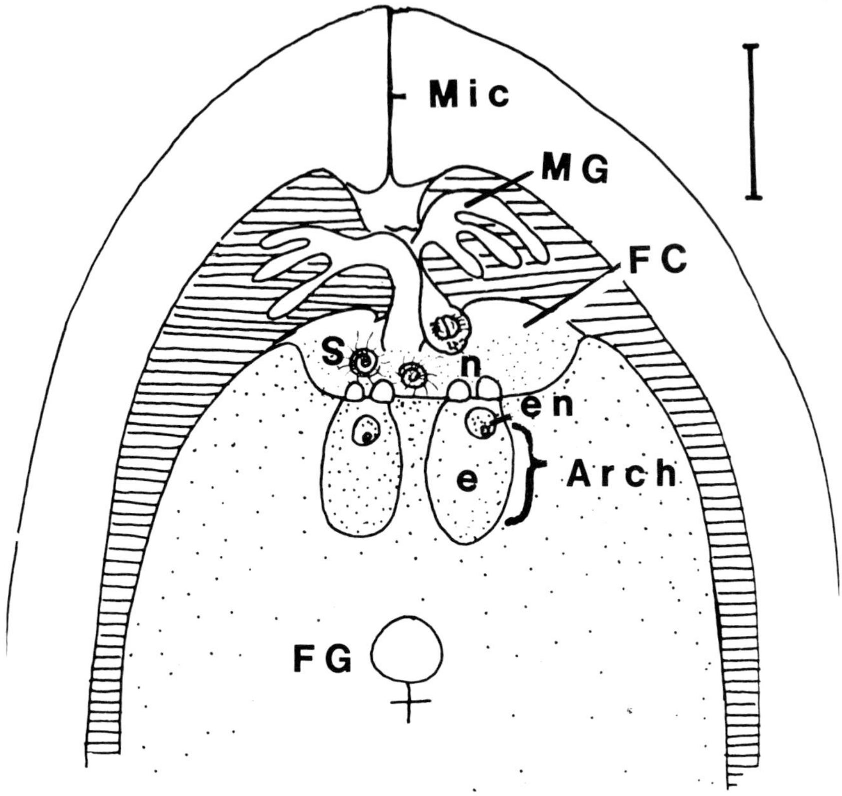

Fig. 10. A diagram illustrating the fertilization of an ovule of *Zamia pumila.* Arch, archegonium with neck cells (n), egg (e), and egg nucleus (en); FC, archegonial chamber; FG, female gametophyte; MG, male gametophytes; Mic, micropyle; S, spermatozoids. Scale bar = 0.1 mm.

organelles that allies the cycad spermatozoid to those of ferns, lycopods, and mosses (Norstog, 1974). We see, therefore, that without the ultrastructural evidence provided by the cycads, and to a lesser degree by *Ginkgo,* an important phylogenetic clue to the origin of seed plants would have escaped us.

Regarding spermatozoids of *Microcycas,* preliminary electron-microscopic studies do not show any startling differences between these cells and the spermatozoids of *Zamia.* However, as noted, the spermatozoid of *Microcycas* is much smaller than those of *Z. pumila* and other cycads, and its spiral band describes only about two turns (gyres) vs. 5–6 in *Zamia.* Also, the structure of flagellar basal bodies of *Microcycas* differs somewhat from those of *Zamia* in that those of *Microcycas* appear to lack a basal procentriolar pattern (Fig. 9B, C). Whether these differences have phylogenetic significance remains to be seen.

The Archegonium, Neck Cells, and Fertilization

The massive female gametophytes of cycads develop relatively slowly, requiring 5–6 months to attain sexual maturity (Fig. 10). In most genera and species, 2 to 5 archegonia are formed, each initially equipped with a pair of neck cells protruding into a circular depression, called the archegonial chamber, which is located at the distal end of the female gametophyte (Fig. 10). *Microcycas* differs from this basic pattern in having as many as several hundred archegonia extending over much of the distal one-third of the gametophyte. The function of this rather extreme redundancy is mystifying in that normally only one egg develops into an embryo.

In *Zamia pumila,* a day or two before the spermatozoids are mature, the archegonial neck cells undergo a last division so that each of the two

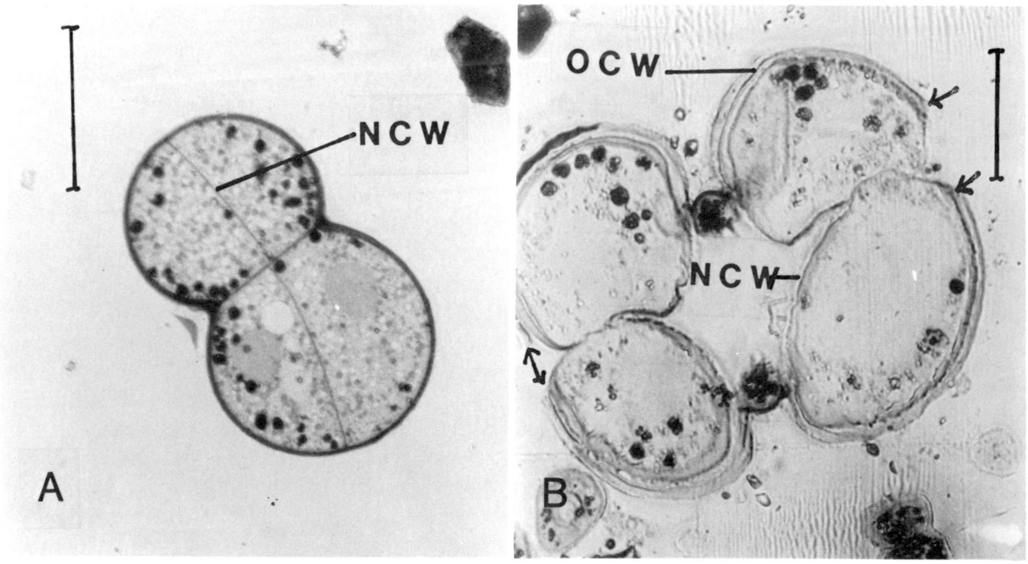

Fig. 11. Development of archegonial neck cells of *Zamia pumila*. **A.** Just prior to fertilization, a division of the paired neck cells occurs, producing thin new cell walls (NCW). Scale bar = 50 μm. **B.** At fertilization, the four neck cells (as seen in Fig. 11A) expand greatly, rupturing the old cell walls (OCW), at arrows, and separating the new cell walls (NCW) at the middle lamellae. This creates an opening through which spermatozoids may enter the archegonium. Scale bar = 50 μm.

neck cells becomes bisected by a delicate new cell wall (Fig. 11A). At fertilization each of the four newly formed neck cells then expands, rupturing the middle lamellae of the new walls and creating an archegonial neck canal (Fig. 11B) through which spermatozoids may pass and fertilize the egg-cell nucleus (Norstog, 1972). In this aspect also, the phylogenetic connection of cycads is seen, for in all other archegoniate plants the neck cells are invariably four-tiered.

Embryogenesis

In *Zamia pumila*, fertilization occurs in late May and is followed by a period of embryo development lasting as late as December, when the female cone dehisces and seeds are released. In early development, the embryo consists of a growing suspensor region bearing a terminal meristematic region called the embryo proper. The embryo proper at this stage is a hemispherical zone of compact meristematic cells, and during embryo development, the suspensor pushes the embryo proper deeper into the flesh of the female gametophyte. However, this action is not solely owing to the elongation of the suspensor, for it has been shown by embryo culture techniques that the suspensor cells are themselves a product of the embryo proper (Monnier & Norstog, 1984). The suspensor grows, however, only by a simple process of cell elongation. Meanwhile the embryo proper develops a pair of cotyledons and, finally, occupies much of the interior length of the female gametophyte.

In vitro cultures of early and late *Zamia* embryos demonstrate that *in ovulo* factors are required for regulation of cotyledonary differentiation. Excised early embryos (early embryo proper stage) do not possess sufficient endogenous support for differentiation to proceed *in vitro*. However, embryos excised 6–8 weeks later (later embryo proper stage), though superficially similar to early embryos, appear committed to cotyledonary differentiation. This is considered to exemplify the developmental phenomenon known as induction (Monnier & Norstog, 1986). Finally, these studies demonstrate the morphogenetic plasticity of cycad embryos, in that 6–8-week-old embryos, when bisected and cultured *in vitro* (later embryo proper stage), can form two wholly new embryos, each with two cotyledons, but older, already differentiated embryos, when bisected, only develop as half-embryos.

Hybridization in the Cycads

From time to time, reports are received of natural and artificial cycad hybrids. In some cases these are of wide crosses, as in the example of the *Ceratozamia* × *Zamia* hybrid described by Chamberlain (1926). More often, suspected natural interspecific hybrids or others produced by artificial hybridizations are described as is the case of the numerous *Encephalartos* crosses reported by Vorster (1986). Since little is known about the genetics of cycads, it is difficult to analyze such hybrids in any satisfactory way, and one of the most important questions regarding them is whether they actually represent interspecific hybridizations. Cycads are such slow-growing and slow-reproducing plants that assembling experimental evidence from several generations of hybridization can represent a major undertaking. To date there are no studies of cycad genetics and cycad hybridization comparable with the many breeding studies accomplished with numerous species of angiosperms.

Methods of Cycad Hybridization

Cross pollination of cycad cones is a relatively simple procedure. At Fairchild Tropical Garden, artificial pollination usually is accomplished by using a syringe to blow dry pollen into a female cone, although sometimes a wet slurry of pollen and water can be successfully employed. The latter method seems to work quite well with *Dioon spinulosum*; the dry method appears appropriate for other cycads. If an interspecific (or intergeneric) cross is to be attempted all that is necessary, in addition, is the protection of the pollinated cone against the chance intrusion of other pollen. In actuality, this probably is only a remote possibility, given the difficulties of pollinating most exotic cycads.

An Experimental Cycad Hybrid, *Zamia furfuracea* × *Zamia spartea*

For the past eight years at Fairchild Tropical Garden, we have been monitoring an interspecific hybridization between cultivated *Z. furfuracea* and *Z. spartea* which in nature are allopatric Mexican species. The parent plants have similar cone characteristics but differ markedly in leaf characters (Fig. 12A). *Zamia spartea* has more than 24 pairs of long–narrow leaflets per frond; *Z. furfuracea* has fewer than 20 short–broad leaflets per frond. In 1979, the pollination of the female of *Z. spartea* with *Z. furfuracea* pollen produced 44 fertile seeds, which, in turn, produced a group of rapidly maturing seedlings. Most of these seedlings produced their first cones in 1984 (very rapid for cycads). The sexually mature hybrids are completely uniform in appearance and quite intermediate between the parental species in terms of leaf characters (Fig. 12A). In 1984, the F_1 plants were interpollinated and produced 9 fertile cones, and these have now produced 228 F_2 progeny, which show a range of leaf morphology ranging between the extremes of the original parental types (Fig. 12B). The variation appears continuous so that one can suspect that some kind of multiple gene inheritance system is operational. This is also evidenced by the observation that none of the second generation plants has leaflets quite as long and narrow as *Z. spartea* nor as broad and short as *Z. furfuracea.* Two-tailed *t*-tests show that differences between the parental types and the first and second generation progeny are highly significant. From these results one can deduce that the original parents are valid species, differing from each other by a number of gene mutations. Also, it seems that the first generation progeny are true Mendelian hybrids, and that the members of the second generation correspond to F_2 progeny of a cross involving several to many genes controlling certain quantitative characters (e.g., leaf length, leaflet pairs, leaflet length and width).

Conclusions and Summary

It would be unwise to predict that hybridizations in other cycad genera and species will necessarily resemble *Z. furfuracea* × *Z. spartea* but some broad inferences perhaps may be drawn. First, if a hybridization is a genuine interspecific one, the first generation progeny ought to be very uniform in appearance. If however, the progeny show considerable variation, then it might be suspected that the parents probably are not "good" species, but are more likely to simply exhibit variations normally expected to occur within the species limits. Secondly, because it seems probable that many of the vegetative characteristics used to define cycad species are the product of quantitative inheritance (multiple gene inheritance), one should not expect the progeny

Fig. 12. *Zamia* hybrids. **A.** Male *Z. furfuracea* parent (left), female *Z. spartea* parent (right), and their F_1 hybrid (center). The F_1 plants are very uniform in appearance and show little statistically significant variation in leaf characters. **B.** Male and female F_1 hybrids (center background) and representative F_2 progeny. Those toward the left tend to resemble *Z. spartea,* while those to the right tend to resemble *Z. furfuracea.*

to closely resemble either original parent. Lastly, it is possible that cycad hybrids may demonstrate hybrid vigor and, as in the case of *Z. furfuracea* × *Z. spartea* also display sexual precocity. Thus, some interesting and valuable ornamental cycads may be produced in the future. Lastly, because cycad breeding and hybridization requires long-term monitoring, it might be wise to concentrate any such research in well-established botanical gardens.

Bringing together and breeding species that normally would not interbreed due to time and distance factors might, as a result of accidental outcrossing pollinations of native species, introduce significant genetic alterations in wild populations. For example, *Zamia pumila,* the native Florida cycad has been shown by Tang (1987) to be pollinated by weevils. Possibly, pollen from introduced species of Caribbean and Central American species of *Zamia* now growing in Florida gardens could be transported by weevils into surrounding wild *Zamia* populations, and thus produce unwanted genetic contaminations. Whether or not this represents a real danger to the gene pools of native cycads is unknown, but even if the wild populations are at only slight risk, more needs to be known about cycad reproduction and, especially, the roles of insects in pollen transport. The involvement of botanical garden personnel in such research is important for the future of both native and introduced species of cycads. Finally, because cycad breeding studies will require long-range planning, the involvement of botanical gardens in such work is a necessity. Only they can guarantee the perpetuation of cycad studies beyond the lifetime of a human individual.

Acknowledgment

Portions of the research reported were funded by a grant BSR 8604221 from the National Science Foundation, Washington, D.C.

Literature Cited

Abraham A. & P. M. Mathew. 1962. Cytological studies in the cycads: Sex chromosomes in *Cycas.* Ann. Bot. (n.s.) **26(102):** 261–266.

Baird, A. M. 1938. A contribution to the life history of *Macrozamia riedlei.* J. Roy. Soc. New South Wales **25:** 153–175.

Chamberlain, C. J. 1926. Hybrids in cycads. Bot. Gaz. **81:** 401–418.

———. 1935. Gymnosperms. Structure and evolution. The University of Chicago Press, Chicago.

Clark, D. A. & D. B. Clark. N.D. Temporal and environmental patterns of reproduction in *Zamia skinneri,* a tropical rain forest cycad. Unpubl. Manuscr., La Selva Biol. Stn., Costa Rica.

Crepet, W. 1979. Insect pollination: A paleontological perspective. BioScience **29:** 102–108.

Crowson, R. A. 1981. The biology of the Coleoptera. Academic Press, New York.

Dilcher, D. L. 1979. Early angiosperm reproduction: An introductory report. Rev. Palaeobot. Palynol. **27:** 291–328.

Faegri, K. & L. van der Pijl. 1971. The principles of pollination ecology. 2nd ed. Pergamon Press, Oxford.

Lawson, A. A. 1926. A contribution to the life history of *Bowenia.* Proc. Roy. Soc. Edinb. **54:** 357–394.

Lee, C. L. 1954. Sex chromosomes in *Ginkgo biloba.* Amer. J. Bot. **41:** 545–549.

Longman, K. A. 1970. Initiation of flowering in first year cuttings of *Metasequoia glyptostroboides.* Nature **227:** 299–300.

Marchant, C. J. 1968. Chromosome patterns and nuclear phenomena in the cycad families Stangeriaceae and Zamiaceae. Chromosoma (Berl.) **24:** 100–134.

Monnier, M. & K. Norstog. 1984. Developmental aspects of immature *Zamia* embryos cultured *in vitro.* Z. Pflanzenphysiol. **113:** 105–115.

——— & ———. 1986. Effect of *in ovulo* period on the differentiation and regulation of immature embryos of *Zamia* cultured *in vitro.* J. Exp. Bot. **184:** 1633–1642.

Newcomer, E. H. 1954. The karyotypes and possible sex chromosomes of *Ginkgo biloba.* Amer. J. Bot. **41:** 542–545.

Newell, S. J. 1985. Intrapopulational variation in leaflet morphology of *Zamia pumila* L in relation to microenvironment and sex. Amer. J. Bot. **72:** 217–221.

Niklas, K. J. & K. Norstog. 1984. Aerodynamics and pollen grain depositional patterns on cycad megastrobili: Implications on the reproduction of three cycad genera (*Cycas, Dioon,* and *Zamia*). Bot. Gaz. **145:** 92–104.

Norstog, K. 1972. Role of archegonial neck cells of *Zamia* and other cycads. Phytomorphology **22:** 125–130.

———. 1974. Fine structure of the spermatozoid of *Zamia*: The *vierergruppe.* Amer. J. Bot. **61:** 449–456.

———. 1977. The spermatozoid of *Zamia chigua* Seem. Bot. Gaz. **61:** 409–412.

———. 1986. The blepharoplast of *Zamia pumila* L. Bot. Gaz. **147:** 40–46.

———, **D. W. Stevenson & K. J. Niklas.** 1986. The role of beetles in the pollination of *Zamia furfuracea* L. f. (Zamiaceae). Biotropica **18:** 300–306.

Ogura, Y. 1967. History of the discovery of spermatozoids of *Ginkgo biloba* and *Cycas revoluta.* Phytomorphology **17:** 109–114.

Rattray, G. 1913. Notes on the pollination of some South African cycads. Trans. Roy. Soc. South Africa **3:** 259–270.

Tang, W. 1985. Studies in cycad reproductive biology. M.S. Thesis. University of Miami, Coral Gables, Florida.

———. 1987. Insect pollination in the cycad *Zamia pumila* (Zamiaceae). Amer. J. Bot. **74:** 90–99.

van Wyk, A. E. & M. I. Classen. 1981. Sex reversal in *Encephalartos umbeluziensis.* Veld & Flora **(Dec.):** 120–122.

Vorster, P. 1986. Hybridization in *Encephalartos.* Excelsa **12:** 101–106.

MARCH 1990 MEMOIRS OF THE NEW YORK BOTANICAL GARDEN 57: 82–88

Micropropagation in Cycads[1]

ROY OSBORNE

Table of Contents

Abstract

OSBORNE, R. (Dept. of Chemistry, University of Natal, King George V Avenue, 4001 Durban, Republic of South Africa). Micropropagation in cycads. Memoirs of the New York Botanical Garden **57:** 82–88. 1990.—A brief review is made of the use of haploid and diploid tissue culture of various cycad species. Callus formation is readily achieved but morphogenesis appears erratic and infrequent. Current work by the author and other workers is described. With primary root tissue of *Stangeria eriopus* callus formation in dark conditions is followed by morphogenesis on transfer to constant light conditions. Further applications of cycad tissue culture work are described.

Resumé

Bref compte rendu de l'utilisation de la culture de tissus haploides et diploides de différentes espèces de cycades. La formation de cals est réalisée sans difficulté, mais la morphogénèse semble être erratique et peu fréquente. Les travaux actuels de l'auteur et d'autres chercheurs y sont décrits. Dans le cas de tissu de racine primaire de la *Stangeria eriopus* la formation de cals dans l'obscurité est suivie de morphogénèse quand on les transfère à la lumière constante. D'autres applications de travaux de culture de tissus de cycades y sont décrites.

Key words: Cycads; Cycadales; tissue culture; micropropagation; *Stangeria*.

Introduction

Plants are typically multiplied either from seeds or vegetatively by division, cuttings, budding and grafting, these methods representing the sexual and asexual routes of propagation respectively. The former route leads to a potential for genetic change while the latter generally guarantees that progeny will be genetically identical to the parent. In this instance all the collectively produced individuals constitute a clone of the original parent.

In cases where there is a limited amount of plant material for vegetative propagation or where the value of the propagules is high, it is clearly an advantage if smaller and smaller portions can be used in vegetative propagation. Recent developments have progressively and dramatically refined micropropagation techniques and have had an enormous impact on the horticultural and agricultural industries. A good illustration of this is found in the orchids where small shoot tips, isolated leaf segments and even young inflorescences can be transferred in aseptic conditions to artificial supporting media where the explant tissues give rise to cellular callus which in turn generates protocorms and eventually plantlets. Morel (1974) has estimated that it is possible to obtain more than one billion plantlets from a single orchid bud in 9 months.

Taking the concept of smaller and smaller tissue units to an extreme, it is possible with several plant groups to use a single cell or even the naked protoplast within that cell to produce a plant identical to its parent. The techniques are highly specialized and not easy; for each success there are dozens and dozens of failures. But the present techniques are constantly being improved and there is now a convincing body of evidence to support the view that, under appropriate conditions, any plant can be propagated by these means.

It has become increasingly apparent that micropropagation can play a vital role in the conservation of endangered plant species. For example, Rubluo et al. (1986) investigated the severely-endangered Mexican plants *Bletia urbana* (Orchidaceae) and *Mammillaria sanangelensis* (Cactaceae); starting with only 5 fruits of the orchid and 5 seeds of the cactus, they managed to produce several thousand plants of each species and both were successfully re-introduced into the environment.

Application of Micropropagation Techniques to the Cycadales

The present-day cycads represent the diverse, modified remnants of a much larger group which flourished in the Mesozoic Era, reaching a zenith in the Jurassic Period about 160 million years ago. The number of currently-valid taxa in the Cycadales is assessed at 132 species in 10 genera (Osborne & Hendricks, 1985), more than half of which have been classified as endangered, vulnerable or rare by the Threatened Plant Unit of the International Union for the Conservation of Nature and Natural Resources (Gilbert, 1984). Confined to tropical and sub-tropical regions, these dioecious plants are generally slow growing, with limited potential for vegetative propagation. Many of the rarer species are so thinly-distributed in their habitat that seed production is restricted and the seeds produced are often non-viable. The potential for tissue culture is thus both obvious and exciting.

Tissue Culture of Cycads—A Review

The life cycle of cycads is often used to illustrate classical botanical principles. Male or female adult plant cells with a diploid chromosome complement undergo meiosis in reproductive organs to give rise ultimately to haploid egg nuclei or haploid microspores, respectively. Pollination leads to fusion of these gametes and the resulting seed contains the zygotic embryo embedded in a well-developed nutritive zone of haploid megagametophytic tissue. Since the latter is somewhat different in origin to the endosperm in seeds of higher plants, much of the early work in cycad tissue culture was carried out to investigate the developmental stages or ontogeny of the material. Only later was the implication made that tissue culture might be a route for cycad propagation.

La Rue (1948) cultured megagametophytes of *Zamia floridana* A. DC. (=*Z. pumila* L.) on simple media and found that 'pustular' growth occurred which, in a small number of cases, led to the formation of roots or shoots or both. Later La Rue (1954) showed that the same tissue material, grown aseptically on moist sand, could sometimes give rise to small detachable balls of meristematic tissue which could be subcultured, eventually with the formation of poorly-vascularized roots and leaves. In the same paper La Rue (1954) reported the production of roots and buds from megagametophytes of *Cycas revoluta* Thunb. after two years in culture. Norstog (1965) extended this work when investigating regeneration from both haploid and diploid tissue of *Zamia integrifolia* Ait. (=*Z. pumila* L.) by culturing megagametophytes and embryos on a variety of media which were modifications of White's (1943) basal formula. Promising results were obtained with medium '8' which contained asparagine, glycine, glutamine and adenine together with the auxin 2,4-dichlorophenoxyacetic acid (2,4-D) and kinetin. In this case, 68% of the haploid explants gave rise to callus within two months, this being followed by some evidence of root and leaf initiation. This type of morphogenesis was promoted by subculture to media without growth factors. Furthermore, Norstog (1965), using a medium devised for the culture of excised barley embryos, was able to obtain adventive embryos ('embryoids') from a few explants of both haploid and diploid *Zamia* tissue; these embryoids seemed analogous to La Rue's pseudobulbils. A further extension of this work was made by Norstog and Rhamstine (1967) using megagametophytes and juvenile embryos from *Zamia pumila* and *Cycas circinalis* L. on a wide variety of stationary and liquid media. Rapid growth of the *Zamia* proembryos was obtained with their medium '59,' a modified Murashige-Skoog (1962) formulation with amino acid and adenine supplements containing 2,4-D and kinetin. On transfer to media without the auxin and cytokinin, adventive embryos were again obtained with some evidence of leaf initiation.

The Italian workers De Luca et al. (1979) investigated the response of haploid explants from megagametophytes of *Ceratozamia mexicana* Brong., *Cycas revoluta* Thunb. and *Encephalartos umbeluziensis* R. A. Dyer to a wide variety of media under different light and temperature conditions. Results varied widely with the plant species and the medium and conditions. Some evidence of somatic embryogenesis was found with *Ceratozamia,* leading in one case to leaf regeneration. With *Cycas* explants, 'pseudobulbils' appeared and these spherical nodules were later shown to be analogous to coralloid root nodules, a normal feature of cycad ontogeny (De Luca & Sabato, 1980). *Encephalartos* megagametophytes produced callus without any subsequent morphogenesis. Similar work with *Cycas revoluta* megagametophytes (De Luca & Sabato, 1980) and with the same explants from *Macrozamia communis* L. Johnson (De Luca et al., 1980b) again led to callusing with subsequent development of coralloid root primordia. This work provided evidence that these organs do not necessarily arise as a response to microbial infection in the formation of a symbiotic process.

Further investigations into regeneration from *Z. pumila* megagametophytes (Rivera Rosa, 1980, unpubl. results) showed that explants on Murashige-Skoog (1962) medium containing 0.1 mg/liter each of 1-naphthaleneacetic acid (NAA) and 6-benzyladenine (BA) led to nodular callus which in turn generated bipolar dicotyledonous embryoids. With 2.0 mg/liter NAA and 0.3 mg/liter BA the callus gave occasional root and shoot formation, but no viable plantlets could be recovered in any of these experiments.

Experiments with the particularly rare Cuban cycad, *Microcycas calocoma* (Miq.) A. DC. showed that callus proliferation, root formation and 'pseudobulbil' development could be obtained from megagametophytic explant of this species on Murashige-Skoog (1962) medium modified by the inclusion of various levels of coconut milk and NAA (Pena & Grillo, 1982). Infertile seeds of *Encephalartos villosus* Lem. on a modified White's (1943) medium gave callus formation with subsequent meristemoidal nodular processes, these being most abundant when the medium was supplemented with 1.0 mg/liter 2,4-D and 0.5 mg/liter kinetin (Laliberte et al., 1983).

Use of the male gametophyte in tissue culture work has been less well explored than that with the female haploid material. *Zamia pumila* microspores have been cultured on White's (1943) medium with a variety of additives (La Rue, 1954). Normal pollen tube extension occurred but growth was limited and the cultures eventually senesced. De Luca and Sabato (1979) observed *in vitro* sperm formation when microsporangia from *Encephalartos altensteinii* Lehm. were cultivated on a medium previously used for pollen studies on *Ginkgo biloba.* In a similar experiment spermatogenesis from microsporangia of *Ceratozamia mexicana* and *Cycas revoluta* led to callus formation from pollen of the former species after 11 months (De Luca et al., 1980a).

While the motivation for much of the work so far described was primarily to investigate aspects of cycad growth and development, certain research has been more specifically directed towards finding a protocol for micropropagation. Early work (Brown & Teas, 1966; Mustoe, 1967) prompted Henson (1980) to carry out a more extensive operation at the micropropagation unit of the Royal Botanic Gardens at Kew, England. Henson (1980) used relatively large explants from leaves, ovules, mega- and micro-sporophylls, coralloid roots and seedling hypocotyls from a large number of species for culture mainly on modified Murashige-Skoog (1962) media supplemented with various concentrations of auxins, cytokinins and gibberellic acid. Good callus growth was obtained from 35 species when 1.0 mg/liter NAA and 0.1 mg/liter BA was added although absolute concentrations were not critical. Callus morphology was variable and the slower-growing, more compact callus, which developed from explants of about half the species tested, gave rise to spherical structures analogous to Norstog's pseudobulbils. Plantlets could not be regenerated in this series of experiments although some root formation was observed on tissue derived from *Zamia* leaves. Stem and root explants from nine species of *Encephalartos* were cultured by Koeleman and Small (1982) on Murashige-Skoog (1962) media containing amino acid and amide supplements together with 10 mg/liter NAA and 15 mg/liter kinetin. Faster growing callus was obtained from mesic species than those from xeric habitats, but again subsequent morphogenesis was disappointing and no plantlets were established.

Extensive use of *in vitro* culture techniques has been made in order to assess the influence of light on root nodule development (Webb, 1981, 1982a, 1982b, 1983, 1984; Webb & De Jesus, 1982; Webb et al., 1984). When embryos from several taxa were cultivated on media containing White's (1943) macronutrients and the Murashige-Skoog (1962) micronutrient formulation, it was found that light stimulated nodule development but inhibited root initiation and development. Using mature embryos from *Zamia pumila*, Webb et al. (1983) found good callus formation on Murashige-Skoog (1962) medium with 1.0 mg/liter NAA and BA. The callus was more compact when the BA level was increased and gave rise to roots, shoots and occasional embryo-like structures not unlike the adventive embryos reported by Norstog and Rhamstine (1967). Although there was considerable leaf elongation *in vitro*, autotrophic plants could not be established. Continuing the studies with *Zamia pumila*, Monnier and Norstog (1984) cultivated immature embryos and their separate embryo, suspensor and egg membrane components, essentially to compare development *in ovulo* and *in vitro*. Maximum growth arose at the embryo base and after two months somewhat callus-like cotyledonary structures were formed.

In summary, therefore, a large amount of experimentation covering many taxa, media and environmental conditions has been carried out. In general, callus formation is quite readily obtained from both haploid and diploid tissue but morphogenesis has been consistently infrequent and has not yet led to the formation of viable cycad plants.

Current Researches into Cycad Tissue Culture

Despite the difficulties experienced so far in attempting to induce morphogenesis in cycads *in vitro*, several workers are convinced that it is possible, and that time and effort will eventually result in workable protocols for cycad micropropagation. Norstog (pers. comm.) is continuing his work with *Zamia pumila*, using primarily megagametophytes and embryos as source material and exploring possible routes to somatic embryogenesis. Litz (pers. comm.), also working in Florida, is using the same plant and also believes that somatic embryogenesis will be a key factor in establishing a protocol for micropropagation. Litz has had some recent success in this direction using nucellar material of *Zamia pumila* and culturing this in liquid media. De Lange (pers. comm.) is working at Kirstenbosch Botanic Garden, Capetown, South Africa with the objective of establishing clonal propagation routes for endangered species among the local flora.

This author (Osborne, 1988) has investigated the response *in vitro* of various explants from South African *Encephalartos* spp. and *Stangeria eriopus* to a number of different media and conditions. Callogenesis was readily accomplished in most instances; good results being obtained when small segments from the primary roots of various species were cultured on the medium of Schenk and Hildebrandt (1972). In general, about two-thirds of the explants gave evidence of callus initiation within 3–4 weeks and this was promoted in the dark. Consistent with the results of Koeleman and Small (1982), the mesic species such as *Encephalartos lebomboensis, E. natalensis* and *E. transvenosus* gave more vigorous callus growth than the xeric species like *E. cycadifolius* and *E. ghellinckii*. Whilst a variety of auxin and cytokinin combinations is possible in inducing this type of growth, a combination of

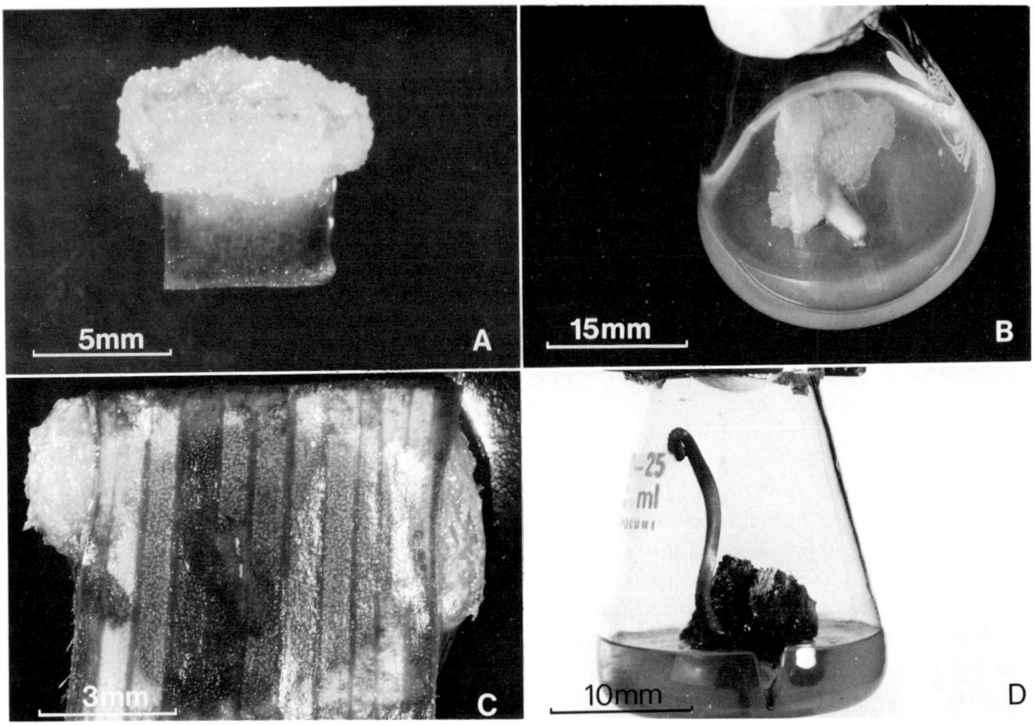

Fig. 1. Callogenesis and morphogenesis from cycad explants. **A.** Callus formation on a segment of primary root tissue from *Encephalartos natalensis* after 37 days in the dark on SH medium supplemented with 1 mg/liter 2,4-D and kinetin. **B.** Callus formation on a mature embryo from *E. villosus* after 40 days in the dark on medium 59 supplemented with 0.3 mg/liter 2,4-D and 3.0 mg/liter BA. **C.** Callus formation on the cut surfaces of a leaflet from *E. woodii* after 37 days in the dark on SH medium supplemented with 10 mg/liter each of NAA and kinetin. **D.** Emergence of a typical circinate leaf from *Stangeria eriopus* root tissue; initial callogenesis was induced in the dark on the medium in A above while morphogenesis followed after subculture and transfer to constant light environment. Culture temperatures were 22–25°C in all cases.

1 mg/liter each of 2,4-D and kinetin seemed to be consistently successful. Pre-treatment with a higher auxin concentration in combination with dimethylsulphoxide as a solvent vehicle appeared beneficial. Figure 1A, B and C shows the type of callus growth obtained from root, embryo and leaflet explants from three different *Encephalartos* species on different media. Despite the comparative ease of production of vigorous, sustained and subcultureable callus, it has not yet been possible to induce any form of morphogenesis with *Encephalartos.*

An encouraging development has been seen with explants from the root of *Stangeria eriopus* (Osborne & van Staden, 1987). Compact callus was obtained in 3–4 weeks on Schenk-Hildebrandt (1972) medium supplemented with 1 mg/liter 2,4-D and kinetin. Optimum callus growth occurs in the dark and may be subcultured on the same medium. On transfer to a medium intensity (100 μE m^{-2} s^{-1}) constant light environment, the cultures gave rise to small green meristematic zones from which buds and leaves arose (Fig. 1D). It may be a general phenomenon that morphogenesis is light-dependent and further work continues in this respect.

Further Applications of Cycad Tissue Culture Work

Although the objective of establishing a viable protocol for cycad micropropagation remains the most important reason for continuing work in this direction, it is by no means the only potential benefit. One of the other exciting prospects lies in the fact that the fast-growing callus, so readily obtained from a variety of cycad tissues, provides excellent material for studies on chromosome karyology. Using callus from the particularly rare species *Encephalartos woodii* (Fig. 1C), a simple aceto-orcein smear preparation shows the chromosome number for the culture is 2n = 18, consistent with the usual diploid number for the genus. This effectively rules out speculation that the solitary original specimen of this taxon may have arisen as a polyploid mutant (Osborne, 1986). Since the undifferentiated callus is perhaps the simplest of all tissues, this may well provide a starting point for detailed investigation into the difference between male and female plants, both in terms of chromosome morphology and biochemically. This information would provide a key to establishing the gender of plants at the seedling stage and to enable possible 'sex change operations' to be devised (Osborne, 1985). *In vitro* cultivation, too, will be vital to understand fully the relation between the cycad and the cyanobacteria in nitrogen-fixing symbiosis, knowledge which may have far-reaching implications in agriculture (Grobbelaar, 1985). Numerous other examples can be drawn from the biochemical and physiological areas, e.g., enzyme polymorphism studies (with possible taxonomic value) and cycad toxin biosynthesis. It is imperative that the present investigations into cycad tissue culture are supported and sustained.

Acknowledgments

Work by the author was supported by a research grant from the University of Natal. This paper was originally read at an International Cycad Conference, Nice, France, my attendance at which was kindly funded by the Alpes-Maritimes Department and by the Naples Botanical Garden.

Literature Cited

Brown C. L. & H. J. Teas. 1966. Cycad tissue cultures on a defined medium. Proc., Assoc. of Southern Agric. Workers, Southern Sect. of Amer. Soc. Plant Physiol., Jackson, Mississippi.

De Luca, P., V. La Valva & S. Sabato. 1980a. Spermatogenesis and tissue formation in cycad pollen grains. Caryologia **33:** 261–265.

———, **A. Moretti & S. Sabato.** 1979. Regeneration in megagametophytes of cycads. Giorn. Bot. Ital. **113:** 129–143.

——— **& S. Sabato.** 1979. *In vitro* spermatogenesis of *Encephalartos* Lehm. Caryologia **32:** 241–245.

——— **& S. Sabato.** 1980. Regeneration of coralloid roots on cycad megagametophytes. Plant Sci. Lett. **18:** 27–31.

———, ———, **A. Balduzzi & R. Nazzaro.** 1980b. Coralloid root regeneration on *Macrozamia* megagametophytes. Giorn. Bot. Ital. **114:** 271–275.

Gilbert, S. 1984. Cycads: Status, trade, exploitation and protection, 1977–1982. World Wildlife Fund, U.S.A.

Grobbelaar, N. 1985. Koraalvormige wortels. Encephalartos **4:** 4–9.

Henson, N. 1980. The formation of organized elements and callus from cycads in culture at Kew. Internal Report of Royal Bot. Gard., Kew, England.

Koeleman, A. & J. G. C. Small. 1982. A note on callus formation by stem and root tissue of some *Encephalartos* species. S. African J. Bot. **1:** 165–166.

Laliberte, S., C. Bertrand & J. Veith. 1983. Callogenèse et degré de différenciation dans des cultures d'endosperm d' *Encephalartos villosus* (Zamiaceae). Rev. Canad. Biol. Exper. **42:** 7–12.

La Rue, C. D. 1948. Regeneration in the megagametophyte of *Zamia floridana.* Bull. Torrey Bot. Club **75:** 597–603.

———. 1954. Studies on growth and regeneration in gametophytes and sporophytes of gymnosperms. Brookhaven Symp. Biol. **6:** 187–208.

Monnier, M. & K. Norstog. 1984. Developmental aspects of immature *Zamia* embryos in culture. Z. Pflanzenphysiol. **113:** 105–116.

Morel, G. M. 1974. Clonal multiplication of Orchids. Pages 169–222 *in* C. L. Withner (ed.), The orchids—Scientific studies. John Wiley.

Murashige, T. & F. Skoog. 1962. A revised medium for rapid growth and bioassays with tobacco tissue cultures. Physiol. Pl. **15:** 473–497.

Mustoe, J. L. 1967. Cultivation of cambial explants of various gymnosperms *in vitro.* M.A. Thesis. University of Georgia, Athens, Georgia, U.S.A.

Norstog, K. 1965. Induction of apogamy in megagametophytes of *Zamia integrifolia.* Amer. J. Bot. **52:** 993–999.

——— **& E. Rhamstine.** 1967. Isolation and culture of haploid and diploid cycad tissue. Phytomorphology **17:** 374–381.

Osborne, R. 1985. Sex-change in cycads—Hope for *woodii*? Encephalartos **2:** 20–22.

———. 1986. Focus on . . . *Encephalartos woodii.* Encephalartos **5:** 4–10.

———. 1988. An investigation into *in vitro* culture and phytochemical aspects of some members of the order Cycadales. Ph.D. Thesis, Univ. of Natal, Pietermaritzburg, South Africa.

——— **& J. G. Hendricks.** 1985. A world list of

cycads. Encephalartos **3:** 13–17, & Supplement 1. Encephalartos **5:** 27.

——— **& J. van Staden.** 1987. *In vitro* regeneration of *Stangeria eriopus.* Hort. Sci. **22:** 1326.

Pena, E. & E. Grillo. 1982. Proliferation of *Microcycas calocoma in vitro.* Rev. Jard. Bot. Nac. (Havana) **3:** 177–196.

Rubluo, A., V. Chavez, O. Martinez & A. Martinez. 1986. The recovery of endangered Mexican plant populations through tissue culture. Abstracts, VI Int. Congress of Plant Tissue & Cell Culture, Minneapolis, Minnesota, U.S.A. Page 428.

Schenk, R. U. & A. C. Hildebrandt. 1972. Medium and techniques for induction and growth of monocotyledonous and dicotyledonous plant cell cultures. Canad. J. Bot. **50:** 199–204.

Webb, D. T. 1981. Effects of light on root nodulation and elongation of seedlings in sterile culture of *Bowenia serrulata.* Phytomorphology **31:** 121–123.

———. 1982a. Importance of the megagametophyte and cotyledons for root growth of *Zamia floridana* DC embryos *in vitro.* Z. Pflanzenphysiol. **106:** 37–42.

———. 1982b. Effect of light intensity on root growth and nodulation of *Zamia floridana* embryos in sterile culture. Phytomorphology **32:** 81–84.

———. 1983. Developmental anatomy of light-induced root nodulation by *Zamia pumila* L. seedlings in sterile culture. Amer. J. Bot. **70:** 1109–1117.

———. 1984. Developmental anatomy and histochemistry of light-induced callus formation by *Dioon edule* (Zamiaceae) seedling roots *in vitro.* Amer. J. Bot. **71:** 65–68.

——— **& S. De Jesus.** 1982. Root nodulation in embryos of *Macrozamia diplomera* in sterile culture. Phytomorphology **32:** 253–256.

———**, M. Nevarez & S. De Jesus.** 1984. Further *in vitro* studies of light-induced nodulation in the Cycadales. Environ. Exp. Bot. **24:** 37–44.

———**, M. E. Rivera, E. Starszak & J. Matos.** 1983. Callus initiation and organized development from *Zamia pumila* embryo explants. Ann. Bot. **51:** 711–717.

White, P. R. 1943. A handbook of plant tissue culture. J. Cattell Press, Lancaster, Pennsylvania, USA.

Conservation Through Cultivation

CYNTHIA GIDDY

Abstract

GIDDY, C. (P.O. Box 45, Umlass Road, 3730 Natal, Rep. of South Africa). Conservation through cultivation. Memoirs of the New York Botanical Garden **57**: 89–93. 1990.—The article examines the conservation status of South African cycads and the threats posed to their natural habitats, the function of the Pollen Bank to ensure pollination of all mature plants in cultivation and the technique of hand pollination are discussed. New advances in the cultivation of artificially produced cycad seedlings are also discussed.

Key words: Pollen bank; conservation; hand pollination; cultivation.

Introduction

Cycads, an ancient group of plants, often called the "coelacanths of the plant world" have changed little since the Mesozoic Age—the time when dinosaurs roamed the earth. Not only are cycads threatened by climatic changes in their natural habitats, but ruthless collecting by avid collectors has brought many of the world's cycads close to extinction. The International Union for Conservation of Nature and Natural Resources (IUCN) has found that 80 of the 96 taxa of cycads of which the status is known, are threatened with extinction. Of the 28 South African species, 1 species, *Encephalartos woodii* from Natal, is "extinct" in nature, 3 are deemed "endangered," while 20 are listed as "vulnerable." This is certainly not a record of which we should be proud.

Threats to the habitat, either climatic in the form of prolonged droughts or man made, e.g., agriculture, afforestation, road and dam development and even the encroachment of urban sprawl, are puny in comparison with those posed by a determined collector with a spade. The collector by selectively removing female plants because the cones are more attractive and command a higher price on the black market, not only reduces the number of plants in the habitat but upsets the ratio of male to female plants. In many instances, natural regeneration in nature is no longer possible. In some localities, male plants outnumber females by 10:1; e.g., *E. eugene-maraisii* in the Middleburg Transvaal district.

The cycad collecting cult has given rise to the development of a thriving national and international market in both legal and illegal plants. Some collectors openly admit that they have no particular interest in the plants themselves and are merely stockpiling them for their investment value.

The Pollen Bank

Several cycad societies exist worldwide where cycad lovers, whether they are collectors, botanists or conservationists, can find common ground. The South African Cycad Society publishes a quarterly newsletter, "Encephalartos," with an informative range of articles from the academic to layman's tips on germination. It has also recently established a Seed Bank and a Pollen Bank to encourage the artificial propagation of seedlings and in this way help to lessen the threat to the habitat. Cycad seedlings at reasonable prices are now commercially available from private nurseries as well as the Department of Nature Conservation in the Transvaal.

Many thousands of cycads produce cones every year in cultivation. Only a few collectors have both a male and female plant of the same species that cone at the same time. Having removed a mature plant from the habitat, let us not compound the error by allowing potentially fertile seed to be wasted. Given the endangered status of our cycads, no female cycad in cultivation should be unproductive. Depending on the species concerned, a single cone produces between 300–450 seeds. A teaspoon of pollen is sufficient to pollinate a cone. With this in mind the South African Cycad Society recently established a Pollen Bank.[1]

Members of the South African Cycad Society are asked on a questionnaire to indicate which mature male and female plants they own or to which they have access. These data are then fed into a computer and categorized into species and areas. As soon as a member notices a female cone emerging in early autumn, they contact the Pollen Bank and attempts are made to locate a pollen source. At present deep frozen pollen is only available to those who live within easy distance of the Pollen Bank in Natal because the pollen deteriorates within 2–3 days. For those further afield a list of possible sources is provided and it is then up to the respective owners to negotiate a time and place and suitable recompense, usually a share of the seeds.

Pollen can however be successfully sent by mail locally within the country if the entire male cone is airmailed just before it commences to shed pollen. The pollen sacs on the undersides of the cone scales progressively mature over a period of several days and fresh pollen can be obtained in this way. At present experiments are being undertaken to determine whether pollen can be stored in liquid nitrogen in the same way that animal semen is sent intercontinentally.

Hand Pollination

The technique of hand pollination is easy to master; the only difficulty lies in the timing of the operation and in acquiring pollen. As soon as the male cone starts shedding pollen, the entire cone can be cut off and placed on a sheet of paper indoors (Fig. 1). This prevents the wind blowing the pollen away and also reduces the danger of cross pollination between species. The pollen sacs will mature over a period of 10–12 days, releasing copious amounts of pollen. In nature pollen is wind borne and settles on the top of the female cone. Subsequent gusts of wind will force the pollen grains between the top scales which open slightly and from where they will gradually sift down spirally.

Hand pollination therefore must attempt to simulate this process. Pollen must be introduced into the top of the cone, as there are no external openings between the tightly packed scales lower down. Determining when the female cone is receptive is not easy. A reasonably reliable guide would be when the male starts shedding pollen, as this would be the normal course of events in nature. Without a male plant of the same species to act as indicator, the female cones should be inspected frequently and as soon as a slight opening of the topmost scales is detected, pollinating should continue for 6–8 days.

It is necessary to place the pollen right inside the cone near to the central axis and to direct it downwards. By using a sharp knife, one of the scales fairly near to the top can be removed. Once the scale is removed the open space around the central stalk is clearly visible. A drinking straw or a kitchen syringe can be used to blow the pollen down into the cone (Fig. 2). Several months will elapse before the seeds are mature and the cone disintegrates. During this time particular care should be given to watering and feeding the mother plant as this will determine how soon that particular female plant will cone again. 'Pregnancy' for a cycad, as for humans and other animals, is a demanding and stressful period.

The South African Cycad Society boasts a membership of more than 500, of whom approximately 10% are overseas. The response to the Pollen Bank from the overseas members has

[1] The Pollen Bank, South African Cycad Society, P.O. Box 45, Umlaas Road, 3730 Natal, South Africa.

Fig. 1. A male cone of *Ceratozamia mexicana* which has been cut off and placed on a sheet of paper indoors. It is a curious phenomenon that such cones are warm to the touch while shedding pollen.

Fig. 2. By means of a syringe the pollen is introduced right into the center of the female cone of *Ceratozamia mexicana* just below the sterile cone scales at the top.

been most enthusiastic as cycad collections are few and far between. However, due to our different summer/winter climatic pattern fresh pollen is not available when they require it in the Northern Hemisphere. We recently had to reluctantly turn down an urgent request from the Botanical Garden of Tenerife in the Canary Islands for pollen of the very rare *Encephalartos laurentianus* from Zaire. Although a male plant was located in South Africa, it did not cone the same season. It is a matter of concern to conservationists that seeds of such an endangered plant group are so difficult to obtain. If every cycad owner were to place his plants at the disposal of a Pollen Bank, we could, through hand pollination of the cycads already in cultivation, ensure the survival of this plant group within a few years.

Recent Advances in the Cultivation of Cycads

Cycads have an undeserved reputation for being slow growing and hence seedlings have in the past had little attraction for collectors or commercial nurseries. Research and recent advances in horticultural technology are, however, proving that cycads grown under optimum conditions can achieve growth rates on a par with other foliage plants such as tree ferns and palms. At our nursery, seeds are placed on a sterile growing medium and kept at a constant 25–28°C by means of soil warming cables. Once germinated, the seedlings are planted into community pots and grown in a plastic tunnel in a controlled environment with a temperature range of 28–30°C and with a relative humidity of 60–70%.

Under these growing conditions with regular weekly applications of liquid fertilizer in the form of N-P-K 3-1-5 in a dilution of 1:1000, tubers weighing 250–300 g and with 6–8 leaves have been obtained within 18 months from the germination. The length of the leaves will of course depend on the species. The nutrients are applied to the root zone. Foliar feeding because of the thick cuticle of cycad leaves is less effective.

Recent experiments in the USA (Dehgan & McConnell, 1985) have shown that more leaves are produced under high light conditions than low light. If young seedlings are subjected to an additional 3–4 hours of artificial light of 250–2000 foot candles per day, the caudex develops rapidly and produces leaves more frequently. To a lesser extent this can also be achieved by planting the caudex well above the level of the growing medium. Higher light intensities (1000 ft c and higher) also prevent the yellowing and premature death of the lower leaves which occurs under natural conditions when the light is reduced by shading from the new crown. Cycads grown under these conditions are more compact and the additional foliage has more appeal.

Cycad seedlings require good drainage. Composted pine bark is put through a 1 cm screen and this as a growing medium gives the necessary air filled porosity and drainage while retaining moisture and nutrients. The pH of the growing medium is in the range of 6.0–6.5 and to date we have observed no species differentiation within *Encephalartos,* with respect to pH requirements. One species, *E. lanatus,* appears to require higher aluminum levels and the addition of aluminum sulphate to the fertilising program for these seedlings has resulted in better leaf color and growth.

The unbranched taproot of *Encephalartos* seedlings is prone to damage on transplanting and such damage can retard growth for several weeks. Dehgan (1983) reports that by cutting the taproot just below the leaf crown, and dipping it in an indolebutyric acid solution, a callus is formed and several branched adventitious roots then develop. These additional roots can then better utilize the water and nutrients, and plant growth is increased considerably. Strict attention must, however, be paid to plant hygiene. A sterile growing medium and regular use of fungicides is necessary, as under the high heat and moisture conditions root rot losses can be high. Previously, the long taproot required very deep containers. The adventitious roots produced under the above method have meant that standard 25 cm nursery containers can now be used for several years growth.

The ultimate conservation strategy for these endangered plants is tissue culture. By making more cycads commercially available the collecting pressure on wild populations could be eased. Also, tissue culture can also be used to select and produce plants with a specific characteristic based on horticultural criteria; e.g., attractive leaf forms, faster maturing clones, and those with cones of an aberrant cone color, as in the yellow-coned form of *E. ferox*.

Conclusion

Protection of their habitats is one key to the future survival of the cycads, another is to closely monitor and control the local and international trade by allowing only the exploitation of common species from wild populations and here there are some species of *Macrozamia* in certain localities that qualify. However, to satisfy the growing demand from an ever increasing number of collectors and gardeners, conservationists are agreed that an increased availability of artificially propagated plants is necessary. Conservation through cultivation will ensure the survival of the cycads for future generations.

Acknowledgments

I am indebted for the details of the conservation status of South African cycads to the Research Officers of the Natal Parks Board and to the report by the World Wildlife Fund entitled "Cycads: Status, Trade, Exploitation and Protection 1977–1982" by Sheryl Gilbert, M.S. Sources not quoted are personal observations and experience gained at our nursery.

Literature Cited

Dehgan, B. 1983. Propagation and growth of cycads—A conservation strategy. Proc. Florida State Hort. Soc. **96:** 137–139.

——— **& D. B. McConnell.** 1985. Cycads with potential as foliage plants. Palm & Cycad Society of Australia. Newsl. 7 (Apr–Jun).

Structural Arrangement of Phycobionts in Coralloid Roots of Cycads

MARIA GRILLI CAIOLA

Abstract

GRILLI CAIOLA, M. G. (University of Rome, via E. Carnevale, 00173 Rome, Italy). Structural arrangement of phycobionts in coralloid roots of cycads. Memoirs of the New York Botanical Garden **57**: 94–103. 1990.—The cycad coralloid root organization has been studied by light and electron microscopy. Gonidial zone sections taken at different levels from the apex to the base were treated by different cytochemical reactions to demonstrate the nature of the components of the mucilage filling the intercellular gonidial layer. The gonidial zone of the coralloids from various cycad species shows a similar structural organization. Phycobionts consist of vegetative cells, heterocysts and akinetes. These latter are abundant in the coralloids cultured for a year at 25°C in the various mineral nutritional media. In the old parts of the coralloids the heterocyst ultrastructure seems to indicate that the heterocysts could divide, originating new vegetative cells or colonies.

Key words: Cycad; coralloid root; phycobionts; ultrastructure.

Introduction

All cycads, in addition to normal roots, carry apogeotropic roots which extend above the ground as numerous dichotomously branched structures. For their resemblance to corals they were named coralloid roots. They are an unusual example of symbiosis between a spermatophyte and the cyanobacteria *Nostoc* or *Anabaena*. The phycobionts occupy a definite layer of the cortical zone, called the "gonidial zone." Special cells (heterocysts) possess a nitrogenase which enables them to fix molecular nitrogen, and the resulting compounds are utilized by the symbiont and the host.

Initial studies regarding the cycad and algal association appeared at the end of the last century (Kny, 1874; Reinke, 1872). At the beginning of this century and more recently, other authors have examined cycad coralloid roots and their phycobionts for different purposes (Pampaloni, 1901; Spratt, 1911, 1915).

So, evolving from Reinke's parasitic concept, the idea of a symbiosis has been reached, with proof forthcoming that the phycobionts are nitrogen fixing (Bergersen et al., 1965; Douin, 1953; Lindblad et al., 1985a, 1985b; Watanabe & Kiyohara, 1963).

In 1963 I (Grilli, 1963) began to examine the coralloid roots of *Cycas revoluta* and, later, numerous other cycads growing in the Botanical Gardens of Naples, with the aim of comparing the structural organization of this association in different cycad species. The main topics I investigated were: structure and ultrastructure of the gonidial zone; the ultrastructure of the phycobiont in the symbiosis; survival of the phycobionts in the coralloid roots and identification of the phycobionts.

To fully investigate these aspects, I extended

my research to various genera and species of cycads that were available in a number of Italian botanical gardens (Grilli Caiola, 1980).

Materials and Methods

For these studies coralloid roots were sectioned and observed both unstained and after treating the sections with Alcian blue at different pH levels (McCracken & Barcellona, 1976), and ruthenium red and iron chloride (Obukowicz et al., 1981), to determine the acid and sulphate polysaccharides and the phenolic compounds, respectively.

Some observations were also made by a Leitz Orthoplan fluorescent microscope equipped with a Ploemopak 2 system and an Osram mercury lamp H BO 100 W with an exciting filter of 546 nm and a barrier filter of 580 nm (Wilde & Fliermans, 1979).

For transmission electron microscopy (TEM), sections of coralloids were routinely fixed in 6% glutaraldehyde in 0.2 M phosphate buffer, pH 6.9, and as follows: a) 3% glutaraldehyde with 0.5% Alcian blue (Grilli Caiola & Pellegrini, 1984); b) 3% glutaraldehyde with 0.1% ruthenium red (de Vecchi & Grilli Caiola, 1986); c) glutaraldehyde with 2% tannic acid for 2 hours at room temperature (Vaara, 1982).

The different fixation and staining techniques have been used in order to reveal the presence of acid and sulphate polysaccharides, and protein. All the samples were postfixed in 2% osmium tetroxide for 1 hour, then dehydrated in a series of graded ethyl alcohols and embedded in Araldite.

Thin and ultrathin sections cut with a diamond knife were observed either stained with uranyl acetate and lead citrate (Reynolds, 1963) using a TEM Philips 400.

Heterocyst frequency was calculated according to the techniques previously reported (Grilli Caiola, 1980).

For the study of survival forms, coralloid roots of *Cycas revoluta* were gathered from a single female plant grown in the Botanical Gardens of Rome (Italy). The material, taken in December, at 10 cm deep, was sterilized by 1% $HgCl_2$, separated in single units and then kept in 1000 ml flasks containing one of the following media: 1) BD (Baslerová & Dvöráková, 1962), nitrogen-free medium; 2) BD with added 0.02 M $NaNO_3$; 3) BD with 0.02 M $NaNO_3$ and 0.002 M NH_4Cl; 4) BD with added 0.002 M NH_4Cl; or 5) BD with added 5% glucose.

The flasks containing the coralloid roots were maintained in continuous light (1000 lux) or in the dark at 25°C or 4°C.

At intervals of 2 to 4 weeks over a period of a year until complete degeneration, the coralloid roots were examined by sectioning them in slices of 3 to 5 nm or by squashing the gonidial zone between slide and cover slide to ascertain the phycobiont surviving the host plant.

A set of coralloid roots was utilized to isolate the phycobionts by the technique used in a previous study (Grilli Caiola, 1980). The isolates were then cultivated in nutritional and environmental conditions similar to those adopted for culturing the coralloids.

Results and Discussion

Structure and Ultrastructure of Gonidial Zone

By comparing numerous thin and ultrathin sections taken at different levels of the coralloid roots from the apex towards the base, it was possible to evidence many aspects common to all the coralloids. In all cases phycobionts are localized intercellularly, rarely intracellularly (Grilli, 1963; Nathanielsz & Staff, 1975), in the gonidial zone. The intercellular spaces are filled with mucilaginous material (Fig. 1) which swells in water or diluted solutions. This material is mainly produced by host cells and only partially by the symbiont. Secretion is evident at the level of the cells of the gonidial layer (Fig. 2). Some of these cells show invaginations typical of "transfer cells." A similar organization has been described in *Zamia skinneri* by Lindblad et al. (1985a) and in *Cycas revoluta* by myself. However, in *C. revoluta* the correlation between the transfer cells and the surrounding host cells is not clear. In many cases the wall invaginations seem to be localized only at the tangential walls, allowing the transport of fixed nitrogen from the gonidial layer towards the other parts of the coralloid root.

The composition of the material filling the gonidial spaces seems to change from the apical zone towards the base of the coralloids. This is evidenced by different staining and cytochemical reactions. At the apical zone the mucilage stains mainly by Alcian blue, toluidine blue and ruthenium red, evidence for sulphate and acid polysaccharides, whereas in the basal zone mainly acid polysaccharides are observed.

After $FeCl_3$ treatment the gonidial zone appears little stained, indicative of the presence of only minimal amounts of phenolic compounds in comparison to the surrounding cortical cells, which are more intensely stained. According to Obukowicz et al. (1981), the presence of phenolic compounds in the mucilage of gonidial cells and of the cortical zone could inhibit the penetration of other microorganisms into the gonidial layer. In fact, a characteristic feature of the gonidial zone is its axenic aspects, i.e., without bacteria, which is not the case in other symbioses with cyanobacteria (i.e., *Azolla–Anabaena*).

Phycobiont Ultrastructure in Symbiosis

Phycobionts in symbiosis are always polymorphic. In all cases vegetative cells and heterocysts are visible (Fig. 3). But in the apical parts vegetative cells are more abundant (Figs. 3–5), whereas in the basal zone heterocysts are more numerous (Figs. 6–8).

Heterocyst frequency (HF) is always high, even though its value changes according to which different host plant species is involved.

Often the heterocysts are double or multiple. From the apex towards the base, besides the change in HF, the structure of the heterocysts also changes, becoming progressively more degenerate and swollen or reduced to a mass of mucilaginous material (Figs. 7, 8), with a small portion of cytoplasm containing some thylakoids as one moves proximally. Among the normal vegetative cells and heterocysts are other cyanobacterial cells. Some of them are smaller and apparently lack all wall. Some others are often irregularly shaped, with a thick cell wall (Figs. 9, 10). It is very difficult to interpret the role and significance of these uncommon cell organizations.

Sometimes many cells, apparently vegetative but smaller and richer in thylakoids, divide, seemingly in a 3-dimensional way, originating a mass of cells.

Another type of phycobiont cell is visible in the basal zone as larger vegetative cells. In the isolated tissue they appear to be dividing and originating spherical colonies containing trichomes of vegetative cells (Fig. 11).

Vegetative phycobiont cell ultrastructure appears very different in various host species. In *Cycas* they show a reticular thylakoidal apparatus (Figs. 2, 5, 6); in *Dioon* they are rich in lipid droplets and cyanophycin; in *Encephalartos* the thylakoids are mainly peripheral; in *Macrozamia* they are very rich in cyanophycin (Grilli Caiola, 1980).

They always contain chlorophyll and phycobiliproteins, as demonstrated by pigment determination in methanolic and buffered phosphate extracts and by their intense autofluorescence released at 580 nm. Phycobilisomes have been described by Lindblad (1985a) in vegetative cells of *Zamia skinneri* phycobionts.

After tannic acid treatment many cells show a very dark and thick cell wall, mainly in the oldest part of the coralloid (Fig. 10).

Survival of the Phycobionts in the Coralloids

Akinetes are rare in the coralloids. They were reported by Grilli Caiola (1974) in *Macrozamia communis* and Lindblad (1985a) in *Zamia skin-*

→

Figs. 1–3. Gonidial zone of the coralloid root. **1.** Phycobionts localized in the intercellular spaces containing mucilaginous material stained by toluidine blue. LM micrographs. **2.** Host cell wall showing a fibrillar mucilaginous production (m) in the gonidial layer (arrow). **3.** Isolated trichomes with vegetative cells (vc) and heterocysts (h) in the apical zone of *Encephalartos* coralloid.

Figs. 4, 5. TEM micrographs of *Cycas revoluta* coralloids. **4.** Trichomes of vegetative cells (vc) and heterocyst (h). **5.** Vegetative cells show an electron transparent sheath. In their cytoplasm thylakoidal apparatus (t), cyanophycin (c) and glycogen (g) granules are visible.

Figs. 6, 7. Basal zone of a *Cycas revoluta* coralloid. **6.** Vegetative cell (vc) and heterocysts (h) embedded in the host mucilage (m). **7.** Among the degenerated heterocysts (h), electron dense granular material (arrows) stained by Alcian blue is evident.

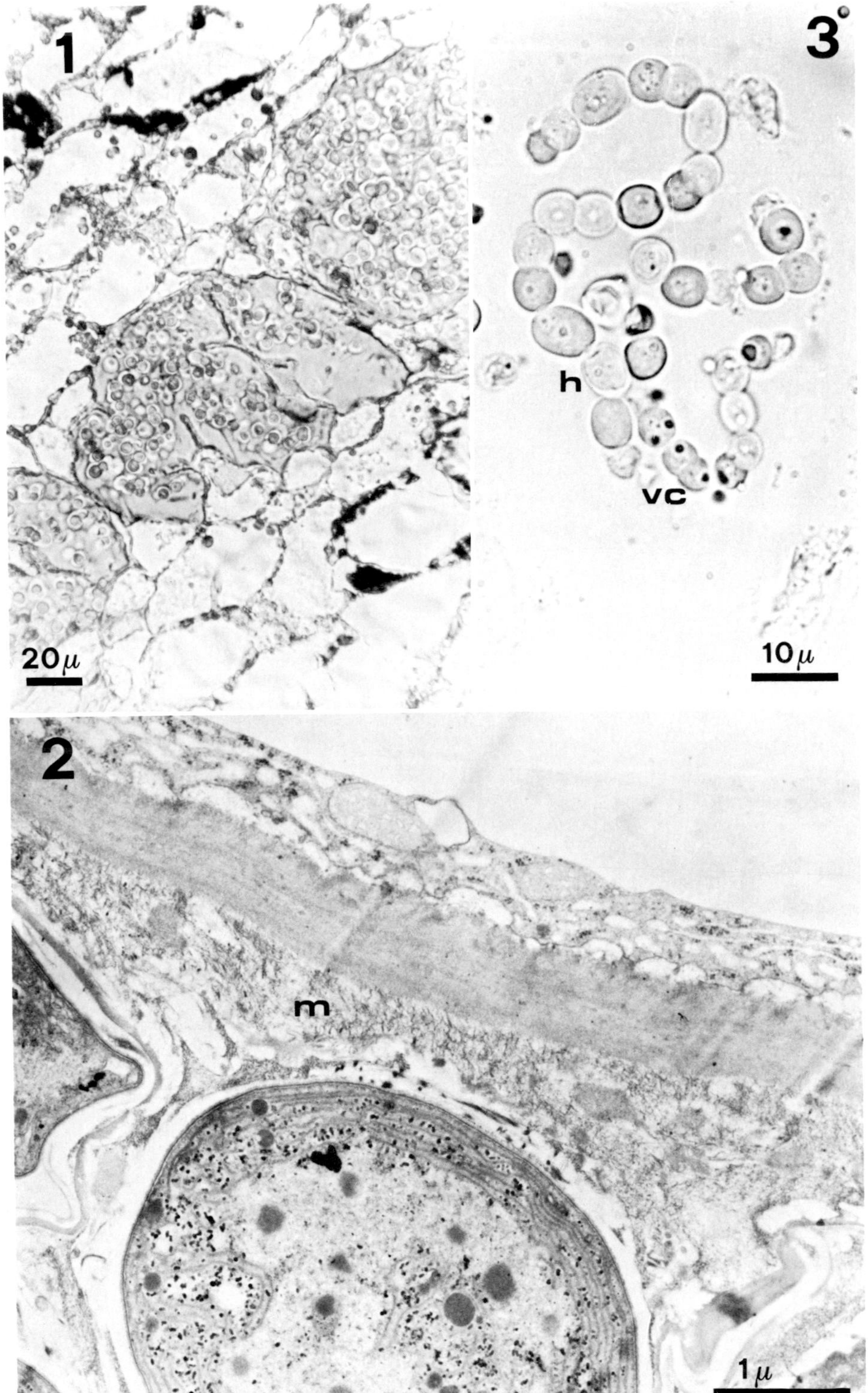
1
20μ
3
h
vc
10μ
2
m
1μ

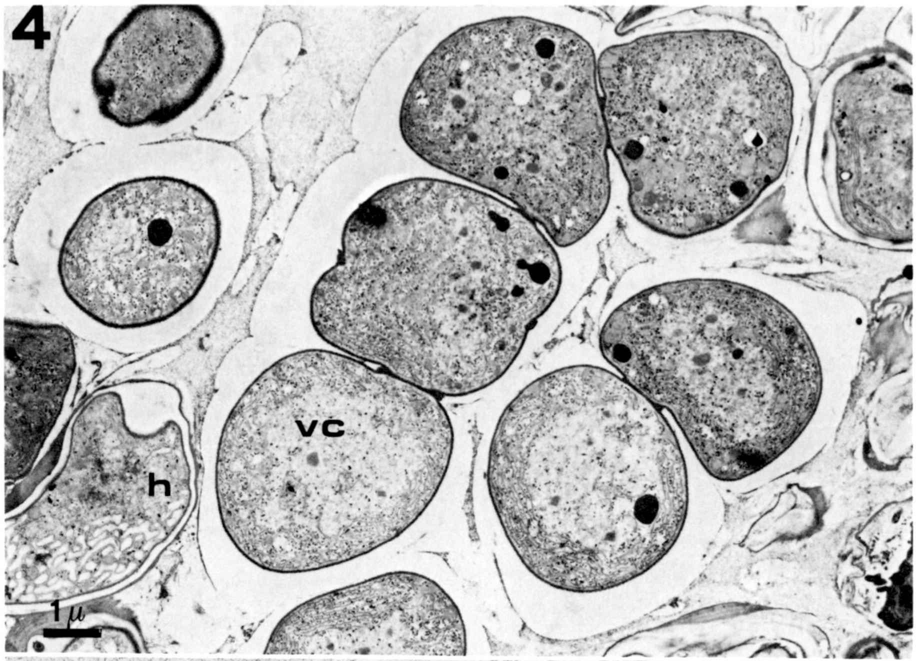
4
vc
h
1μ

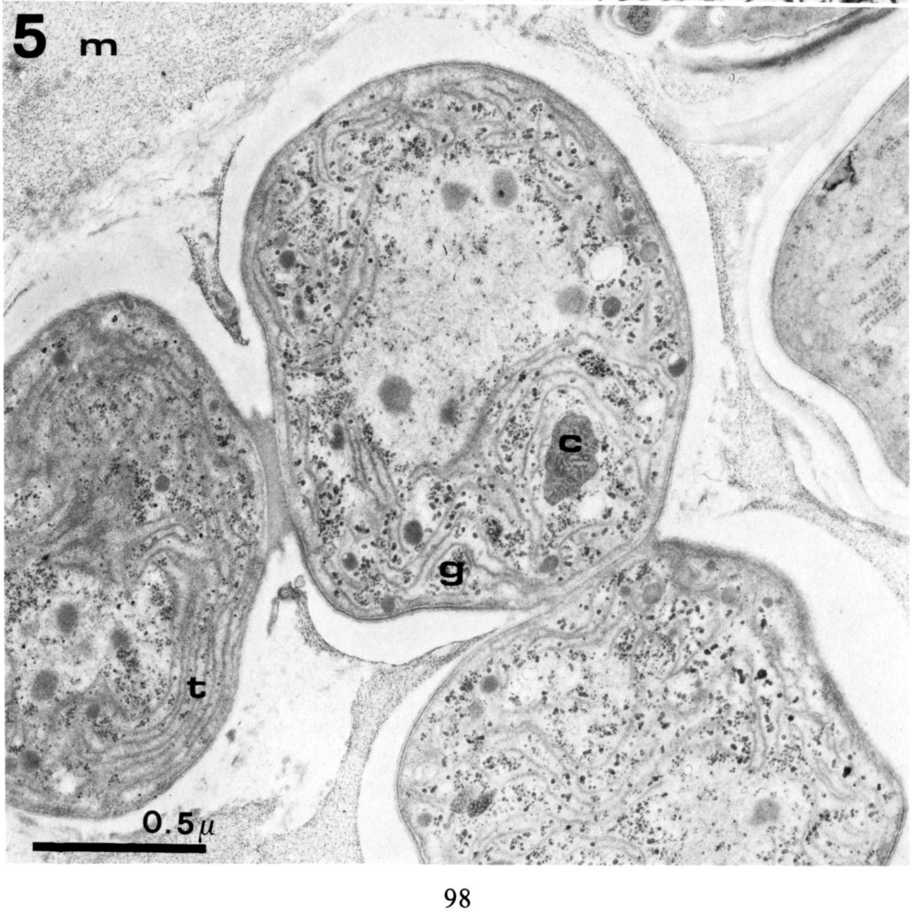
5
m
c
g
t
0.5μ

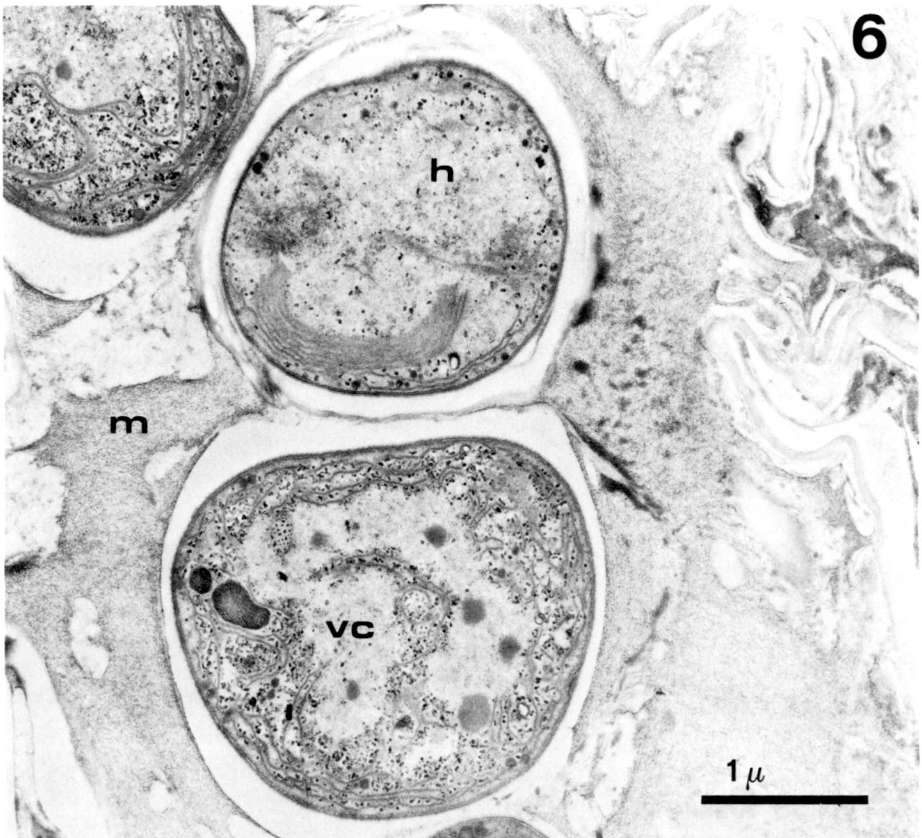
6
h
m
vc
1 μ

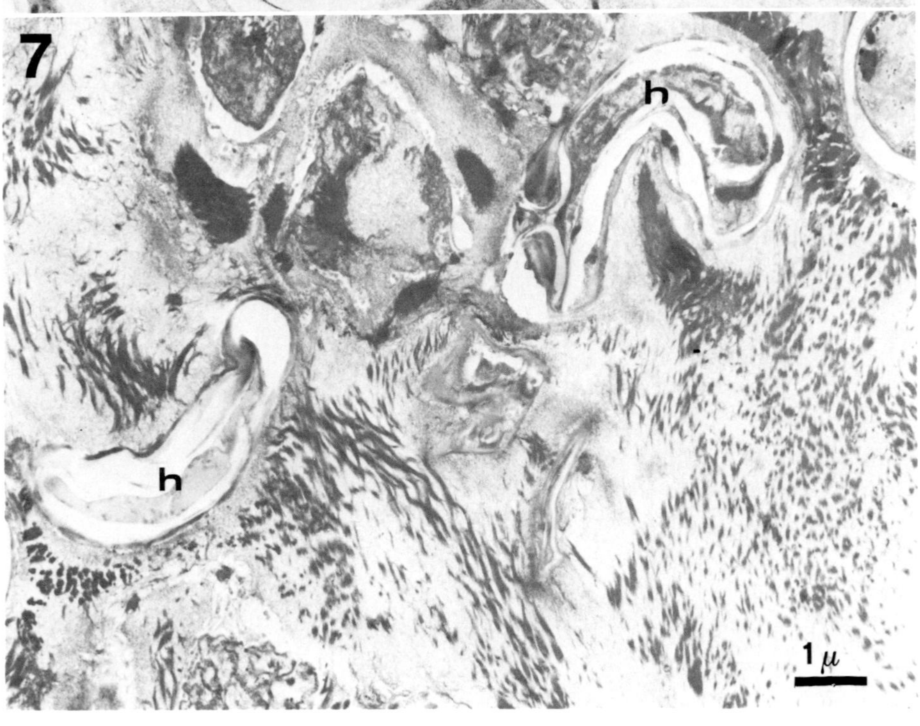
7
h
h
1 μ

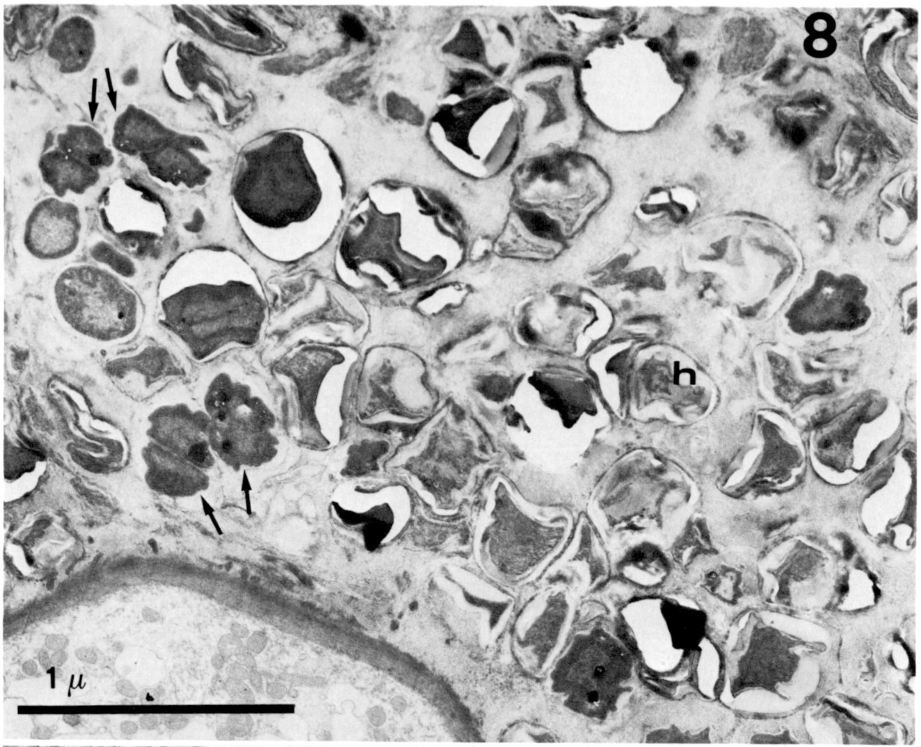
8
h
1 μ

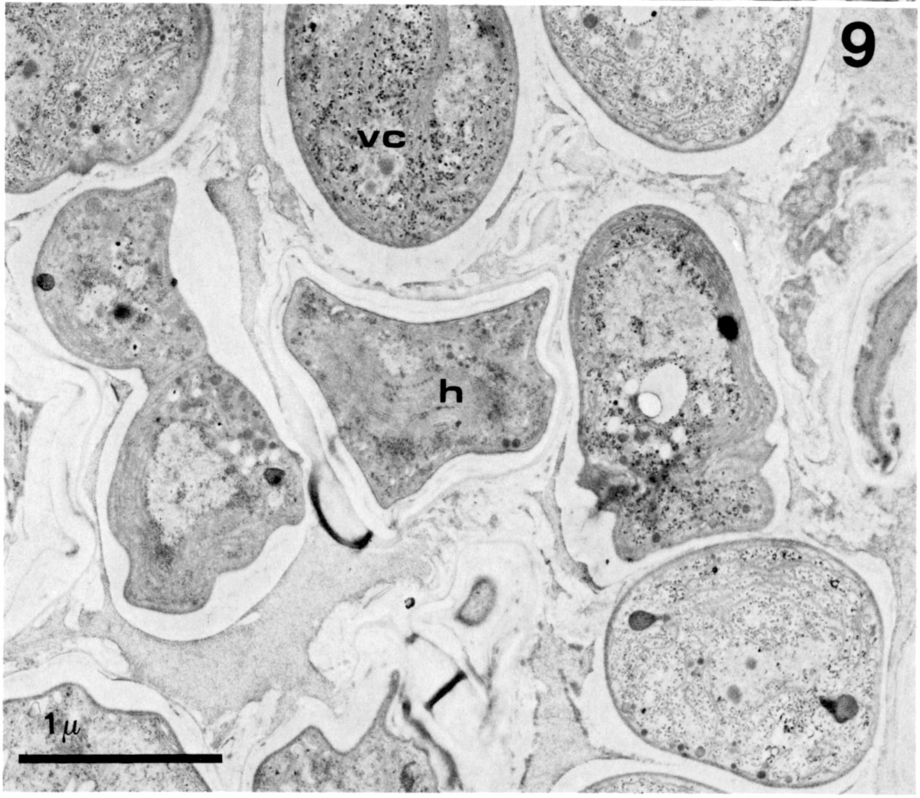
9
vc
h
1 μ

neri, but they are not common to all the coralloids, not even to the oldest ones (Grilli Caiola, 1980). However, their absence could depend on the timing of the harvesting and observation of the material, or their differentiation could be very rapid and of short duration.

On this matter I experimented with *Cycas revoluta,* using coralloids taken during the winter and grown for a year in mineral media, in different light and temperature conditions. From these experiments I have obtained two kinds of information: 1) about akinete differentiation, and 2) about heterocyst evolution. With regard to the akinetes I noted that they did not appear in the coralloids kept at 4°C, but only in those kept at 25°C, independent of the light/dark regime or the nutritional medium. In fact, akinetes were observed in the coralloids grown in nitrogen-free medium, illuminated (Fig. 12) or in the dark. In addition, akinetes were noted in the coralloids grown in the nitrogen-free medium with nitrate added and illuminated. In all the experiments the akinetes were recorded in degenerated coralloids, at the end of the experiments. From these results it seems that temperature was a more important factor than either light or nitrogen compounds for akinete differentiation. This observation has been confirmed also by experiments on *Nostoc* from the same coralloids isolated and grown in cultures at different temperature, light intensity and nutritional conditions. Also in the *Nostoc* from cycad coralloid roots, isolated and grown in culture, it has been observed that the akinete differentiation anticipates (Grilli Caiola & Forni, 1984) or occurs only at the higher temperature (25°C), though other factors, such as ammonia and the light cycle, could influence the process.

Low temperature could inhibit akinete differentiation in the natural habitat if the soil does not reach a temperature similar to that in the laboratory. In cycad coralloid roots, the phycobiont biology appears influenced as much by the external factors as by the host plants.

Other information concerns heterocyst evolution. Comparison of the heterocysts and gonidia in the different zones of the coralloids convinces me that many heterocysts can dedifferentiate and divide, originating the small cells visible in the thin and ultrathin sections of the coralloids. It appears that the heterocysts at one stage of aging can lyse their envelopes, reorganize their thylakoidal apparatuses and form new cell walls, originating new cells capable of rapid and multiple division (Fig. 11). These irregularly shaped cells resemble those observed in the different zones of the coralloids (Fig. 8). In the oldest coralloids these cells are visible only in the apical zone, but in the earlier stages they appear numerous in the peripheral and basal zones of gonidial layer. These results differ from those of Spratt (1911) because that author hypothesized that the heterocysts germinated directly, not after redifferentiation into vegetative cells.

In my previous paper (Grilli Caiola, 1980) I affirmed the impossibility of demonstrating the presence of heterocyst germination and akinete differentiation in isolated gonidial tissue. This impossibility could be attributed to the lack of an appropriate nutritional medium, as well as to inappropriate temperature and light of cultural conditions. Spratt (1911) observed heterocyst germination in the gonidial tissue; my previous researches were carried out with tissue squashed in mineral media. This suggests the possibility of mixotrophic metabolism in the gonidial layer and the capacity of the heterocysts to divide, thus originating new vegetative cells.

Nitrogen fixation determinations (Lindblad et al., 1985b) suggest the median zone is more nitrogen-fixing than the basal one, although the HF was higher in the latter. The explanation of this apparent contradiction lies in the different structure of the heterocysts, which in the oldest part of coralloids degenerate, losing their contents, or dedifferentiating. The host plant could utilize the nitrogen compounds released in the gonidial layer.

This hypothesis is in accordance with the results of Lindblad et al. (1985b) and also with

←

Figs. 8, 9. Basal zone of a *Cycas revoluta* coralloid. **8.** Among the degenerated heterocysts (h) numerous small irregularly shaped cells (arrows) and dividing cells are noticeable in the oldest part of the coralloid. **9.** Aspects of a heterocyst (h) and an irregularly shaped dividing cell. In both cells the cytoplasm is more electron dense and possesses a peripheral thylakoidal system.

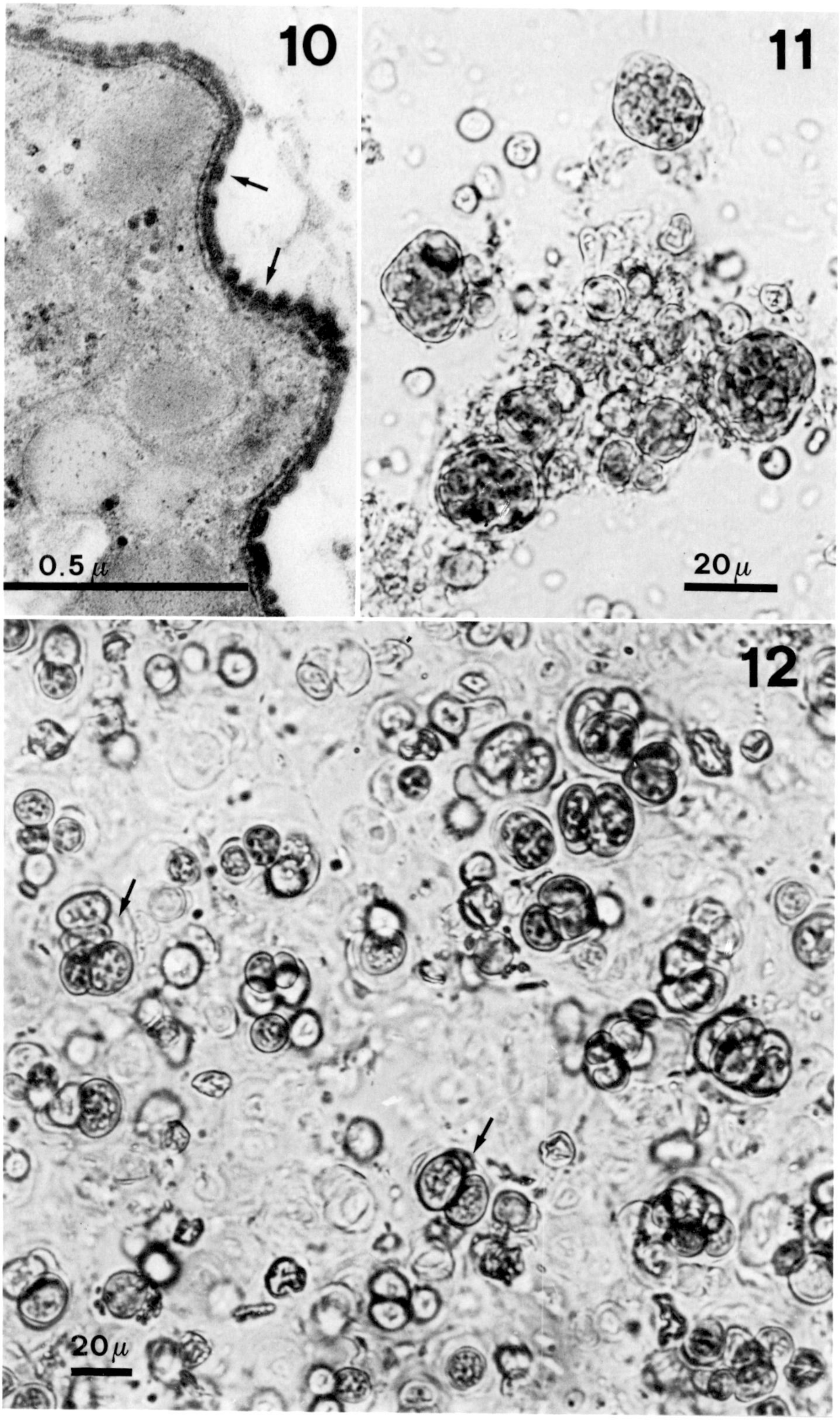
10
0.5 μ
11
20 μ
12
20 μ

those of Perraju et al. (1986), indicating that the enzymes for nitrogen fixation change from the apex toward the base of the coralloids.

Acknowledgments

I thank Prof. P. De Luca, Director of Botanical Garden of Naples (Italy) very much for the supply of the material used in the present study. This research was supported by CNR, Italy, Special grant I.P.R.A. Sub-project 1 Paper no. 1407.

Literature Cited

Baslerová, M. & J. Dvŏráková. 1962. *Algarum, Hepaticarum Muscorumque* in culturis collectio. Nakladatelství Ceskoslovenské Akademie věd., Praha. Pages 7–8.

Bergersen, F. J., G. S. Kennedy & W. Wittmann. 1965. Nitrogen fixation in the coralloid roots of *Macrozamia communis* (L.) Johnson. Austral. J. Biol. Sci. **18:** 1135–1142.

de Vecchi, L. & M. Grilli Caiola. 1986. An ultrastructural and cytochemical study of *Anabaena* sp. (Cyanophyceae) envelopes. Phycologia **25:** 415–422.

Douin, M. R. 1953. Sur la fixation de l'azote libre per les Myxophytes des Cycadacéés. C. R. Acad. Paris **236:** 956–958.

Grilli, M. 1963. Ricerche al microscopio elettronico sulle cianoficee viventi nei tubercoli raicali di *Cycas revoluta.* Ann. Microbiol. **13:** 109–124.

Grilli Caiola, M. 1974. A light and electron microscopic study of the blue-green algae living either in the coralloid roots of "*Macrozamia communis*" or isolated in culture. Giorn. Bot. Ital. **108:** 161–173.

———. 1980. On the phycobionts of cycad coralloid roots. New Phytol. **85:** 537–544.

——— **& C. Forni.** 1984. Study on the cycad phycobionts in culture. Third Int. Symp. on "Nitrogen fixation with non legumes," 2–8 September 1984, Helsinki.

——— **& S. Pellegrini.** 1984. Lysis of *Microcystis aeruginosa* (Kuts) by *Bdellovibrio*-like bacteria. J. Phycol. **20:** 471–475.

Kny, P. 1874. Eine grüne parasitische Alge. Sitzungsber. Ges. naturf. Freunde Berlin. *In* A. Schneider "Mutualistic Symbiosis of Algae and Bacteria with *Cycas revoluta.*" Bot. Gaz. **19:** 25–32.

Lindblad, P., B. Bergman, A. V. Hofsten, L. Hällbom & J. E. Nylund. 1985a. The cyanobacterium–*Zamia* symbiosis: An ultrastructural study. New Phytol. **101:** 707–716.

———**, L. Hällbom & B. Bergman.** 1985b. The cyanobacterium–*Zamia* symbiosis: C_2H_2—Reduction and heterocyst frequency. Symbiosis **1:** 19–28.

McCracken, M. D. & W. J. Barcellona. 1976. Electronhistochemistry and ultrastructural localization of carbohydrate-containing substances in the sheath of *Volvox.* J. Histochem. Cytochem. **24:** 663–673.

Nathanielsz, C. P. & I. A. Staff. 1975. On the occurrence of intracellular blue-green algae in cortical cells of the apogeotropic roots of *Macrozamia communis* (L.) Johnson. Ann. Bot. **39:** 363–368.

Obukowicz, M., M. Schaller & G. S. Kennedy. 1981. Ultrastructure and phenolic histochemistry of the *Cycas revoluta–Anabaena* symbiosis. New Phytol. **87:** 751–759.

Pampaloni, L. 1901. *Nostoc punctiforme* nei suoi rapporti coi tubercoli radicali delle Cicadee. Nuovo Giorn. Bot. Ital., n.s. **8:** 626–632.

Perraju, B. T. V. V., A. N. Rai, A. P. Kumar & H. N. Singh. 1986. *Cycas circinalis–Anabaena cycadeae* symbiosis: Photosynthesis and the enzymes of nitrogen and hydrogen metabolism in symbiotic and cultured *Anabaena cycadeae.* Symbiosis **1:** 239–250.

Reinke, J. 1872. Parasitische *Anabaena* in Wurzelen der Cycadeen. Gottingen Nachrichten, 107.

Reynolds, E. S. 1963. The use of lead citrate at high pH as an electronopaque stain in electron microscopy. J. Cell Biol. **17:** 208–212.

Spratt, E. R. 1911. Some observations on the life-history of *Anabaena cycadeae.* Ann. Bot. **25:** 369–380.

———. 1915. The root-nodules of the Cycadaceae. Ann. Bot. **2:** 619–626.

Vaara, T. 1982. The outermost surface structures in chroococcacean cyanobacteria. Canad. J. Microbiol. **28:** 929–941.

Watanabe, A. & T. Kiyohara. 1963. Symbiotic blue-green algae of lichens, liverworts and cycads. Microalgae & photosynthetic bacteria. Pages 189–196 *in* Jap. J. Pl. Physiol. Special volume.

Wilde, E. W. & C. B. Fliermans. 1979. Fluorescence microscopy for algal studies. Trans. Amer. Microsc. Soc. **98:** 96–102.

←

Figs. 10–12. *Cycas revoluta* coralloid. **10.** Many cells show the cell wall with a thickened and electron dense outer membrane. **11.** Gonidial tissue with new colonies. **12.** Squashed gonidial tissue from a year-old coralloid in culture. Numerous germinating akinetes are visible (arrows).

Nitrogen and Carbon Metabolism in Coralloid Roots of Cycads

PETER LINDBLAD

Table of Contents

Abstract

LINDBLAD, P. (Department of Physiological Botany, University of Uppsala, Box 540, S-751 21 Uppsala, Sweden). Nitrogen and carbon metabolism in coralloid roots of cycads. Memoirs of the New York Botanical Garden **57**: 104–113. 1990.—Filamentous heterocystous cyanobacteria of the genus *Nostoc* are found in a zone between two cortex layers within cycad coralloid roots. The zone is traversed by elongated cycad cells, supposedly involved in the transfer of metabolites between the symbionts. Nitrogen fixation was maximal in the growing tip and declined towards basal parts of the root. In contrast, the heterocyst frequency gradually increased towards basal, older parts and formation of multiple heterocysts commenced simultaneously. The inverse relationship between nitrogen fixation and total heterocyst frequency indicates a high frequency of inactive heterocysts in the multiple complexes, especially since nitrogenase activity correlates well with the frequency of single heterocysts. As in free-living cyanobacteria, nitrogenase occurred in the heterocysts only. The activity and cellular/subcellular localization of the primary ammonia assimilating enzyme, glutamine synthetase, was comparable to that of free-living cyanobacteria. Moreover, glutamate synthase showed full *in vitro* activity, suggesting that the nitrogen metabolite transferred from the cyanobiont to the cycad may be a product of ammonia assimilation. The light harvesting phycobiliprotein phycoerythrin was present in the vegetative cells, although the cyanobionts occur in darkness. The absence of light was, however, manifested by no *in vivo* CO_2-fixation. However, both ribulose 1,5-bisphosphate carboxylase and phosphoribulokinase activities were found in cell-free extracts.

Key words: CO_2-fixation; coralloid root; cyanobacteria; cycad; N_2-fixation; *Nostoc*; symbiosis.

Introduction

Cycads produce at least three different types of roots; a tap root, lateral roots and dichotomously branched, coral-like clusters, referred to as coralloid roots. The coralloid roots are produced by the cycad as axenically grown cycad seeds still form these structures (see Webb et al., 1984). Furthermore, the megagametophyte may produce coralloid roots (DeLuca & Sabato, 1980; DeLuca et al., 1980).

Coralloid Roots

Coralloid roots have been found in all cycads examined so far. The following description of cycad coralloid roots is based on field-material from the Costa Rican cycad *Zamia skinneri* (Fig. 1). When cross- or longitudinal-sectioning an infected coralloid root, a distinct, dark-green cyanobacterial zone is seen (Fig. 1C). The zone is built up by elongated *Zamia* cells, interconnecting two cortex layers (Fig. 1D).

Ultrastructurally, these host cells showed the characteristics of transfer cells: considerably expanded/invaginated cell walls and numerous mitochondria and amyloplasts (Pate & Gunning, 1972). It was suggested that these elongated cells are involved in the transfer of solutes between the host and the cyanobiont (Lindblad et al., 1985a). They resemble the glands (hairs) or fingerlike protrusions, of host origin, which frequently develop into the cyanobacterial zone of most other symbioses known to contain cyanobacteria, such as *Anthoceros* (Duckett et al., 1977), *Azolla* (Duckett et al., 1975; Neumuller & Bergman, 1981), *Blasia* (Duckett & Prasad, 1974; Duckett et al., 1977; Rodgers & Stewart, 1977) and *Gunnera* (Silvester & McNamara, 1976). The cyanobacteria were only found between the elongated cells, i.e., intercellularly, contrasted with the reports of intracellular occurrence in *Cycas revoluta* (Grilli, 1963; Obukowicz et al., 1981) and *Macrozamia communis* (Nathanielsz & Staff, 1975).

The Cyanobacterium

The cyanobacterium has the characteristics of the genus *Nostoc,* according to the criteria given by Rippka et al. (1979); i.e., spherical vegetative cells and hormogonia are produced. *Nostoc* also seems to be the cyanobiont found in six other genera: *Dioon, Macrozamia* (Grilli Caiola, 1980), *Cycas* (Grilli Caiola, 1980; Florenzano et al., 1985), *Encephalartos* (Grilli Caiola, 1980; Grobbelaar et al., 1987), *Bowenia* (Bowyer & Skerman, 1968) and *Stangeria* (Grobbelaar et al., 1987). However, *Anabaena* has been reported in *Cycas revoluta* (Obukowicz et al., 1981; Zhu, 1982), both *Anabaena* and *Nostoc* in *Macrozamia communis* (Nathanielsz & Staff, 1975), and *Calothrix* in *Encephalartos* (Grobbelaar et al., 1987).

The *Nostoc* filaments consist of vegetative cells and heterocysts, as well as a few akinetes (spores) (Fig. 1E). Ultrastructurally, the photosynthetic apparatus in the cyanobiont of *Zamia skinneri* appears to be well developed since the vegetative cells contained carboxysomes and thylakoids with distinct phycobilisomes, harboring phycobiliproteins. Furthermore, the cyanobionts retain their chlorophyll *a* and contain both nitrogen (cyanophycin granules) and carbon (glycogen granules) reserve material (Lindblad et al., 1985a).

Heterocyst Frequency

As in other cyanobacteria-eukaryotic plant symbioses, an increased heterocyst frequency, compared to the 5–6% found in free-living nitrogen-fixing cyanobacteria, is evident in symbiotic cyanobacteria of cycads (Grilli Caiola, 1975, 1980; Zhu, 1982; Lindblad, 1984; Lindblad et al., 1985b; Kumar et al., 1986; Lindblad & Bergman, 1986). A dramatic increase in the heterocyst frequency was observed, from 17% in the growing tip to 46% in basal, older parts in *Zamia skinneri* coralloid roots (Lindblad et al., 1985b). A more detailed study of the tip region revealed a heterocyst frequency as low as 5.4% (Lindblad & Bergman, 1986); a frequency similar to that observed in free-living nitrogen-fixing cyanobacteria.

There is not only an increase in the total heterocyst frequency but also a change in the heterocyst spacing pattern within the filaments. From being mainly single heterocysts, an increasing proportion of multiple (double, Fig. 1F, to quadruple) heterocysts are formed. In *Zamia skinneri,* the relative amount of single heterocysts decreased from 86% to 31% (Lindblad et al., 1985b). Simultaneously, double heterocysts in-

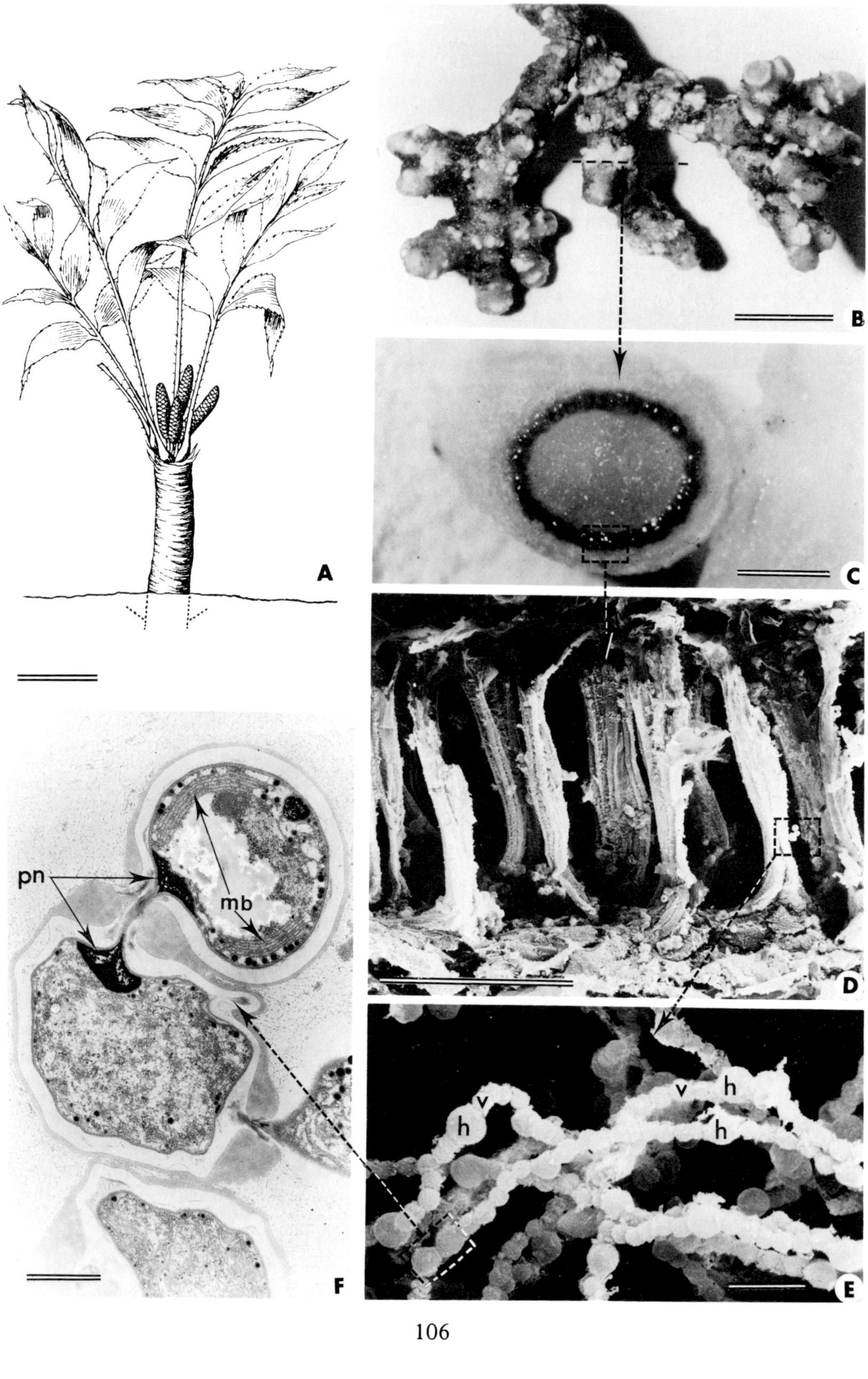

A
B
C
D
E
F
pn
mb
h
v

creased from 14 to 47%, triple from 0 to 20% and even quadruple heterocysts were occasionally formed.

Nitrogen Fixation

Nitrogen fixation occurs in all cycads examined so far. An annual nitrogen fixation capacity of 19 kg N/ha^{-1}/year^{-1} has been reported for two natural *Macrozamia riedlei* populations in Western Australia (Halliday & Pate, 1976). When comparing nitrogen fixation in symbiotic cyanobacteria of cycads with that found in free-living isolates, higher fixation rates have been found in the symbiotic stage (Kumar et al., 1986; Perraju et al., 1986). However, our own results showed similar or lower rates in the symbiotic stage (Lindblad et al., 1985b; Lindblad & Bergman, 1986), when compared with the free-living form(s).

A gradient in nitrogen fixation (C_2H_2-reduction) along coralloid roots of *Cycas revoluta* and *Zamia skinneri* has been observed, being maximal in the growing tip and declining towards the basal, older parts (Lindblad et al., 1985b; Lindblad & Bergman, 1986), while the heterocyst frequency showed the opposite pattern. In *Zamia skinneri,* there was an inverse relationship between nitrogen fixation and the percentage of multiple heterocysts, indicating a high frequency of inactive heterocysts in the multiple complexes (Lindblad et al., 1985b). However, a good correlation was found between nitrogen fixation and frequency of single heterocysts (Lindblad et al., 1985b).

Using antibodies against the Fe-protein of the nitrogenase complex in the cyanobiont of *Cycas revoluta,* the Fe-protein was found to be localized in the heterocysts only, subcellularly evenly distributed throughout the cytoplasm (Bergman et al., 1986). At least two possibilities may explain the inverse correlation between frequency of multiple heterocysts and nitrogenase activity: 1) absence of nitrogenase in multiple heterocysts and 2) lack of fixed carbon from distant vegetative cells. Inasmuch double as well as triple heterocysts contain nitrogenase (Bergman et al., 1986), the second explanation is favored.

Assimilation of Fixed Nitrogen

Free-living cyanobacteria assimilate the ammonia produced by nitrogen fixation primarily by the enzyme system glutamine synthetase (GS)–glutamate synthase (GOGAT) (Dharmawardene et al., 1973; Stewart & Rowell, 1975; Thomas et al., 1975), while the activity of ammonia assimilating glutamate dehydrogenase (NADPH-dependent GDH) is very low (Haystead et al., 1973). Symbiotic cyanobacteria of *Cycas revoluta, Ceratozamia mexicana* and *Zamia skinneri* all showed GS activities, using both the transferase and the coupled biosynthetic assay, similar to that found in free-living cyanobacteria and in cyanobacteria isolated from cycads (Table I; Lindblad & Bergman, 1986). Furthermore, by using antibodies directed towards cyanobacterial GS, a cellular/subcellular distribution of the protein similar to that of free-living cyanobacteria was found (Lindblad & Bergman, 1986). The GS protein was found to be evenly distributed in both heterocysts and vegetative cells, with labelling 30–40% higher in heterocysts compared to vegetative cells (Lindblad & Bergman, 1986). In accordance with the gradient in nitrogen fixation (Lindblad et al., 1985b; Lindblad & Bergman, 1986), GS activity declined towards the basal, older parts of the coralloid roots (Table I; Lindblad & Bergman, 1986).

These data are in contrast to the majority of studies on GS activities and protein levels in other symbioses involving cyanobacteria (Stewart & Rowell, 1977; Ray et al., 1978; Haselkorn et al., 1980; Orr & Haselkorn, 1982; Stewart et al., 1983; Meeks et al., 1985; Rowell et al., 1985). However, the finding that the GS activities and

←

Fig. 1. Coralloid root of the Costa Rican cycad *Zamia skinneri* Warsc. **A.** *Zamia skinneri,* with four male cones. Bar = 25 cm. **B.** Coralloid root, collected 10 cm below the soil surface (Lindblad et al., 1985a). Bar = 5 mm. **C.** Transverse section showing the cyanobacterial zone. Bar = 1 mm. **D.** Scanning electron micrograph showing the elongated cycad cells transversing the cyanobacterial zone. Most of the cyanobacteria and mucilage are washed away (Lindblad et al., 1985a). Bar = 100 μm. **E.** Scanning electron micrograph showing cyanobacterial filaments with both heterocysts (h) and vegetative cells (v) (Lindblad et al., 1985a). Bar = 10 μm. **F.** Transmission electron micrograph showing a double heterocyst with polar nodules (pn) and membranes (mb) (Lindblad et al., 1985a). Bar = 1.0 μm.

Table I

In vitro glutamine synthetase activity in (a) symbiotic cyanobacteria freshly isolated from different sections, each about 2 mm thick, of coralloid roots of *Cycas revoluta, Ceratozamia mexicana* and *Zamia skinneri,* and in (b) free-living (*Anabaena cylindrica* and *Nostoc muscorum*) and isolated cyanobacteria (*Nostoc* PCC 7422, originally isolated from *Cycas* sp.). The coralloid roots were cut into about 2 mm thick sections and numbered 1–5 starting at the growing tip of the coralloid root

Cycad species		Sections of coralloid roots				
		1	2	3	4	5
(a)						
Cycas revoluta[a]	Transf.[b]	617 ± 198	585 ± 147	400 ± 121	292 ± 118	85 ± 10
	Biosynth.[c]	76	56	35	22	N.D.[d]
Ceratozamia mexicana	Transf.	651 ± 187	540 ± 226	447 ± 218	314 ± 295	251 ± 79
	Biosynth.	59	52	32	27	5
Zamia skinneri[a]	Transf.	868 ± 22	833 ± 64	617 ± 229	489 ± 156	N.D.
	Biosynth.	60	74	59	24	N.D.
(b)						
Anabanea cylindrica (CCAP 1403/2a)	Transf.	647 ± 123				
	Biosynth.	65 ± 10				
Nostoc muscorum (CCAP 1453/12)	Transf.	523 ± 112				
	Biosynth.	55 ± 11				
Nostoc 7422 (PCC 7422)	Transf.	685 ± 208				
	Biosynth.	62 ± 13				

[a] Data from Lindblad & Bergman, 1986.
[b] Transferase activity; nmol γ-glutamylhydroxamate formed·mg protein^{-1}·min^{-1}.
[c] Coupled biosynthetic activity; nmol NADH oxidized·mg protein^{-1}·min^{-1}.
[d] N.D. = Not determined.

protein levels in older parts of the coralloid roots are significantly reduced agrees with the trend observed in *Peltigera* (Rowell et al., 1985) and *Azolla* (Haselkorn et al., 1980), although in the latter cases the total reduction was much more pronounced.

The *in vitro* GOGAT activity in the cyanobiont of *Cycas revoluta* was also similar to that of free-living cyanobacteria, while ammonia-assimilating GDH was undetectable in both (Lindblad et al., 1987). Similar results have been reported for the cyanobiont of *Cycas circinalis* (Perraju et al., 1986).

Using $^{15}N_2$ it is known that the nitrogen fixed is transferred to the cycad (Bergersen et al., 1965; Renaut et al., 1975). However, the form in which the fixed nitrogen is released by the cyanobiont is not known. The fact that, unlike other cyanobacterial symbioses, the cyanobiont of *Cycas revoluta* possesses full *in vitro* GS-GOGAT activities suggests that the nitrogen metabolite transferred may be a product of ammonia assimilation (Lindblad & Bergman, 1986; Lindblad et al., 1987). Recently, citrulline and glutamine have been identified as the principal N-solutes translocated in xylem from the coralloid root to the rest of the cycad in *Macrozamia, Lepidozamia* and *Encephalartos* (all members of Zamiaceae) (Pate et al., 1988). Citrulline and glutamine were readily labeled in coralloid roots of *M. riedlei* by $^{15}N_2$, and citrulline and aspartic acid with $^{14}CO_2$ (Pate et al., 1988). In contrast, citrulline was not found in extracts from coralloid roots or in xylem exudate of species of *Cycas* (Cycadaceae) and *Bowenia* (Boweniaceae), and the principal product of $^{15}N_2$-fixation by *C. revoluta* coralloid roots was glutamine.

Carbon Fixation

The cyanobacteria in cycad coralloid roots are often living beneath the soil surface, i.e., not reached by light. As a consequence, a heterotrophic carbon metabolism has been suggested (Lindblad & Bergman, 1986; Lindblad et al., 1987), with the fixed carbon being provided by the cycad. *Nostoc* cells freshly isolated from *Mac-*

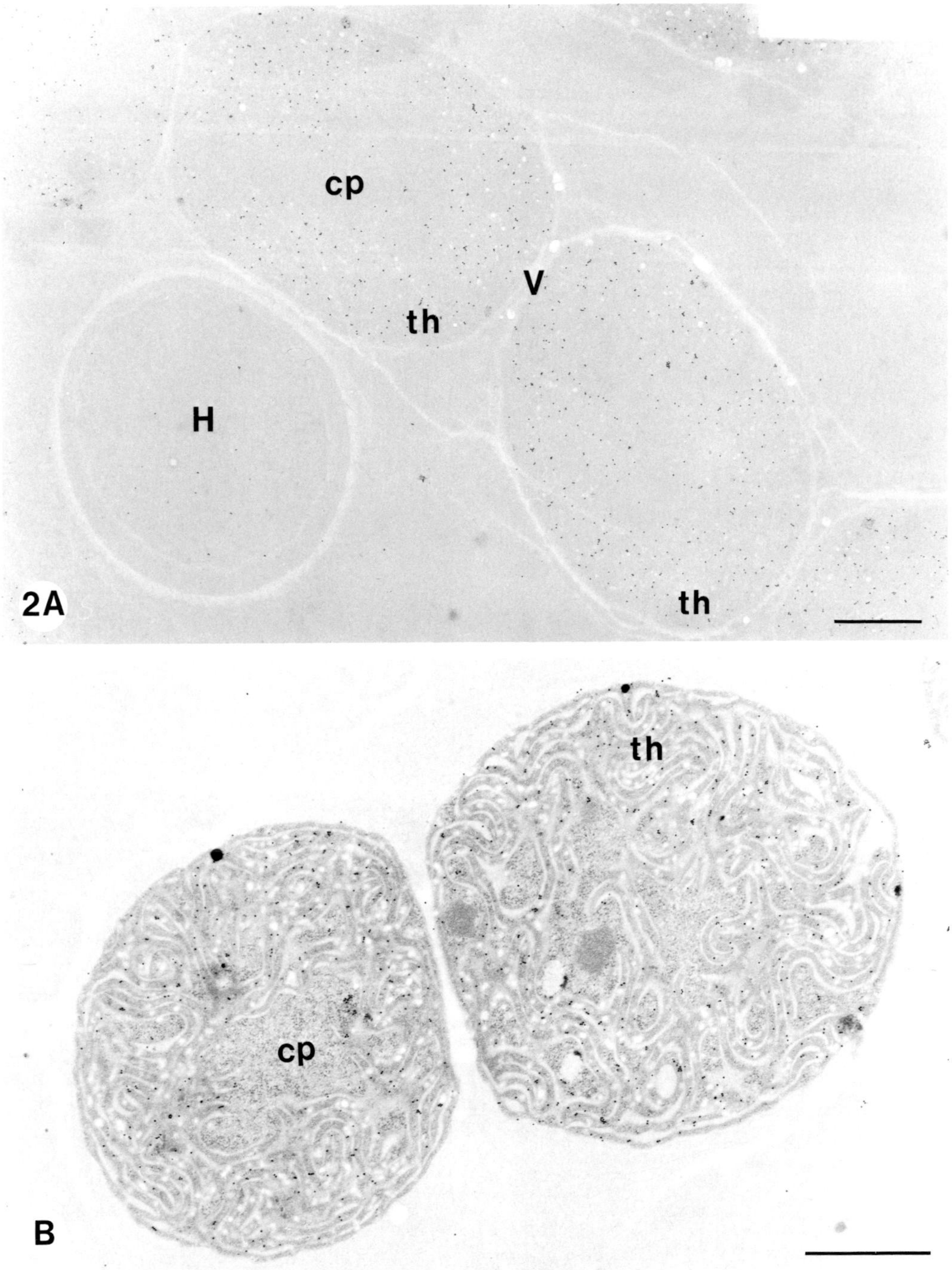

Fig. 2. Immunocytochemical localization of phycoerythrin in the free-living cycad isolate *Nostoc* 7422, originally isolated from *Cycas* sp. Fixation, embedding, immuno-gold labelling, control of the specificity of the labelling and post-staining were all performed as described in Bergman et al., 1985. The primary antibody used was raised against phycoerythrin, isolated and purified from the cyanobacterium *Phormidium persicinum* (Guard-Friar et al., 1986). Bar = 1.0 μm. **A.** Heterocyst (H) and vegetative cells (V). Not post-stained. Phycoerythrin is predominantly found in the thylakoid regions (th) of the vegetative cells, absent from the centroplasm (cp) of the vegetative cells as well as from the heterocyst. **B.** Vegetative cells. Post-stained. Note the localization of phycoerythrin to the photosynthetic thylakoids (th) and the absence from the centroplasm (cp).

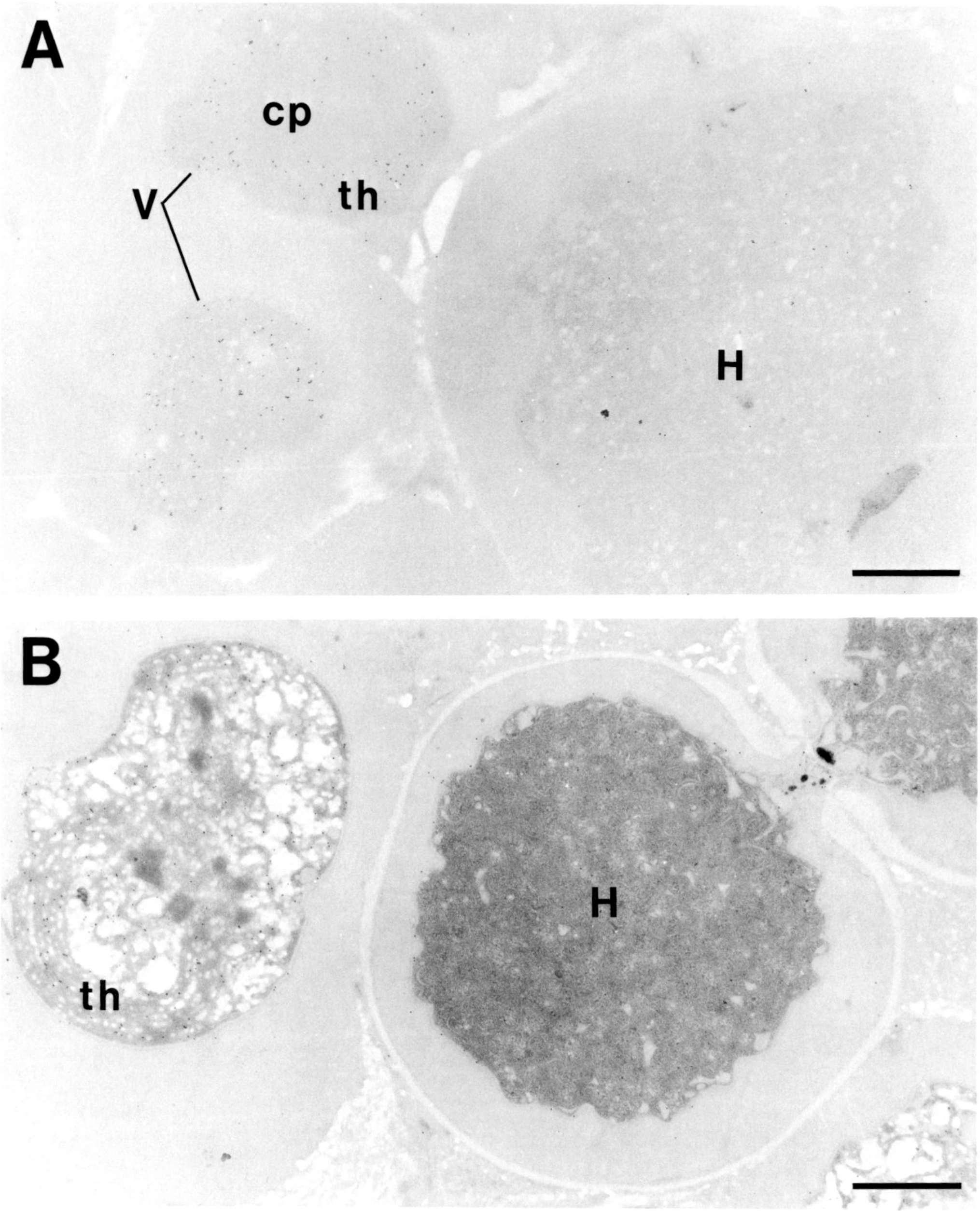

Fig. 3. Immunocytochemical localization of phycoerythrin in symbiotic *Nostoc* of coralloid roots of *Cycas revoluta.* Method see text to Figure 2. Bar = 1.0 μm. **A.** Heterocyst (H) and vegetative cells (V). Not post-stained. Phycoerythrin is predominantly found in the thylokoid regions (th) of the vegetative cells, absent from the centroplasm (cp) of the vegetative cells as well as from the heterocysts. Note the difference in labelling in sections going through, and not through, the central plane of vegetative cells. **B.** Heterocyst (H) and vegetative cell (V). Post-stained. Note the localization of phycoerythrin to the vegetative cells, predominantly in the thylakoid regions (th).

rozamia lucida are capable of heterotrophic growth in complete darkness (Hoare et al., 1971).

In our ultrastructural study of the cyanobiont of *Zamia skinneri* (Lindblad et al., 1985a), no subcellular modifications were observed in the *Nostoc* cells due to a symbiotic mode of life in the absence of light. In *Z. skinneri,* thylakoids with distinct phycobilisomes as well as carboxysomes were present. By immunohistochemistry, phycoerythrin was found in vegetative cells, subcellularly closely associated to the thylakoid membranes, while absent from the nitrogen-fixing heterocysts in *Nostoc* of *Cycas revoluta* and in the free-living isolated *Nostoc* PCC 7422 (Figs. 2, 3; Lindblad & Bergman, 1989). Using other techniques, phycocyanin, but not phycoerythrin, was found in the endophyte of *Cycas circinalis* (Perraju et al., 1986), while phycobilisomes were not observed in the phycobiont of *Macrozamia communis* (Grilli Caiola, 1974). Moreover, cyanobacteria in symbiosis with liverworts (Rodgers & Stewart, 1977) and *Gunnera* (Silvester, 1976) apparently lack such proteins, while phycobiliproteins are reported as present in lichens (Stewart et al., 1981; Bergman & Hällbom, 1982) and in *Azolla* (see Kaplan et al., 1986).

The undetectability of *in vivo* CO_2-fixation in the cyanobiont of *Cycas revoluta* (Lindblad et al., 1987) is probably due to the absence of light. However, both ribulose 1,5-bisphosphate carboxylase and the ribulose 1,5-bisphosphate-generating phosphoribulokinase showed similar *in vitro* activities in symbiotic as in free-living/isolated *Nostoc* (Table II; Lindblad et al., 1987). Also, in the cyanobionts of *Anthoceros* (Rodgers & Stewart, 1977) and *Gunnera* (Silvester, 1976), the CO_2-fixation is virtually absent. In contrast, cyanobacteria in *Azolla* fix atmospheric carbon (Ray et al., 1979; Peters et al., 1982; Peters & Calvert, 1983) though the mRNA level for RuBisCO is reduced (Nierzwicki-Bauer & Haselkorn, 1986). Other CO_2-fixing enzymes, such as PEP-carboxylase, PEP-carboxykinase, PEP-carboxytransphosphorylase and pyruvate carboxylase, showed low *in vitro* activities in the cyanobiont of *Cycas revoluta* (Table II; Lindblad et al., 1987).

Table II

In vitro activities of ribulose 1,5-bisphosphate (RuBis) carboxylase, phosphoribulokinase (PRK), PEP carboxylases and pyruvate carboxylase in symbiotic *Nostoc* freshly isolated from *Cycas revoluta* (a) and in the free-living *Cycas* isolate *Nostoc* 7422 (b). Data from Lindblad et al., 1987

	Activity (nmol product formed·mg protein^{-1}·min^{-1})	
Enzymes	Symbiotic *Nostoc* (a)	*Nostoc* 7422 (b)
RuBis carboxylase	279.4	321.7
PRK	153.2	163.6
PEP carboxylase	<1.1	3.7
PEP carboxykinase	N.D.	<1.9
PEP carboxytransphosphorylase	<1.8	N.D.
Pyruvate carboxylase	3.5	4.2

Mean of duplicate experiments.
N.D. = not detectable.

Acknowledgments

It is a pleasure to acknowledge the advice and cooperation of Dr. B. Bergman. This work was supported by the Swedish Natural Science Research Council and the Swedish Agency for Research Cooperation with Developing Countries.

Literature Cited

Bergersen, F. J., G. S. Kennedy & W. Wittman. 1965. Nitrogen fixation in the coralloid roots of *Macrozamia communis* L. Johnson. Austral. J. Biol. Sci. **18:** 1135–1142.

Bergman, B. & L. Hällbom. 1982. *Nostoc* of *Peltigera canina* when lichenized and isolated. Canad. J. Bot. **60:** 2092–2098.

———, **P. Lindblad, A. Pettersson, E. Renström & E. Tiberg.** 1985. Immuno-gold localization of glutamine synthetase in a nitrogen-fixing cyanobacterium (*Anabaena cylindrica*). Planta **166:** 329–334.

———, ——— **& A. N. Rai.** 1986. Nitrogenase in free-living and symbiotic cyanobacteria: Immunoelectron microscopic localization. FEMS Microbiol. Lett. **35:** 75–78.

Bowyer, J. W. & V. B. D. Skerman. 1968. Production of axenic cultures of soil-borne and endophytic blue-green algae. J. Gen. Microbiol. **54:** 299–306.

DeLuca, P. & S. Sabato. 1980. Regeneration of coralloid roots on cycad megagametophytes. Plant Sci. Lett. **18:** 27–31.

———, ———, **A. Balduzzi & R. Nazzaro.** 1980. Coralloid root regeneration on *Macrozamia* megagametophites. Giorn. Bot. Ital. **114:** 271–275.

Dharmawardene, M. W. N., A. Haystead & W. D. P.

Stewart. 1973. Glutamine synthetase of the nitrogen-fixing alga *Anabaena cylindrica.* Arch. Microbiol. **90:** 281–295.

Duckett, J. G. & A. S. K. Prasad. 1974. Ultrastructural studies on the *Nostoc* colonies associated with *Blasia pusilla* and *Anthoceros* spp. J. Bryol. **8:** 181–182.

———**, A. K. S. K. Prasad, D. A. Davies & S. Walker.** 1977. A cytological analysis of the *Nostoc*–Bryophyte relationship. New Phytol. **79:** 349–362.

———**, R. Toth & S. L. Soni.** 1975. An ultrastructural study of the *Azolla, Anabaena azollae* relationship. New Phytol. **75:** 111–118.

Florenzano, G., M. R. Tredici, L. Giovannetti, L. Tomaselli & M. C. Margheri. 1985. Metabolismo mixotrofico ed azotofissazione in *Nostoc* sp. da noduli coralloidi di *Cycas circinalis* L. Ann. Microbiol. **35:** 129–141.

Grilli, M. 1963. Ozzervazioni sulle strutture e la infrastrutture delle cellule dei tubercoli radicali di *Cycas revoluta.* Ann. Fac. Agr. **3:** 467–481.

Grilli Caiola, M. 1974. A light and electron microscopic study of the blue-green algae living either in the coralloid roots of *Macrozamia communis* or isolated in culture. Giorn. Bot. Ital. **108:** 161–173.

———. 1975. A light and electron microscopic study of the blue-green algae growing in the coralloid roots of *Encephalartos altensteinii* and in culture. Phycologia **14:** 25–33.

———. 1980. On the phycobionts of the cycad coralloid roots. New Phytol. **85:** 537–544.

Grobbelaar, N., W. E. Scott, W. Hattingh & J. Marshall. 1987. The identification of the coralloid root endophytes of the southern African cycads and the ability of the isolates to fix dinitrogen. S. African J. Bot. **53:** 111–118.

Guard-Friar, D., B. L. Eisenberg, M. R. Edwards & R. MacColl. 1986. Immunochemistry on cryptomonad biliproteins. Pl. Physiol. **80:** 38–42.

Halliday, J. & J. S. Pate. 1976. Symbiotic nitrogen fixation by coralloid roots of the cycad *Macrozamia riedlei*: Physiological characteristics and ecological significance. Austral. J. Pl. Physiol. **3:** 349–358.

Haselkorn, R., B. Mazur, J. Orr, D. Rice, N. Wood & R. Rippka. 1980. Heterocyst differentiation and nitrogen fixation in cyanobacteria. Pages 259–278 *in* W. E. Newton & H. Orme-Johnson (eds.), Nitrogen fixation. Vol. 2. Symbiotic associations and cyanobacteria. University Park Press, Baltimore, USA.

Haystead, A., M. W. N. Dharmawardene & W. D. P. Stewart. 1973. Ammonia assimilation in a nitrogen-fixing blue-green alga. Plant Sci. Lett. **1:** 439–445.

Hoare, D. S., L. O. Ingram, E. L. Thurston & R. Walkup. 1971. Dark heterotrophic growth of an endophytic blue-green alga. Arch. Microbiol. **78:** 310–321.

Kaplan, D., H. E. Calvert & G. A. Peters. 1986. The *Azolla-Anabaena azollae* relationship. XII. Nitrogenase activity and phycobiliproteins of the endophyte as a function of leaf age and cell type. Pl. Physiol. **80:** 884–890.

Kumar, A. P., B. T. V. V. Perraju & H. N. Singh. 1986. Carbon nutrition and the regulation of uptake hydrogenase activity in free-living and symbiotic *Anabaena cycadeae.* New Phytol. **104:** 115–120.

Lindblad, P. 1984. Diversion between C_2H_2-reduction and heterocyst frequency in a cycad root. Page 511 *in* C. Veeger & W. E. Newton (eds.), Advances in nitrogen fixation research. Nijhoff/Junk, The Hague, The Netherlands.

——— **& B. Bergman.** 1986. Glutamine synthetase: Activity and localization in cyanobacteria of the cycads *Cycas revoluta* and *Zamia skinneri.* Planta **169:** 1–7.

——— **&** ———. 1989. Occurrence and localization of phycoerythrin in symbiotic *Nostoc* of *Cycas revoluta* and in the free-living isolated *Nostoc* 7422. Pl. Physiol. **89:** 783–785.

———, ———**, A. v. Hosten, L. Hällbom & J. E. Nylund.** 1985a. The cyanobacterium–*Zamia* symbiosis: An ultrastructural study. New Phytol. **101:** 707–716.

———**, L. Hällbom & B. Bergman.** 1985b. The cyanobacterium–*Zamia* symbiosis: C_2H_2-reduction and heterocyst frequency. Symbiosis **1:** 19–28.

———**, A. N. Rai & B. Bergman.** 1987. The *Cycas revoluta–Nostoc* symbiosis: Enzyme activities of nitrogen and carbon metabolism in the cyanobiont. J. Gen. Microbiol. **133:** 1695–1699.

Meeks, J. C., C. S. Enderlin, C. M. Joseph, J. S. Chapman & M. W. L. Lollar. 1985. Fixation of $[^{13}N]N_2$ and transfer of fixed nitrogen in the *Anthoceros–Nostoc* symbiotic association. Planta **164:** 406–414.

Nathanielsz, C. P. & I. A. Staff. 1975. On the occurrence of intracellular blue-green algae in cortical cells of the apogeotropical roots of *Macrozamia communis* (L.) Johnson. Ann. Bot. **39:** 363–368.

Neumuller, M. & B. Bergman. 1981. The ultrastructure of *Anabaena azollae* in *Azolla pinnata.* Physiol. Pl. **51:** 69–76.

Nierzwicki-Bauer, S. A. & R. Haselkorn. 1986. Differences in mRNA levels in *Anabaena* living freely or in symbiotic association with *Azolla.* EMBO J. **5:** 29–35.

Obukowicz, M., M. Schaller & G. S. Kennedy. 1981. Ultrastructure and phenolic histochemistry of the *Cycas revoluta–Anabaena* symbiosis. New Phytol. **87:** 751–759.

Orr, J. & R. Haselkorn. 1982. Regulation of glutamine synthetase activity and synthesis in free-living and symbiotic *Anabaena* spp. J. Bacteriol. **152:** 626–635.

Pate, J. S. & B. E. S. Gunning. 1972. Transfer cells. Ann. Rev. Plant Physiol. **23:** 173–196.

———**, P. Lindblad & C. A. Atkins.** 1988. Pathways of assimilation of fixed nitrogen in coralloid roots of cycad–*Nostoc* symbioses. Planta **176:** 461–471.

Perraju, B. T. V. V., A. N. Rai, A. P. Kumar & H. N. Singh. 1986. *Cycas circinalis–Anabaena cycadeae* symbiosis: Photosynthesis and the enzymes of nitrogen and hydrogen metabolism in symbiotic and cultured *Anabaena cycadeae*. Symbiosis **1:** 239–250.

Peters, G. A. & H. E. Calvert. 1983. The *Azolla–Anabaena* symbiosis. Pages 109–145 *in* L. J. Goff (ed.), Algal symbiosis. Cambridge University Press, USA.

———, ———, **D. Kaplan, O. Ito & R. E. Toia, Jr.** 1982. The *Azolla–Anabaena* symbiosis: Morphology, physiology and use. Israel. J. Bot. **31:** 305–323.

Ray, T. B., B. C. Mayne, R. E. Toia, Jr. & G. A. Peters. 1979. *Azolla–Anabaena* relationship. VIII. Photosynthetic characterization of the association and individual partners. Pl. Physiol. **64:** 791–795.

———, **G. A. Peters, R. E. Toia, Jr. & B. C. Mayne.** 1978. *Azolla–Anabaena* relationship. VII. Distribution of ammonia-assimilating enzymes, proteins, and chlorophyll between host and symbiont. Pl. Physiol. **62:** 463–467.

Renaut, J., A. Sasson, H. W. Pearsson & W. D. P. Stewart. 1975. Nitrogen-fixing algae in Morocco. Pages 229–246 *in* W. D. P. Stewart (ed.), IBP, Nitrogen fixation by free-living micro-organisms. Cambridge University Press, USA.

Rippka, R., J. Deruelles, J. B. Waterbury, M. Herdman & R. Y. Stanier. 1979. Generic assignments, strain histories and properties of pure cultures of cyanobacteria. J. Gen. Microbiol. **111:** 1–61.

Rodgers, G. A. & W. D. P. Stewart. 1977. The cyanophyte–hepatic symbiosis. I. Morphology and physiology. New Phytol. **78:** 441–458.

Rowell, P., A. N. Rai & W. D. P. Stewart. 1985. Studies on the nitrogen metabolism of the lichens *Peltigera aphthosa* and *Peltigera canina*. Pages 145–160 *in* D. H. Brown (ed.), Lichen physiology and cell biology. Plenum Press, New York, USA.

Silvester, W. B. 1976. Endophyte adaptation in *Gunnera–Nostoc* symbiosis. Pages 521–538 *in* P. Nutman (ed.), IBP. Vol. 7. Nitrogen fixation by symbiotic organisms. Cambridge University Press, USA.

——— **& P. J. McNamara.** 1976. The infection process and ultrastructure of the *Gunnera-Nostoc* symbiosis. New Phytol. **77:** 135–141.

Stewart, W. D. P., A. N. Rai, R. H. Reed, E. Creach, G. A. Codd & R. Rowell. 1981. Studies on the N_2-fixing lichen *Peltigera aphthosa*. Pages 237–243 *in* A. H. Gibson & W. E. Newton (eds.), Current perspectives in nitrogen fixation. Australian Acad. Sciences, Canberra, Australia.

——— **& P. Rowell.** 1975. Effects of L-methionine-DL-sulphoximine on the assimilation of newly fixed NH_3, acetylene reduction and heterocyst production in *Anabaena cylindrica*. Biochem. Biophys. Res. Comm. **65:** 846–856.

——— **&** ———. 1977. Modifications of nitrogen-fixing algae in lichen symbioses. Nature **265:** 371–372.

———, ——— **& A. N. Rai.** 1983. Cyanobacteria–eukaryotic plant symbioses. Ann. Microbiol. (Inst. Pasteur) **134B:** 205–228.

Thomas, J., C. P. Wolk, P. W. Shaffer, S. M. Austin & A. Galonsky. 1975. The initial organic products of fixation of ^{13}N-labelled nitrogen gas by the blue-green alga *Anabaena cylindrica*. Biochem. Biophys. Res. Comm. **67:** 501–507.

Webb, D. T., M. Nevarez & S. de Jesus. 1984. Further *in vitro* studies of light-induced root nodulation in the cycadales. Env. Exp. Bot. **24:** 37–44.

Zhu, C. 1982. Fine structure of blue-green algae and the cells lined along the endophyte cavity in the coralloid roots of *Cycas*. Acta Bot. Sin. **24:** 109–114.

MARCH 1990 MEMOIRS OF THE NEW YORK BOTANICAL GARDEN **57:** 114–122

Cytotaxonomy of Cycads

ALDO MORETTI

Table of Contents

Abstract

MORETTI, A. (Dipartimento di Biologia vegetale, Facoltà di Scienze, Università di Napoli, via Foria 223, Napoli, Italy) Cytotaxonomy of cycads. Memoirs of the New York Botanical Garden **57:** 114–122. 1990.—Nine genera of cycads have been examined from a cytotaxonomic point of view. Karyotypic data, most of which are newly described, are figured and discussed from a phylogenetic point of view. Robertsonian changes, which involve centric fission or fusion, chromosome loss and chromosome deletion, are proposed to explain karyotype differentiation in *Zamia,* the only genus with interspecific chromosome variation. On the basis of these mechanisms, a scheme showing the evolution of karyotypes in *Zamia* is presented. The scheme appears realistic enough even though it does not show correlation with morphological and taxonomic characters of *Zamia* species. Two cases of infraspecific chromosome variation; *Z. roezlii* with 2n = 22, 24, 25, 26 and *Z. paucijuga* with 2n = 19, 23, 24, 25, 26, 27, 28, are also discussed. The Robertsonian changes involved in the karyotypic differentiation of nine genera of cycads would be the loss of chromosomes, the rearrangement of large genomic segments, centric fission, and the deletion of chromosome pieces or arms. The different rates of change determined by such mechanisms in the karyotype of each genus are used to single out four generic groupings: *Bowenia*; *Stangeria*; *Encephalartos, Lepidozamia, Macrozamia* and *Dioon*; *Zamia, Ceratozamia* and *Microcycas*. Thus, groupings of genera based on karyotypic characters parallel those of families and tribes of the order as established by morphological criteria.

Key words: Cycads; cytotaxonomy; karyotypic evolution; Robertsonian changes.

Introduction

The origin of the cycads and the phylogenetic relationships of their genera probably represent the most enigmatic aspect in the study of these plants. On account of their very ancient origin, dating back to the Paleozoic Era, and from their morphology and distribution, the extant cycad genera appear to be relic forms with many lost relatives.

Over the last few years, I have been studying cycad genera from a cytotaxonomic point of view

Table I
Chromosome numbers and karyotypes of African and Australian cycad genera

Species	2n	Karyotype	Source
Bowenia serrulata (W. Bull) Chamberlain	18	12M+6S	a
B. spectabilis Hook. f.	18	12M+6S	a
Encephalartos lehmannii (Jacq.) Lehm.	18	8M+10S	a
Lepidozamia hopei Regel	18	8M+8S+2T	a
L. peroffskyana Regel	18	8M+8S+2T	a
Macrozamia communis L. Johnson	18	8M+8S+2T	b
M. diplomera (F. Muell.) L. Johnson	18	8M+8S+2T	b
M. heteromera C. Moore	18	8M+8S+2T	b
M. miquelii (F. Muell.) A. DC.	18	8M+8S+2T	b
M. moorei F. Muell.	18	8M+8S+2T	b
M. pauli-guilielmi W. Hill & F. Muell.	18	8M+8S+2T	b
M. riedlei (Fisch. ex Gaudich.) C. A. Gardn.	18	8M+8S+2T	b
M. secunda C. Moore	18	8M+8S+2T	b
M. spiralis (Salisb.) Miq.	18	8M+8S+2T	b
Stangeria eriopus (Kunze) Baill.	16	10M+4S+2T	a

[a] A. Moretti (unpubl.); plants examined are cultivated at the Naples Botanical Garden.
[b] Moretti (1982).

(Moretti, 1982; Moretti & Sabato, 1984) with the aim to improve knowledge of cycad karyotypes and, in addition, by their comparative analysis, to perchance learn something more about the phylogenetic affinities and divergences among the genera.

Unfortunately, as with other ancient groups, the attempt to correlate chromosome variation with morphological differentiation in cycads is very often unsuccessful. Typically, various cycad characteristics evolve at different rates. Thus, it happens that we find at one time and in the same individual, for example in *Cycas,* derived (leaflet type and platispermic seed) and primitive (unreduced megasporophyll) characteristics. Similarly, in *Microcycas,* in which possession of a very high number of sperms, considered a primitive characteristic, coexists with very reduced sporophylls, sessile ovules, articulated leaflets and stipulate leaves in the seedlings, which are considered derived characteristics (S. Sabato, pers. comm.). Such situations, therefore, make the correlation between chromosomal and morphological evolution in cycads very confusing.

Despite these difficulties, in my opinion the cytotaxonomic study of cycads can be very useful; it is a research field that has been little explored. In this regard, it will surely be useful to extend such a study at a molecular level, by using techniques of chromosome banding and DNA analysis. Indeed, these two kinds of approach are currently in progress in my laboratory.

In the present paper, I present results obtained from my study on the cytotaxonomy of nine genera of cycads. I have no new data on *Cycas,* whose karyological examination is programmed for future study. The phylogenetic implications of the results are discussed at both the infrageneric and intergeneric levels. Some preliminary results achieved by using the fluorochrome quinacrine technique are also discussed.

Chromosome numbers and karyotypes were determined using both Feulgen and banding techniques. Part of the results presented here have already been published (Moretti, 1982; Moretti & Sabato, 1984). Species whose karyotypic data are reported for the first time here are listed in Tables I and II. For the examination of these last species I used the methods of Moretti and Sabato (1984).

Literature Data on the Karyology of Cycads

Most of the literature data on the karyology of cycads is reported by Johnson (1963), Marchant (1968), Resende and Rijo (1948), and Sax and Beal (1934). Genera studied at that time had 2n = 16 or 18, except for *Cycas* and *Microcycas* which were very different from the others, with

Table II
Chromosome numbers and karyotypes of American cycad genera

Species	2n	Karyotype	Source
Ceratozamia hildae Landry & Wilson	16	12M+2S+2T	a
C. kuesteriana Regel	16	12M+2S+2T	a
C. matudae Lundell	16	12M+2S+2T	a
C. mexicana Brongn.	16	12M+2S+2T	a
C. microstrobila Vovides & Rees	16	12M+2S+2T	a
C. norstogii Stevenson	16	12M+2S+2T	a
Dioon purpusii Rose	18	8M+8A+2T	a
D. edule Lind.	18	8M+8A+2T	a
D. holmgrenii De Luca, Sabato & V. Torres	18	8M+8A+2T	a
D. caputoi De Luca, Sabato & V. Torres	18	8M+8A+2T	a
D. merolae De Luca, Sabato & V. Torres	18	8M+8A+2T	a
D. tomasellii De Luca, Sabato & V. Torres	18	8M+8A+2T	a
D. spinulosum Dyer	18	8M+8A+2T	a
D. mejiae Standl. & Williams	18	8M+8A+2T	a
D. rzedowskii De Luca, Moretti, Sabato & V. Torres	18	8M+8A+2T	a
D. califanoi De Luca & Sabato	18	8M+8A+2T	a
Microcycas calocoma (Miq.) A. DC.	26	2M+2A+22T	a
Zamia fischeri Miq.	16	12M+4S	b
Z. pygmaea Sims emend. Schuster	16	12M+4S	a
Z. amblyphyllida D. Stevenson	16	12M+4S	a
Z. furfuracea L. f.	18	10M+2S+2A+4T	b
Z. loddigesii Miq.	18	10M+2S+2A+4T	b
Z. manicata Linden ex Regel	18	10M+2S+2A+4T	a
Z. paucijuga Wieland	19	9M+2S+2A+6T	a
Z. paucijuga Wieland	23	5M+2S+2A+14T	b
Z. paucijuga Wieland	24	4M+2S+2A+16T	a
Z. paucijuga Wieland	25	3M+2S+2A+18T	b
Z. paucijuga Wieland	26	2M+2S+2A+20T	b
Z. paucijuga Wieland	27	1M+2S+2A+22T	b
Z. paucijuga Wieland	28	2S+2A+24T	b

[a] Moretti (in prep.).
[b] Moretti & Sabato (1984).

2n = 22 and 26, respectively. In addition, only *Zamia* had two chromosome numbers (2n = 16 and 18). More recently, and especially thanks to Norstog's work in the 1980's, the interest in cytotaxonomy of cycads has been renewed. Norstog has made a very useful study of *Zamia,* in which he has examined several species. Among other things, he showed that the infrageneric variation of *Zamia* is very wide, finding plants with 2n = 16, 18, 22 and 24 (Norstog, 1980). In addition, he showed that infraspecific variation occurs in *Z. roezlii*. This species shows four chromosome numbers: 2n = 22, 24, 25 and 26 (Norstog, 1981).

Cytotaxonomy of African and Australian Genera

With reference to the African (*Encephalartos, Stangeria*) and Australian (*Bowenia, Lepidozamia, Macrozamia*) genera, I have reexamined the relative chromosome numbers and karyotypes from several species (Table I). My data generally confirmed earlier reports in the literature (see above). Some discrepancies appeared in the morphological determination of acrocentric and submetacentric chromosomes.

Cytotaxonomy of American Genera

The major part of my work on cytotaxonomy of cycads was devoted to four American genera: *Ceratozamia, Dioon, Microcycas* and *Zamia*. The published karyological information on these genera was very limited and only a few species had been examined.

With my colleagues of Naples and with Drs. Mario Vazquez Torres from the Universidad Veracruzana, Xalapa, Mexico and Dennis Wm. Stevenson from the New York Botanical Garden, I spent a long time in the field in order to study the American genera. In the course of several trips we collected many plants, most of which are now cultivated in the greenhouses at the Naples Botanical Garden, Italy. The results of karyological studies of these four genera are reported in Table II.

All examined species of *Ceratozamia* have the same number and karyotype, 2n = 16 with 12M+2S+2T. I examined all the known species of *Dioon*. All of them have the same chromosome number and karyotype, 2n = 18 with 8M+8A+2T. *Microcycas calocoma,* the only species in the genus, has 2n = 26 with 2M+2A+22T. The examination of *M. calocoma* proved to be very difficult and I am not fully confident about this karyotype. In some preparations, a pair of submetacentric chromosomes seems to occur instead of acrocentrics.

The examination of the genus *Zamia* proved to be more interesting. The study confirmed that the variation in chromosome number in this genus is indeed very wide (Table II). But more interesting was the discovery in *Z. paucijuga* of a very clear infraspecific variation like that occurring in *Z. roezlii*. *Zamia paucijuga* has 2n = 19, 23, 24, 25, 26, 27 and 28 (relative karyotypes are reported in Table II). Indeed, 2n = 28 is the highest chromosome number found in cycads. *Zamia paucijuga* is distributed along 2000 km of the Pacific coast of Mexico. In the southern part of the distribution, cytotypes have relatively low chromosome numbers, 2n = 19 to 26; in the northern part, higher chromosome numbers occur, 2n = 27 and 28. In Table II, we can see that the cytotypes differ substantially in the number of metacentric and telocentric chromosomes. Such variation was interpreted as being the result of the phenomenon of centric fission of metacentric chromosomes (Moretti & Sabato, 1984). Let us examine this phenomenon, for example, in the change from 2n = 26 with 2M+2S+2A+20T to 2n = 27 with 1M+2S+2A+22T. Then, one metacentric was involved in the fission mechanism, producing two telocentrics. The arrangement of the seven cytotypes (Table II) shows a disjunction between 2n = 19 and 2n = 23. I cannot exclude the possibility of completing the series when the examination of all the *Z. paucijuga* samples in my possession is complete. The presence of 2n = 19 suggests that a 2n = 18 cytotype might exist. My hypothesis, if confirmed, would mean that *Z. paucijuga* has some relationship with the 2n = 18 group of Mexican Zamias.

In Table III are listed all *Zamia* chromosome numbers whose karyotypes are of certain or almost certain determination. Some karyotypes given in Table III, determined in different ways by various authors, were modified to reflect my interpretation of their data and to give uniformity to the terminology of the chromosome morphology.

Karyotype Evolution in *Zamia*

Unlike the other American genera (*Ceratozamia, Dioon* and *Microcycas*) whose karyotypes show a constant infrageneric uniformity, *Zamia,* with its high rate of infrageneric chromosome variation, is suitable for a discussion of the mechanisms involved in karyotype differentiation. Specifically, karyotype analysis of all *Zamia* species would show that the centric fission of metacentric chromosomes is heavily involved in the karyotype differentiation in this genus. In addition, when the karyotype differentiation is not explicable on the basis of this mechanism, two further mechanisms may be involved. These are chromosome loss and chromosome deletion. With these three mechanisms in mind, we can now examine all the karyotypes of the genus *Zamia*.

In Figure 1, a simplified scheme with chromosome numbers (circled) and karyotypes (in rectangles) of *Zamia* is represented. The scheme was constructed on the basis of the data reported in Table III.

In the scheme the karyotype sequence from the top to the bottom generally reflects an increasing asymmetry in the karyotypes. In fact, we can see a dominant presence of metacentrics and submetacentrics towards the top and the gradual appearance of acrocentrics and telocentrics towards the bottom. The presence of telocentric chromosomes becomes dominant in the two cytotype series represented at the bottom in the scheme. If we suppose that the most symmetrical karyotypes are most primitive in cycads, then we can deduce that the 12M+4S karyotype is the most ancient and that all the other *Zamia* karyotypes have originated from it by means of various Robertsonian changes. Let us now examine each step.

Step from 12M+4S to 10M+4S+2T Karyo-

Table III
Chromosome numbers and karyotypes in *Zamia*

Species	2n	Karyotype	Source
Z. angustifolia Jacq.	16	12M+4S	a
Z. fischeri Miq.	16	12M+4S	b, c
Z. floridana A. DC.	16	12M+4S	b
Z. media Jacq.	16	12M+4S	d
Z. portoricensis Urban	16	12M+4S	a
Z. purpurea Vovides, Rees & Vazquez Torres	16	12M+4S	e
Z. pygmaea Sims emend. Schuster	16	12M+4S	a, f
Zamia sp. nov. (D. Wm. Stevenson, in prep.)	16	12M+4S	f
Z. chigua Seem.	16	10M+4S+2T	a
Z. lindenii Regel	16	10M+4S+2T	a
Z. pseudoparasitica Yates	16	10M+4S+2T	a
Z. tuerckeimii J. D. Smith	16	10M+4S+2T	a
Z. inermis Vovides, Rees & Vazquez Torres	16	12M+2S+2A	e
Z. splendens Schutzman	16	12M+2S+2A	h
Z. furfuracea L. f.	18	10M+2S+2A+4T	a, c
Z. loddigesii Miq.	18	10M+2S+2A+4T	a, c, e
Z. manicata Linden ex Regel	18	10M+2S+2A+4T	f
Z. obliqua A. Braun	18	10M+2S+2A+4T	a
Z. skinneri Warsz.	18	10M+2S+2A+4T	a
Z. roezlii Linden	22	4M+4S+2A+12T	g
Z. roezlii Linden	24	2M+4S+2A+16T	g
Z. roezlii Linden	25	1M+4S+2A+18T	g
Z. roezlii Linden	26	4S+2A+20T	g
Z. paucijuga Wieland	19	9M+2S+2A+6T	f
Z. paucijuga Wieland	23	5M+2S+2A+14T	c
Z. paucijuga Wieland	24	4M+2S+2A+16T	f
Z. paucijuga Wieland	25	3M+2S+2A+18T	c
Z. paucijuga Wieland	26	2M+2S+2A+20T	c
Z. paucijuga Wieland	27	1M+2S+2A+22T	c
Z. paucijuga Wieland	28	2S+2A+24T	c

[a] Norstog (1980); [b] Marchant (1968); [c] Moretti and Sabato (1984); [d] Sax and Beal (1934); [e] Vovides (1983); [f] Moretti (in prep.); [g] Norstog (1981); [h] Schutzman (1984).

type.—Two mechanisms are involved: centric fission of two metacentric chromosomes, and the subsequent loss of two of the four resulting telocentrics. The number of submetacentric chromosomes remains the same. In this step, the opposite change from the 10M+4S+2T to 12M+4S karyotype by a mechanism of "centric fusion" is not possible, as shown by counting the number of arms in each karyotype.

Step from 12M+4S to 12M+2S+2A.—Here we do not have centric fissions, but rather the conversion of two submetacentric chromosomes into two acrocentric chromosomes by deletion of terminal parts of the chromosomes. One of the two submetacentric pairs is conserved; metacentrics remain the same. Some species of the 12M+4S karyotype, namely *Z. fischeri, Z. floridana* (=*Z. integrifolia*) and *Z. purpurea,* have secondary constrictions on the short arms of one submetacentric pair. Therefore, the possibility exists that just at this point a fracture and then the deletion occurred.

Step from 12M+2S+2A to 10M+2S+2A+4T.—This is a very clear case of centric fission involving two metacentric chromosomes with the production of four telocentrics. Submetacentrics and acrocentrics are unchanged.

Examination of Z. paucijuga.—The cytotypes of this species (series at the bottom right of Fig. 1) are compatible with their derivation from the 10M+2S+2A+4T karyotype, once more on the basis of the fission mechanism.

Zamia roezlii.—*Zamia roezlii* (cytotype series at the bottom left of Fig. 1) would be derived from the 12M+2S+2A karyotype. Here we have repeated centric fissions and deletion of chro-

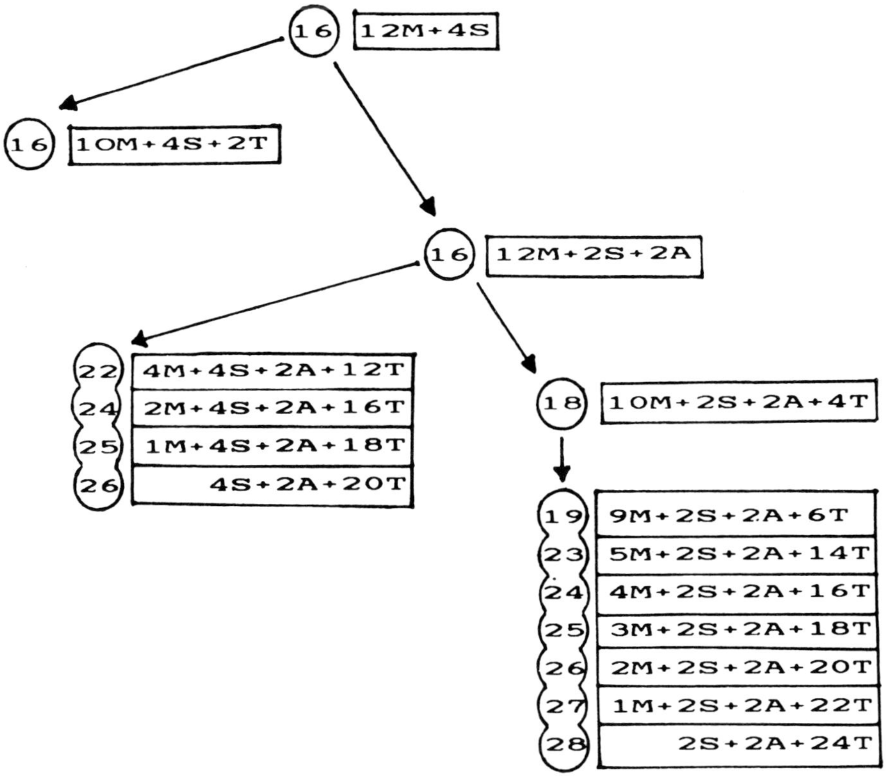

Fig. 1. Karyotype evolution in *Zamia*.

mosomal pieces. Let us examine the consequences of the two mechanisms. Six metacentric chromosomes produce 12 telocentrics by centric fission. Two additional metacentric chromosomes lose terminal parts of their arms to become two submetacentrics. In this way, the submetacentric number increases from two to four. Acrocentric number is the same. While the first cytotype is formed by these mechanisms, all other ones are derived by repeated fissions of metacentric chromosomes.

Correlation between Karyological and Morphological Characters in *Zamia*

So far, we have examined *Zamia* species on a karyological basis exclusively. If we try to find some correlation between these data and morphological ones, our work becomes very complicated, as already stated in the introduction. In this regard, *Z. roezlii* is a good example. In the scheme of Figure 1, *Z. roezlii* is represented as a derived species and not as a primitive one. In contrast, some of its morphological features (very large trunk, female cone and seeds) suggest that it can be considered a primitive species. To explain such a discrepancy, we cannot exclude that the karyotypic differentiation of this species happened by a mechanism of centric fusion rather than of fission. The fusion mechanism causes a reduction in the chromosome number that leads to an evolutionary sequence which has an opposite direction to that represented in the scheme of Figure 1.

In conclusion, from the examination of *Zamia* karyotypes, we can see that there exist steps that are still not explained. The scheme of Figure 1 cannot be considered a definitive result of the *Zamia* study but, rather, as a "working hypothesis." Only by the acquisition of new data and by using new techniques of studying chromosomes (e.g., Q, C and DNase banding, and *in situ* hybridization) can we see whether this hypothesis is correct.

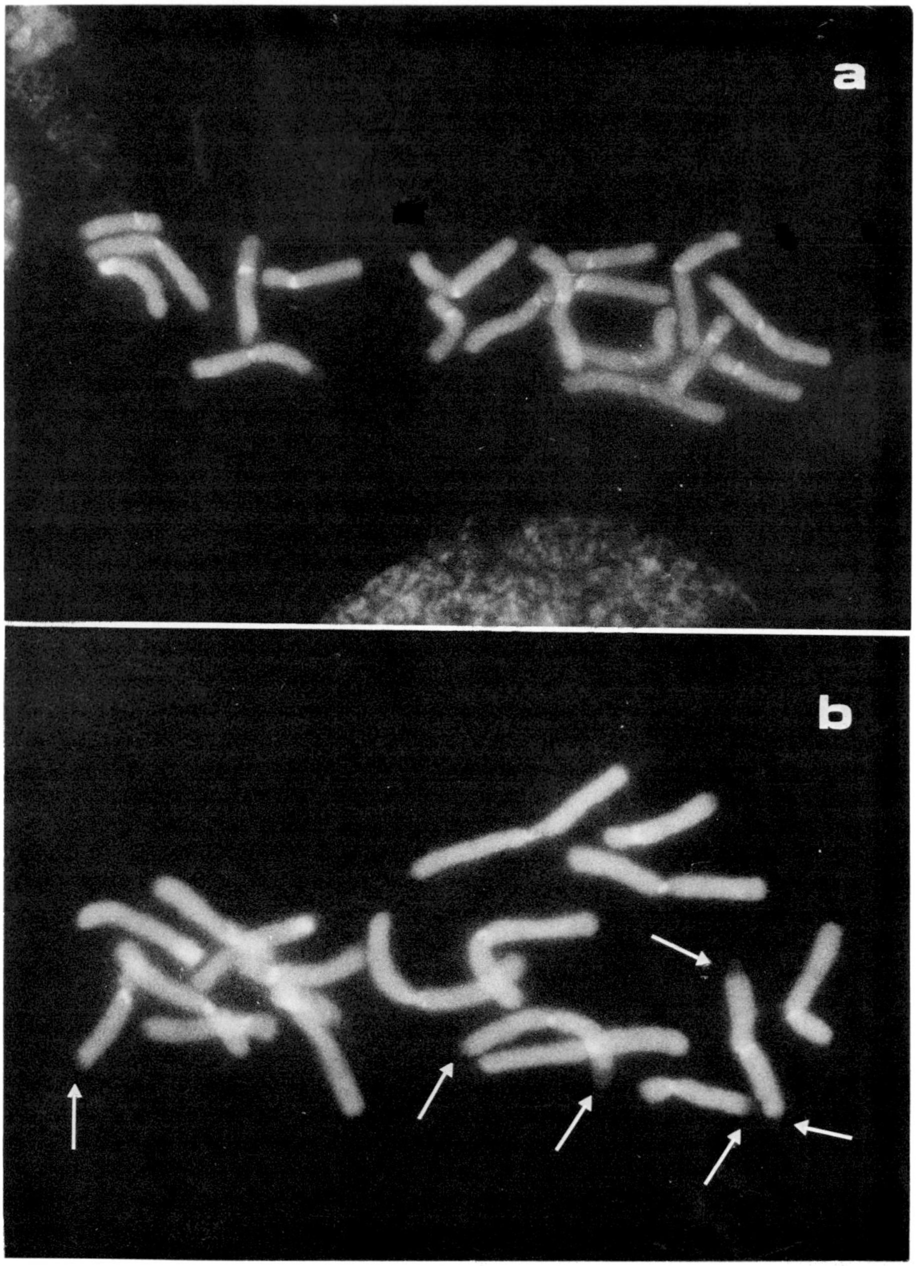

Fig. 2. **a.** The quinacrine stained chromosomes of *Macrozamia secunda* showing enhanced fluorescence at the centromeric regions (×1100). **b.** The quinacrine stained chromosomes of *Macrozamia secunda*. Besides enhanced fluorescence at centromeric regions, note on the distal regions of some chromosomes the dark bands (arrowed) of reduced fluorescence (×1300). (After Moretti, 1982.)

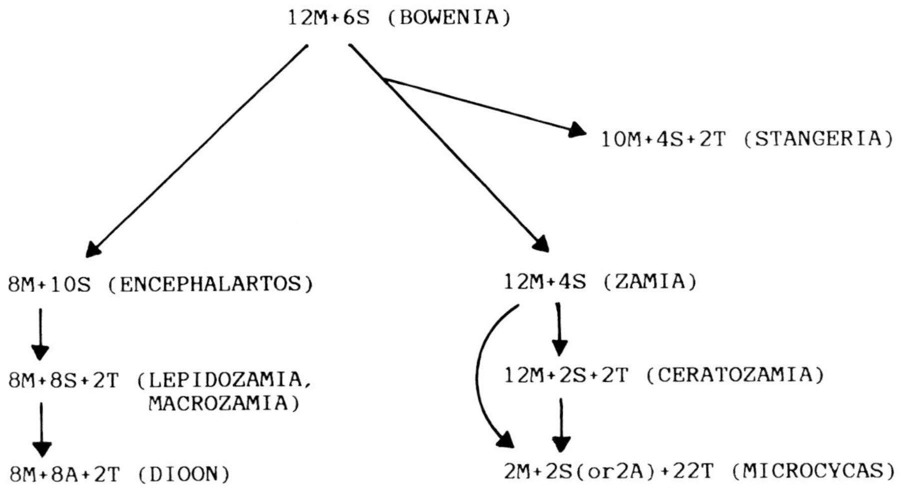

Fig. 3. Karyotype evolution of nine cycad genera.

Karyotypic Study of Cycads by Banding Techniques

Among the new cytotaxonomic techniques that I am adopting in my laboratory, the one which has given the most interesting results is the use of fluorochromes for studying heterochromatin. I have obtained data from the genus *Macrozamia* (Moretti, 1982) and some species of *Dioon* (A. Moretti, unpubl.). For example, using the fluorochrome quinacrine, it has been possible to obtain banding of the chromosomes of all the species of *Macrozamia*. An enhanced fluorescence at the level of the centromeric heterochromatin has been seen with all the chromosomes (Fig. 2a). The pattern of banding differs in some species, where the fluorescence is lacking on the telocentric chromosomes. Moreover, it has been possible to observe reduced fluorescence at the level of the terminal heterochromatic regions (Fig. 2b).

Karyotype Evolution in Cycads

In conclusion, I present a scheme summarizing the karyological data for the nine genera examined by me (Fig. 3).

At the top is *Bowenia*, with its highly symmetric karyotype. I suggest that all the other karyotypes are derived from such a karyotype. Robertsonian mechanisms involved in these steps are certainly more complicated than those proposed for *Zamia*. Let us examine which mechanisms could be involved in each step.

From Bowenia to Encephalartos.—Four metacentric chromosomes change into submetacentric ones by either deletion or large rearrangement of parts of the chromosomes.

From Encephalartos to Lepidozamia and Macrozamia.—*Lepidozamia* and *Macrozamia* have the same karyotype. They have two telocentric chromosomes which could be derived from two submetacentric chromosomes of *Encephalartos* by deletion of the short arms.

From Lepidozamia and Macrozamia to Dioon.—This change involves the formation of eight acrocentrics from eight submetacentrics by either deletion or rearrangement of a very small part of the submetacentric chromosomes.

From Bowenia to Zamia.—(Among the different karyotypes of *Zamia* I have chosen that with 12M+4S because it is the most symmetrical and perhaps the most primitive.) The number of metacentric chromosomes is the same. The submetacentrics go from six to four. This implies that a pair of chromosomes is lost.

From Zamia to Ceratozamia.—The number of metacentrics is the same. In *Ceratozamia* there are two telocentric chromosomes while the number of submetacentrics is reduced by two. The mechanism involved here is a deletion of the short arms of a pair of submetacentrics.

From Ceratozamia to Microcycas.—The number of metacentric and telocentric chromosomes both change. These changes in number can readily be explained by the centric fission of the 10 metacentric chromosomes. However, this step is not as simple as it appears in the scheme. The karyotype of *Microcycas* is not known with certainty. As already mentioned, in some preparations *Microcycas* appears to have acrocentric chromosomes. If that is so, the origin of its karyotype becomes more complicated. It could have originated from various karyotypes of *Zamia,* including the cytotypes of *Z. paucijuga.*

From Bowenia to Stangeria.—This passage involves several complicated mechanisms, like centric fission and simultaneous loss of chromosome pairs or arms. Depending upon which chromosomes are involved in centric fission, there would be various kinds of loss. With fission of two metacentrics there would be loss of two telocentrics and two submetacentrics; with fission of two submetacentrics there would be loss of two metacentric chromosomes and the short arms of two submetacentric chromosomes. It is worthwhile to point out that there are some species of *Zamia* which have the same karyotype as *Stangeria,* but I cannot accept that *Stangeria* is derived from *Zamia,* because a careful karyotype comparison of the two shows a clear difference in the dimensions of the chromosomes. In the scheme, however, I have taken into consideration such a hypothesis by indicating *Stangeria* as originating from a hypothetical precursor whose karyotype would be between those of *Bowenia* and *Zamia.*

Correlation between Karyological and Morphological Characters in Cycads

Although the scheme of Figure 3, like the one for *Zamia* (Fig. 1), is only a "working hypothesis," I believe that it allows some taxonomic and phylogenetic considerations. Some of the mechanisms proposed for the evolutionary changes described above cause huge changes at the genomic level because they either significantly change the karyotype or they cause a major loss of genome as large chromosome pairs (metacentrics or submetacentrics) are lost. These mechanisms include those involved in the steps from *Bowenia* to *Encephalartos, Bowenia* to *Zamia* and *Bowenia* to *Stangeria.* The other mechanisms involved in the remainder of the steps cause less drastic changes because they simply lead either to the rearrangement of genetic material or to the loss of small pieces of chromosomes.

The scheme reflects these two kinds of changes, so that on the basis of whichever change is involved, four groups of genera can be established: *Bowenia/Stangeria/Encephalartos, Lepidozamia, Macrozamia* and *Dioon/Zamia, Ceratozamia* and *Microcycas.*

If one accepts this proposal, it is possible to see some correspondence between karyotypic differentiation on the one hand and morphology and taxonomy on the other hand. The grouping of genera based on karyological data reflects some of the morphological and taxonomic characters of the order. In fact, *Bowenia* is placed in the family Boweniaceae, and *Stangeria* in the family Stangeriaceae. The genera of the other two groupings are all in the family Zamiaceae, but *Zamia, Ceratozamia* and *Microcycas* are in the tribe Zamieae and *Encephalartos, Lepidozamia, Macrozamia* and *Dioon* in the tribes Encephalarteae and Dioeae.

Literature Cited

Johnson, L. A. S. 1963. Cytological and taxonomic notes on Zamiaceae. Contr. New South Wales Natl. Herb., Fl. New South Wales **3(4):** 235–240.

Marchant, C. J. 1968. Chromosome patterns and nuclear phenomena in the cycad families Stangeriaceae and Zamiaceae. Chromosoma **24:** 100–134.

Moretti, A. 1982. Quinacrine fluorescence analysis of the chromosomes of *Macrozamia* Miq. (Cycadales, Zamiaceae). Delpinoa **24:** 129–136.

——— **& S. Sabato.** 1984. Karyotype evolution by centromeric fission in *Zamia* (Cycadales). Pl. Syst. Evol. **146:** 215–223.

Norstog, K. 1980. Chromosome numbers in *Zamia* (Cycadales). Caryologia **33(3):** 419–428.

———. 1981. Karyotypes of *Zamia chigua* (Cycadales). Caryologia **34(3):** 255–260.

Resende, F. & L. Rijo. 1948. Structure of chromosomes as observed in root tips. V. Olisterochromatin, chromatic agglutination and mutations. Port. Acta Biol. **2:** 117–147.

Sax, K. & J. M. Beal. 1934. Chromosomes of the Cycadales. J. Arnold Arb. **15:** 255–258.

Schutzman, B. 1984. A new species of *Zamia* L. (Zamiaceae, Cycadales) from Chiapas, Mexico. Phytologia **55(5):** 299–304.

Vovides, A. P. 1983. Systematic studies on the Mexican Zamiaceae. I. Chromosome numbers and karyotypes. Amer. J. Bot. **70(7):** 1002–1006.

Table III
Macrozamin and cycasin percentage in ripe seeds of cycads[a]

Species	Macrozamin % (fr. wt)[b]	Cycasin % (fr. wt)[b]
Cycadaceae		
Cycas basaltica C. A. Gardner	0.29	0.12
C. cairnsiana F. Muell.	0.20	0.10
C. circinnalis L.	0.42	0.10
C. lane-poolei C. A. Gardner	0.45	0.72
C. pruinosa Maconochie	0.33	0.10
C. revoluta Thunb.	0.26	0.28
C. thouvarsii R. Brown	0.31	0.06
Stangeriaceae		
Stangeria eriopus (Kunze) Baill.	4.70	0.02
Zamiaceae		
Tribe Dioeae		
Dioon edule var. *angustifolium* Miq.	0.64	0.12
Dioon sp. (Guerrero, Mex.)	0.62	0.13
Dioon sp. (Nayarit, Mex.)	0.65	0.01
Dioon sp. (Sonora, Mex.)	0.68	0.01
Tribe Encephalarteae		
Encephalartos altensteinii Lehm.	2.11	0.07
E. ferox Bertol. f.	2.10	0.05
E. lebomboensis Verdoorn	2.16	0.05
E. umbeluziensis R. A. Dyer	2.86	0.08
E. villosus Lem.	2.09	0.08
Lepidozamia peroffskyana Regel	1.11	0.21
Macrozamia diplomera (F. Muell.) L. Johnson	2.41	0.16
M. fawcettii C. Moore	2.49	0.06
M. miquelii (F. Muell) A. DC.	3.88	0.09
M. moorei F. Muell.	3.72	0.08
Tribe Zamieae		
Bowenia serrulata (W. Bull) Chamberlain	4.33	0.06
B. spectabilis Hook. f.	5.04	0.42
Ceratozamia matudai Lundell	1.05	0.08
C. mexicana Brongn.	1.01	0.02
Ceratozamia (Hidalgo, Mex.)	1.06	0.01
Ceratozamia sp. (Oaxaca, Mex.)	1.02	0.03
Microcycas calocoma (Miq.) A. DC.	0.13	0.07
Zamia latifoliolata Prenl.	1.02	0.02
Zamia sp. (Nayarit, Mex.)	1.25	0.01
Zamia sp. (Oaxaca, Mex.)	1.16	0.16
Zamia sp. (Oaxaca, Mex.)	1.23	0.15
Zamia sp. (Veracruz, Mex.)	1.01	0.28

[a] After Moretti et al., 1983.
[b] Each datum is the average of three measurements.

lysis and identified the acidic components and the sugars linked thereto. On the basis of these data, I analyzed by a gas-chromatographic method the composition of the mucilage in individuals of *Encephalartos longifolius* of different age, sex, and growth conditions (Siniscalco Gigliano, 1980). Rhamnose, fucose, arabinose, mannose, and galactose were identified in the polysaccharide obtained by precipitation with 95% ethanol. Our results showed that the amounts of each monosaccharide obtained on hydrolysis of the mucilage are constant in individuals of *E. longifolius* irrespective of age, sex, and growth conditions (Table IV).

These findings, suggesting that the mucilage composition could be significant from a chemotaxonomical point of view, have led me to analyze the monosaccharides of hydrolyzed muci-

Table IV

Monosaccharide composition of the mucilage in a number of individuals belonging to the species *Encephalartos longifolius* Lehm.[a]

		117[b]	118[b]	126[b]	128[b]	129[b]
R	mg/g	95.95	96.63	96.00	95.63	95.94
	mmol/g	0.5845	0.5825	0.5847	0.5825	0.5844
F	mg/g	49.28	48.88	49.35	49.30	49.91
	mmol/g	0.3002	0.2977	0.3006	0.3003	0.2979
A	mg/g	196.08	195.80	196.00	194.51	195.88
	mmol/g	1.3061	1.3042	1.3055	1.2956	1.3047
M	mg/g	8.30	8.40	8.26	8.39	8.30
	mmol/g	0.0461	0.0466	0.0458	0.0466	0.0461
G	mg/g	255.30	254.74	256.71	257.24	256.86
	mmol/g	1.4170	1.4140	1.4249	1.4278	1.4257

[a] After Siniscalco Gigliano, 1980.
[b] Catalog number of specimen.
R = Rhamnose, F = Fucose, A = Arabinose, M = Mannose, G = Galactose.

lages in a number of species of *Encephalartos,* representative of the whole geographical distribution range of this genus (Moretti et al., 1981b). Galactose, arabinose, and rhamnose are the most abundant monosaccharides in all the examined species. They constitute a very high proportion of the total polysaccharide, whereas fucose, mannose, and methylrhamnose represent a very small proportion of the total. The monosaccharide distribution patterns are similar in all the species examined. Figure 1, reporting monosaccharide amounts in millimoles, supports this finding to

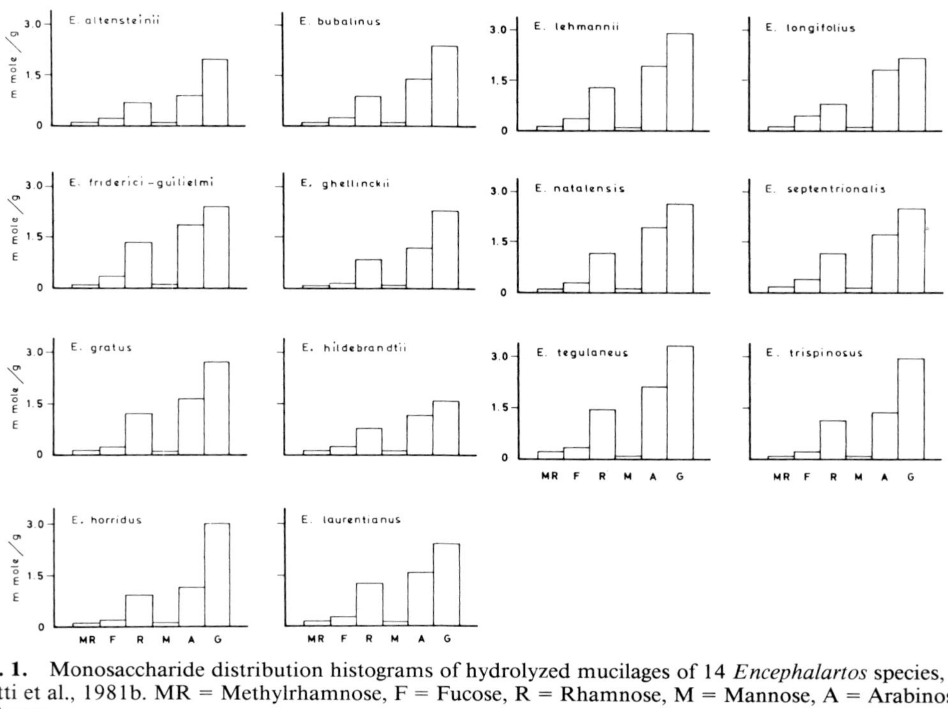

Fig. 1. Monosaccharide distribution histograms of hydrolyzed mucilages of 14 *Encephalartos* species, after Moretti et al., 1981b. MR = Methylrhamnose, F = Fucose, R = Rhamnose, M = Mannose, A = Arabinose, G = Galactose.

Table IV
Extended

273[b]	278[b]	Average
95.91	95.94	96.99 ± 0.28
0.5842	0.5844	0.5839 ± 0.0009
49.28	49.30	49.33 ± 0.28
0.3002	0.3003	0.2996 ± 0.0011
196.54	195.52	195.76 ± 0.58
1.3091	1.3023	1.3039 ± 0.0039
8.40	8.33	8.34 ± 0.05
0.0466	0.0462	0.0463 ± 0.0003
256.10	256.96	256.27 ± 0.86
1.4215	1.4263	1.4224 ± 0.0048

a greater extent. The monosaccharide ratios appear similar among the species even if the total monosaccharide amount in the mucilages is different. The monosaccharide composition of the mucilages showed that differences may occur at a specific level while the composition in additional individuals examined appeared constant. However, the most striking feature about monosaccharide composition of the mucilages within the examined *Encephalartos* species is the occurrence of a common pattern of monosaccharide distribution.

Later, the mucilage of representative species of all ten genera were analyzed in order to determine the distribution of the monosaccharide (De Luca et al., 1982). Figure 2 shows the monosaccharide content for one or two species in each genus. Arabinose, fucose, galactose, mannose, rhamnose, and methylrhamnose are the monosaccharides of the polysaccharide in *Bowenia, Ceratozamia, Cycas, Dioon, Encephalartos, Lepidozamia, Macrozamia, Microcycas,* and *Zamia*. However, *Ceratozamia* contains galactose only in traces. The only genus which lacks some of the monosaccharides is *Stangeria,* in which there is a complete absence of methylrhamnose and rhamnose.

The monosaccharides occur in the mucilages in quantities varying between 32% (w/w) in *Dioon mejiae* Standl. & Williams and 94% (w/w) in *Macrozamia secunda* C. Moore. Fucose is the most abundant monosaccharide in *Dioon*; rhamnose in *Ceratozamia* and *Zamia*; galactose in *Encephalartos, Lepidozamia,* and *Microcycas*; arabinose in *Bowenia, Cycas, Macrozamia,* and *Stangeria.*

The monosaccharide compositions of the mucilages show that differences occur at generic and higher levels, while the monosaccharide distribution patterns are characteristic at the generic level. *Stangeria* (of the monotypic family Stangeriaceae) is the most distinct, whereas *Lepidozamia* and *Encephalartos* are the only two genera to show similar patterns.

The polysaccharides of the African, Australian and Australasian genera contain mainly arabinose and galactose, whereas those of the American genera contain higher proportions of fucose and galactose in *Dioon*; rhamnose and arabinose in *Ceratozamia*; rhamnose, fucose and methylrhamnose in *Zamia*; galactose and methylrhamnose in *Microcycas*. The Stangeriaceae are well separated chemically from the other families.

The American genera of the Zamiaceae may be distinguished from the other members of the family by their different monosaccharide composition. This may indicate an evolutionary divergence at an early stage in the history of this taxon. On the other hand, however, the monosaccharide distribution patterns of *Cycas,* included in the Cycadaceae, and *Bowenia,* included in the Boweniaceae, appear to be very similar to that of *Macrozamia,* included in the Zamiaceae.

Among the Zamiaceae, the data regarding *Lepidozamia, Macrozamia, Microcycas,* and *Zamia* appear of systematic interest. *Lepidozamia,* a genus established in the past century by Regel (1857) and later included by most taxonomists in *Macrozamia* (Schuster, 1932), appears well separated from *Macrozamia* on the basis of their different monosaccharide compositions. *Microcycas,* morphologically not well separated from *Zamia,* nonetheless appears well separated from *Zamia* on the basis of their different monosaccharide compositions. In contrast, the monosaccharide composition of the mucilages of *Zamia* does not support the subdivision of this genus into *Zamia* and *Aulacophyllum* as proposed by Regel (1876), because the monosaccharide composition is identical in species attributed both to *Zamia* (*Z. furfuracea* L. f., *Z. floridana* A. DC., *Z. manicata* Linden ex Regel) and to *Aulacophyllum* (*Z. lindenii* Regel ex Andrè, *Z. wallisii* hort. Veitch ex A. Braun).

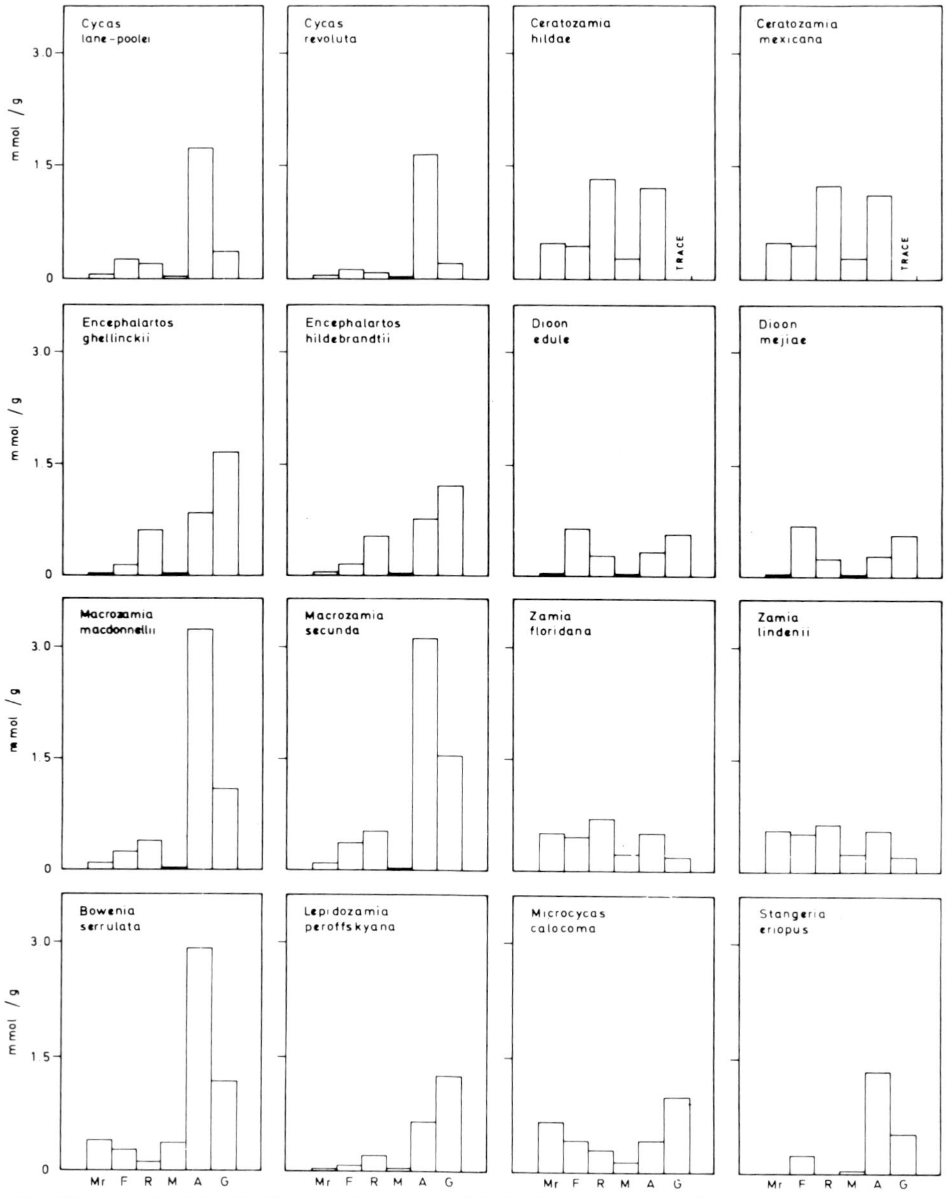

Fig. 2. Monosaccharide distribution histograms of hydrolyzed cycad mucilage, after De Luca et al., 1982. MR = Methylrhamnose, F = Fucose, R = Rhamnose, M = Mannose, A = Arabinose, G = Galactose.

Conclusion

The analyses for MAM glycosides have revealed them to be a very important tool in understanding the relationships among cycads and between cycads and other seed plants. In particular the exclusive occurrence of MAM glycosides in the cycads supports the concept that they are a monophyletic group.

Cycasin, simple ovulate cones and girdling leaf-traces are unique synapomorphies which distinguish cycads from other seed plants (Crane, 1985).

Macrozamin probably represents the most primitive MAM glycoside. Cycasin and neocycasins whose carbohydrate component consists of glucose units may be considered derived. On the basis of these characters, the cycads appear only very anciently related to other seed plants; they probably diverged at an early time in seed plant evolution.

The mucilage composition patterns also revealed useful trends in substantiating evolutionary relationships and/or divergences among the cycad genera. The monosaccharide composition pattern may be considered a marker at the generic level. *Zamia,* the most widespread and diversified American genus, shows a similar pattern both in specialized and primitive species, i.e., *Zamia manicata* Linden ex Regel and *Zamia vallisii,* that from their leaflet morphology may appear to belong to separate genera, show monosaccharide composition patterns identical to those of other species of *Zamia.*

The discovery of the new genus *Chigua* in Northern Colombia is a very exciting scientific event. The analyses of MAM glycosides and mucilages from this new plant may, in my opinion, elucidate its systematic position and help in understanding the evolutionary history of the cycads.

Literature Cited

Cooper, J. M. 1940. Isolation of a toxic principle from the seeds of *Macrozamia spiralis.* J. Roy. Soc. New South Wales **74:** 450–454.

Crane, P. R. 1985. Phylogenetic analysis of seed plants and the origin of angiosperms. Ann. Missouri Bot. Gard. **72:** 716–793.

Dastur, D. K. & R. S. Palekar. 1966. Effect of boiling and storing on cycasin content of *Cycas circinalis* L. Nature **210:** 841–843.

De Luca, P., A. Moretti, S. Sabato & G. Siniscalco Gigliano. 1980. The ubiquity of cycasin in cycads. Phytochemistry **19:** 2230–2231.

———, ———, ——— & ———. 1982. A comparative study of cycad mucilages. Phytochemistry **21:** 1609–1611.

Gasco, A., A. Serafino, V. Mortarini & E. Menziani. 1974. An antibacterial and antifungal compound from *Calvatia lilacina.* Tetrahedron Letters **38:** 3431–3432.

Johnson, L. A. S. 1959. The families of cycads and the Zamiaceae of Australia. Proc. Linn. Soc. New South Wales **84:** 64–71.

Laquer, G. L. & M. Spatz. 1968. Toxicology of cycasin. Cancer Res. **28:** 2262–2267.

Lythgoe, B. & N. V. Riggs. 1949. Macrozamin. Part I. The identity of the carbohydrate component. J. Chem. Soc. **4:** 2716–2718.

Matsumoto, H. & F. M. Strong. 1963. The occurrence of methylazoxymethanol in *Cycas circinalis* L. Arch. Biochem. Biophys. **101:** 299–310.

Moretti, A., S. Sabato & G. Siniscalco Gigliano. 1981a. Distribution of macrozamin in australasian cycads. Phytochemistry **20:** 1415–1416.

———, ——— & ———. 1981b. Monosaccharide composition of the mucilages in *Encephalartos* Lehm. (Zamiaceae). Giorn. Bot. Ital. **115:** 291–297.

———, ——— & ———. 1983. Taxonomic significance of methylazoxymethanol glycosides in the cycads. Phytochemistry **22:** 115–117.

Nagahama, T. 1964. Studies on neocycasins, new glycosides of cycads. Bull. Fac. Agric. Kagoshima Univ. **14:** 1–50.

Nishida, K., A. Kobayashi & T. Nagahama. 1955. Studies on cycasin, a new toxic glycoside of *Cycas revoluta* Thunb. Agric. Chem. Soc. **19:** 77–83.

Pant, D. D. 1973. *Cycas* and Cycadales. Central Book Dept., Allahabad, India.

Regel, E. 1857. Zwei neue Cycadeen. Bull. Soc. Nat. Mosc. **1:** 163–182.

———. 1876. Cycadearum generum specierumque revisio. Acta Horti petrop. **4:** 275–320.

Schoental, R. 1969. Carcinogenic action of elaiomycin in rats. Nature **221:** 765–766.

Schuster, J. 1932. Cycadaceae. *In* Engler, Das Pflanzenreich **99(IV. 1):** 1–168.

Siniscalco Gigliano, G. 1980. Analisi gascromatografica dei monosaccaridi delle mucillagini di *Encephalartos longifolius* Lehm. (Zamiaceae). Delpinoa **21:** 63–70.

Stephen, A. M. & D. C. De Bruyn. 1967. The gum exudate of *Encephalartos longifolius* Lehm. (female) (family Cycadaceae). Carbohyd. Res. **5:** 256–265.

Stevenson, D. Wm. 1981. Observation on ptyxis, phenology, and trichomes in the Cycadales and their systematic implication. Amer. J. Bot. **68:** 1104–1114.

Flavonoid Chemistry and the Taxonomy of Cycads

P. MICK RICHARDSON

Table of Contents

Abstract

RICHARDSON, P. M. (New York Botanical Garden, Bronx, New York 10458, USA) Flavonoid chemistry and the taxonomy of cycads. Memoirs of the New York Botanical Garden **57**: 132–141. 1990.—The flavonoid chemistry of the cycads is comprehensively reviewed from a historical and taxonomic viewpoint. The chemistry and phylogeny of lower plants is discussed with a view to choice of an outgroup for phylogenetic analysis of the cycads based upon their flavonoid chemistry, using mainly *C*-glycosylflavones, biflavones and proanthocyanidins. Lack of variation within genera limits the resolution of resultant cladograms. Literature reports of other phenolic substances such as cinnamic acids and lignins are also discussed.

Key words: Cycads; biochemical systematics; flavonoids; tannins; lignins; cladistics.

Introduction

A variety of compounds are used in the study of the biochemical systematics of plants. The compounds in plants can be classified in various ways: either into macromolecules and micromolecules or into primary and secondary compounds. Macromolecules are large compounds such as DNA and proteins; micromolecules are compounds with a small molecular weight; primary compounds are chemicals involved in the primary metabolism of the plant and generally occur in all plant species, e.g., chlorophyll; secondary compounds are not involved in primary metabolism and characteristically have a restricted distribution. It is this restricted distribution which has made secondary compounds the most favored ones for biochemical systematics studies. Secondary compounds are themselves classified into groups of compounds based upon their biosynthetic pathways—terpenes, phenolics, nitrogen compounds, etc. The phenolic compounds can be further divided into smaller groups, one of which is the group known as flavonoids.

Flavonoids are a group of plant pigments, some of which are familiar to most of us in the form of reddish autumn leaves or red and blue flowers. However, it is only a minority of the compounds which absorb light in the visible part of the spectrum, while the majority absorb only in the ultraviolet region. Thus, they remain invisible without the use of special techniques such as ultraviolet spectrophotometry, even though they occur in the cell vacuoles of all parts of the plant. Flavonoids are related by their common biosynthetic pathway and are all related to the parental molecule flavone. This molecule contains 15 carbon atoms in the form of two benzene rings linked by 3 carbon atoms which form a third ring system. Flavone is a most unusual flavonoid compound because there are no hydroxyl groups attached to the molecule. Generally, flavonoids have various numbers of hydroxyl groups attached and in many cases there are sugars or methyl groups attached to the hydroxyl groups. Thus we can have a variety of oxygenation, glycosylation and methoxylation patterns. It is the occurrence of these patterns, or various combinations of them, which provide the characters when flavonoids are used in biochemical systematics.

Flavonoids and Taxonomy

Flavonoids are divided into various groups more or less based on the degree of oxidation of the central ring. The commonest types of flavonoids found in plants are glycosides of flavones, flavonols and flavanones. Such compounds are called monomers because they contain only one 15-carbon flavonoid molecule. However, compounds other than monomers also occur. These can be either dimers which consist of two flavonoid molecules attached to one another or they can be oligomers where several flavonoid molecules are attached to one another. The commonest dimers are biflavones, consisting of two flavone molecules, but "hybrid" molecules can also occur, e.g., a flavone linked to a flavanone. The dimers are generally referred to as biflavonoids (although the term biflavanoid is perhaps more correct, and neither term should be confused with bioflavonoids, which is the layman's term for flavonoids in general). The oligomers are generally termed proanthocyanidins because, although they themselves are colorless, they react with hot acid to form reddish-colored anthocyanidins. Older names for proanthocyanidins include condensed tannins and leucoanthocyanidins.

For years phytochemists have investigated the occurrence of the different types of flavonoids in various plant groups. The resulting data have generally been interpreted in a narrative fashion (Humphries & Richardson, 1980). Two questionable generalizations have come to exist in the flavonoid literature. The first of these is that chemical complexity is equated with evolutionary advancement; in other words, the compound at the end of a biosynthetic pathway is evolutionarily more advanced than any of its precursors. In the biosynthetic pathway A → B → C → D, compound D would be considered more advanced than compound C. The second generalization is that primitive compounds occur in primitive plants and there is a general increase in flavonoid complexity from the algae to the angiosperms. However, both of these generalizations are generally invalid (Gornall & Bohm, 1978).

In the author's opinion, the only way to determine whether a compound is advanced or primitive is to examine its occurrence in the nearest relatives of the plant in question, a process called out-group analysis. The occurrence of advanced flavonoid features can then be used to determine the phylogeny of the members of a plant group. The species (or genera) are grouped together on the basis of the advanced features which they share. Advanced compounds which occur more widely form larger taxonomic groups. A detailed example concerning *Solanum* section *Androceras* has already been published (Richardson, 1982), and in this case the cladogram based upon flavonoid features compared favorably with a cladogram based upon morphological, geographical and ecological features.

Knowing that I was interested in such cladistic studies of chemical features, Dennis Stevenson asked me to attempt a similar analysis of the flavonoids of the cycads. First, it was essential to examine the occurrence of flavonoids in plant groups which are more primitive than the cycads. Then, using a suitable outgroup, the advanced flavonoid features of the cycads could be defined and a cladogram of the cycad genera could be constructed. These three areas are described in more detail below.

Flavonoid Chemistry of Land Plants

A brief review of the chemistry of the lower land plants will allow us to discern any flavonoid compounds which are unique to the gymnosperms. The plant groups are discussed in the order in which they appear on a cladogram of the green plants in a recent publication by Bremer (1985).

Flavonoids first appear in the plant kingdom in the algal genera *Chara* and *Nitella,* which contain only *C*-glycosylflavones. The next group in the cladogram is the liverworts, and these too contain *C*-glycosylflavones. However, primitive though they are, the liverworts not only have *C*-glycosylflavones but also contain numerous other types of flavonoid compounds, including flavone *O*-glycosides, flavonols, flavanones, chalcones, aurones and dihydrochalcones. Liverworts also possess several flavonoid features which were earlier considered to be unique to higher plants, e.g., *O*-methylation, complex glycosylation, acylation and 8-hydroxylation. Interestingly, the primitive genus *Takakia* contains the same compounds as members of the advanced Marchantiaceae. The occurrence of such a wide range of flavonoid compounds in this early group of plants leaves little room for new diversity in the remainder of the plant kingdom. The next group on the Bremer cladogram is the mosses. They contain all the types of flavonoids found in the liverworts and they also have some novel compounds, e.g., a biflavonoid and 3-deoxyanthocyanins as well as 8-hydroxylation.

The next few groups of plants on the cladogram (*Psilotum,* lycopods, *Equisetum, Ophioglossum, Marattia* and true ferns) form a paraphyletic group, the pteridophytes. They contain few new compounds. *Equisetum* is the first group to contain proanthocyanidins and some of the true ferns contain anthocyanins.

Next we come to another paraphyletic group, the gymnosperms, and it is in this group that we find the cycads and the outgroup (the group to be used to determine the derived flavonoid compounds in the cycads). The cycads contain no novel types of flavonoids, possessing only biflavones, *C*-glycosylflavones and proanthocyanidins. *Ginkgo* contains only biflavones, flavonols and proanthocyanidins. The conifers are the first group to contain isoflavones and they also contain biflavones, *C*-glycosylflavones, flavanones, flavonols, anthocyanins and proanthocyanidins.

Finally we come to the angiosperms and these have several novel types of compound as well as containing all the previous types of flavonoids. This brief overview of flavonoid distribution in the plant kingdom shows us that the flavonoids in the cycads are not remarkable and are actually rather similar to those found in other gymnosperms. However, a more detailed examination of the actual compounds in the cycads and their sister group is necessary in order to decide which compounds can be considered as advanced characters and used to reconstruct a possible phylogeny.

The Chemistry of the Outgroup

The choice of an outgroup for the cycads is discussed at length in recent papers on the phylogeny of the land plants (Crane, 1985; Doyle & Donoghue, 1986). There is some disagreement among the authors as to which group is the actual outgroup and at the moment it is only possible to say that the outgroup is either *Ginkgo* or the conifers or both *Ginkgo* and the conifers. The flavonoid chemistry of the cycads, conifers and *Ginkgo* is rather similar in many respects. All three groups contain biflavonoids and proanthocyanidins. In addition, cycads contain *C*-glycosylflavones, which are also found in conifers but not in *Ginkgo*. The presence of flavonols in *Ginkgo* and conifers, and of flavanones, anthocyanins and isoflavones in conifers is immaterial to the current question because these compounds are absent in cycads and can thus give us no taxonomic information about the group. The compounds which potentially contain information of taxonomic value are *C*-glycosylflavones, biflavonoids and proanthocyanidins.

Chemotaxonomy of Cycads

C-Glycosylflavones

C-Glycosylflavones have been reported to occur in several genera of cycads. The compounds were first found in *Dioon spinulosum* by Carson and Wallace (1972). Compounds based on both apigenin and luteolin were present, although the latter was not fully identified. Later, Niemann and Miller (1975) reported the presence of two apigenin-based compounds in *Cycas taiwania* and the absence of *C*-glycosylflavones in *Ceratozamia mexicana, Encephalartos lehmannii, Dioon edule* and another species of *Dioon*. Lebreton (1980) mentions the presence of unidentified *C*-glycosylflavones in *Cycas circinalis, C. revoluta* and *Stangeria paradoxa* and the absence of the compounds in *Dioon edule, Ceratozamia mexicana, C. longifolia, Encephalartos lehmannii, E. horridus, E. longifolius* and *Macrozamia diplomera*. In summary, *C*-glycosylflavones have been reported from *Dioon, Cycas* and *Stangeria*. Also, the compounds are not present in *Ginkgo* but are found in some genera of conifers.

Using *Ginkgo* as an outgroup would suggest that *C*-glycosylflavones are an advanced feature in cycads and this would link *Cycas, Stangeria* and *Dioon* as a taxonomic group. However, there is a problem here in that we do not yet know the true distribution of these compounds in the cycads. Using the other outgroups (conifers or conifers plus *Ginkgo*), the presence of *C*-glycosylflavones in some genera of conifers (Lebreton et al., 1978; Niemann & Miller, 1975) means that we cannot say whether the compounds are a primitive or an advanced feature in the cycads. It is possible that these compounds will prove to be of taxonomic interest within genera even if they are of no value above the generic level.

Biflavonoids

The biflavonoids are the characteristic flavonoid compounds of the cycads. They are also characteristic of *Ginkgo* and the conifers except for the Pinaceae, where they have only been reported once (Chatterjee et al., 1984). There have been numerous reports of the compounds in cycads during the past 30 years. The first discovery was by Kariyone and Sawada (1958) who reported sotetsuflavone (amentoflavone 7′′-methyl ether) from *Cycas revoluta*. The same compound from the same plant is also mentioned in Sawada (1958), Kawano and Yamada (1960, as cited by Carson & Wallace, 1972; Locksley, 1973; and Dossaji et al., 1975, although a plant source is not mentioned in Chemical Abstracts) and Nakajima (1964). In the 1970's things moved more rapidly and less repetitively. Geiger and Pfleiderer (1971) found amentoflavone, sotetsuflavone, 2,3-dihydroamentoflavone, hinokiflavone and 2,3-dihydrohinokiflavone in *Cycas revoluta*. Handa et al. (1971) reported amentoflavone, sequoiaflavone (amentoflavone 7-methyl ether), bilobetin (amentoflavone 4′-methyl ether), ginkgetin (amentoflavone 7,4′-dimethyl ether), sciadopitysin (amentoflavone 7,4′, 4′′′-methyl ether) and amentoflavone 7,7′′,4′,4′′′-tetramethyl ether in *Zamia angustifolia*. Several *Dioon* species were found to contain amentoflavone and several of its methyl ethers (Dossaji et al., 1973). Varshney et al. (1973) reported both old and new compounds from three *Cycas* species. In 1975 Dossaji published the results of his Ph.D. studies (Dossaji et al., 1975), reporting the distribution of amentoflavone and hinokiflavone, their methyl ethers and 2,3-dihydroderivatives in 82 species from all 10 genera. The significance of these results will be discussed in the next section. There have been few recent reports on biflavonoids in cycads. Gadek (1982) examined the distribution of these compounds in the seed testa, reporting *Cycas* to contain amentoflavone and its 4′′′-monomethyl ether and *Macrozamia* to contain amentoflavone and several of its mono-, di- and trimethyl ethers as well as cupressuflavone and a mono- and a dimethyl ether. The cupressuflavones have not previously been known to occur in the cycads. Mohammad et al. (1983) examined three species of *Encephalartos* and reported the presence of amentoflavone as well as a mono-, a di-, a tri- and a tetramethyl ether. Curiously, the methyl ethers were not fully named even though the authors appeared to have fully characterised the compounds by comparison with authentic samples. Gadek et al. (1984) used aluminum chloride fluorescence to locate phenolic compounds in several gymnosperms, including cycads, and claimed that the fluorescence in the cuticle was due to biflavonoids. They did not explain why the fluorescence could not be due to other phenolic compounds such as *C*-glycosylflavones. Reviews on the occurrence of biflavonoids in cycads and other plants include those of Locksley (1973), Geiger and Quinn (1975, 1982), Lebreton (1980) and Hegnauer (1986).

The most significant study has been the work of Dossaji, whose comprehensive findings were published in 1975 (Dossaji et al., 1975). Dossaji interpreted his results as confirming the classification of Johnson (1959): the Stangeriaceae lacked biflavonoids; the Zamiacae had only non-hinokiflavone biflavonoids; and the Cycadaceae had both hinokiflavone and non-hinokiflavone biflavonoids. Furthermore, Dossaji interpreted the absence of biflavones as an advanced character but he did not comment on the evolutionary polarity of hinokiflavone. I might mention here that such a narrative interpretation of the data would not support Stevenson's (1981) concept of the family Boweniaceae, for *Bowenia* has a biflavonoid chemistry identical to that of *Encephalartos*. I will return to this topic in a later section.

A review by Geiger and Quinn (1982) reiterates Dossaji's narrative interpretation of the biflavonoid data. In contrast, an earlier review by Lebreton (1980) boldly reinterpreted Dossaji's results, Lebreton claiming that the Cycadaceae were defined by the presence of biflavanones, hinokiflavone and sotetsuflavone and that the Zamiaceae were defined by sciadopitysin. However, a closer examination of Lebreton's tabulation of the data reveals that he only included 7 of the 10 genera studied by Dossaji. When the missing genera are added, the situation changes dramatically: *Microcycas* contains sotetsuflavone but no sciadopitysin and this causes Lebreton's chemical definitions of the Cycadaceae and the Zamiaceae to fall apart. Indeed, the Zamiaceae contain no unique compounds to define them as a group; if they did, surely Dossaji would have pointed out this fact. Furthermore, Lebreton claims the occurrence of biflavanones in *Cycas*. As far as I am aware these compounds, which would be composed of two flavanone molecules linked together, have not yet been reported from the cycads, being known so far only from the angiosperms (Geiger & Quinn, 1982). Presumably, the compounds to which Lebreton is referring are 2,3-dihydroamentoflavone and 2,3-dihydrohinokiflavone, both of which are biflavonoids which consist of a flavone molecule and a flavanone molecule linked together. Thus they are neither biflavones nor biflavanones but, rather, a hybrid molecule. In the cycads these flavone-flavanone compounds are unique to *Cycas,* as is hinokiflavone. It should be noted that in general it is easier to find a chemical character to define a group consisting of a single genus than it is to find a character which occurs in every member of a large group, e.g., the eight genera of the Zamiaceae. However, the presence of two unique groups of compounds in the Cycadaceae does set them apart from the Zamiaceae. Furthermore, *Cycas* contains only mono- and dimethyl ethers of amentoflavone and lacks the more highly methylated compounds which are found in the Zamiaceae, where tetramethylated compounds are found in five of the eight genera and trimethylated compounds are lacking only in *Microcycas*. It would be possible to talk of a trend towards an increase or decrease in the amount of methylation in the cycads but it is perhaps more constructive to discuss actual distributions rather than trends.

A cladistic analysis of the biflavonoid data was attempted in order to see if any more information could be gleaned from Dossaji's data. This data set was chosen for the analysis for several reasons. It was the most complete set of information available. The one report which conflicts with Dossaji's data has been ignored because Dossaji's data contains no conflicts with other data from several groups of workers. Finally, as discussed by Geiger and Quinn (1975: 726 and 1982: 530) differences in biflavonoid data are minimized when the plants are screened by the same group of workers.

The 15 molecular features (hydroxylation, methylation, linkage of the two flavonoid molecules, etc.) of the biflavonoids in the cycads are shown in Table I. The various cycad compounds and the molecular features contained in each are shown in Table II. From this we can tabulate the molecular features present in each genus (Table III). Most of the molecular features are universal (except in *Stangeria*) and thus of little taxonomic value within the group. This eliminates 9 of the 15 features, leaving only 6 features with which to work. A fully resolved cladogram for the ten genera would require a minimum of nine characters, so by definition any cladogram produced

Table I
The 15 molecular features which occur in the biflavonoids of the cycads

1	5-OH
2	7-OH
3	4′-OH
4	5″-OH
5	7″-OH
6	4‴-OH
7	5-OCH_3
8	7-OCH_3
9	4′-OCH_3
10	5″-OCH_3
11	7″-OCH_3
12	4‴-OCH_3
13	2,3-dehydrogenation
14	3′,8″-linkage
15	4′,6″-ether linkage

from the six characters can only be partially resolved. Which of these six characters are unique to cycads and potentially useful as characters within the group?

The occurrence of biflavonoids in the plant kingdom is fairly well known (Geiger & Quinn, 1982), the compounds being found in the mosses, *Psilotum* and *Tmesipteris, Selaginella, Osmunda* (Okuyama et al., 1979), the cycads, *Ginkgo,* the conifers and the angiosperms. Table IV shows the biflavonoid molecular features found in non-angiospermous groups other than the cycads. Several of these groups were tried as outgroups for the cycads and the results were as follows.

Using *Osmunda* as the outgroup suggests that features 3, 6, 7, 10, 13 and 15 are advanced features, while using mosses through to *Osmunda* suggests only 7, 10 and 13 as being advanced. *Ginkgo* suggests features 3, 7, 10, 11, 13 and 15 are advanced, while conifers give only 7, 10 and 13 as advanced, as would using conifers and *Ginkgo* together. Using any of these groups as the outgroup gives a poorly resolved cladogram but all of the cladograms would have some features in common. Methylation of the 5 and 5″ hydroxyl groups is unique to *Dioon* and 2,3-dehydrogenation is unique to *Cycas*. In addition, the *Gingko*-based tree suggests that the nine biflavonoid-containing genera are grouped together on the basis of 4′-hydroxylation, that the 4′,6‴-ether linkage is unique to *Cycas* and that loss of 7″-methylation is unique to *Macrozamia*.

Despite the limited amount of data available

Table II
Cycad biflavonoids and the molecular features which occur in each compound. The notations 8A.1, 8A.2, etc. refer to the review by Geiger and Quinn (1982). The numbers 1 through 15 refer to the molecular features in Table I

	Compound	Molecular features
8A.1	Amentoflavone	1, 2, 3, 4, 5, 6, 14
8A.3	Sequoiaflavone	1, 3, 4, 5, 6, 8, 14
8A.4	Bilobetin	1, 2, 4, 5, 6, 9, 14
8A.5	Sotetsuflavone	1, 2, 3, 4, 6, 11, 14
8A.6	Podocarpus flavone A	1, 2, 3, 4, 5, 12, 14
8A.7	Ginkgetin	1, 4, 5, 6, 8, 9, 14
8A.11	Isoginkgetin	1, 2, 4, 5, 9, 12, 14
8A.14	Sciadopitysin	1, 4, 5, 8, 9, 12, 14
8A.17	Amentoflavone tetramethyl ether	1, 4, 8, 9, 11, 12, 14
8A.18	Amentoflavone hexamethyl ether	7, 8, 9, 10, 11, 12, 14
8B.1	2,3-dihydro-amentoflavone	1, 2, 3, 4, 5, 6, 13, 14
10A.1	Hinokiflavone	1, 2, 4, 5, 6, 15
10B.1	2,3-dihydrohino-kiflavone	1, 2, 4, 5, 6, 13, 15

from biflavonoids we can conclude the following: 1) *Cycas* and *Dioon* both contain unique compounds; 2) the absence of biflavonoids in *Stangeria* indicates that it is different from the other genera; 3) there are no biflavonoid characters to link the genera in either Johnson's or Stevenson's Zamiaceae; and 4) there is no evidence to either support or controvert Stevenson's Boweniaceae (the exact similarity of *Bowenia* and *Encephalartos* chemistry could readily be the retention of a suite of primitive characters common to almost all cycads).

By knowing the distribution of biflavonoids in the cycads and related groups, it is possible to comment on the presence or absence of the compounds in the ancestor of the cycads. An ancestor containing biflavonoids is the most parsimonious explanation for both the cladogram of Stevenson (this volume) and three of the four cladograms based on Johnson's three family classification. (In the fourth case, where Stangeriaceae would be the sister group to Cycadaceae and Zamiaceae, it is equally parsimonious for the ancestor to possess or lack biflavonoids.) Generally, the loss of biflavonoids in *Stangeria*

Table III

The flavonoid features of the biflavonoids which occur in the cycad genera. The numbers refer to the features in Table I. Note that *Stangeria* lacks biflavonoids completely

Genus	Molecular features
Cycas	1, 2, 3, 4, 5, 6, –, 8, 9, –, 11, –, 13, 14, 15
Stangeria	– – – – – – – – – – – – – – –
Lepidozamia	1, 2, 3, 4, 5, 6, –, 8, 9, –, 11, 12, –, 14, –
Macrozamia	1, 2, 3, 4, 5, 6, –, 8, 9, –, –, 12, –, 14, –
Encephalartos	1, 2, 3, 4, 5, 6, –, 8, 9, –, 11, 12, –, 14, –
Dioon	1, 2, 3, 4, 5, 6, 7, 8, 9, 10, 11, 12, –, 14, –
Ceratozamia	1, 2, 3, 4, 5, 6, –, 8, 9, –, 11, 12, –, 14, –
Zamia	1, 2, 3, 4, 5, 6, –, 8, 9, –, 11, 12, –, 14, –
Bowenia	1, 2, 3, 4, 5, 6, –, 8, 9, –, 11, 12, –, 14, –
Microcycas	1, 2, 3, 4, 5, 6, –, 8, 9, –, 11, –, –, 14, –

is a more simple explanation than the gain and subsequent loss of the compounds or their parallel gain in two separate cycad lines. However, if *Stangeria* is the sister group of the remainder of the cycads it is equally parsimonious to have an ancestor with or without biflavonoids.

Similarly, an ancestor for cycads, conifers and *Ginkgo* would be expected to contain biflavonoids. Otherwise, biflavonoids would have had to evolve independently in three separate lines. The loss of biflavones in the Pinaceae would be equivalent to their loss in *Stangeria*. In the angiosperms biflavonoids have been explained as either the retention of a primitive character (Geiger & Quinn, 1975) or the result of independent origin in each group in which they occur (Geiger & Quinn, 1982).

PROANTHOCYANIDINS

Proanthocyanidins are one group of tannins which occur in plants (Swain, 1979). In a phylogenetic sense the compounds are first found in *Equisetum* and they are characteristic of ferns, gymnosperms and woody angiosperms.

Bate-Smith and Lerner (1954) reported leucoanthocyanidins in *Dioon* and *Macrozamia* and their absence in *Stangeria*. The compounds were also present in *Ginkgo* and various conifers. It was many years before a more intensive investigation was undertaken (Lebreton et al., 1980). The compounds were reported from *Cycas, Stangeria, Dioon, Ceratozamia, Encephalartos* and *Macrozamia*. The authors also estimated the relative amounts of procyanidin and prodelphinidin. Based on their examination of seven species in six genera, they concluded that the Cycadaceae were rich in prodelphinidin and that this compound almost disappeared from *Stangeria* and was completely absent from the Zamiaceae. In a second paper, Lebreton (1980) added a further three species to the analysis (but no new genera) and the conclusion remained unchanged.

Preliminary results from my laboratory (Richardson et al., unpubl.) based solely on color produced in boiling acid indicate that the compounds are present in *Macrozamia, Encephalartos, Dioon, Microcycas* and *Ceratozamia*. Perhaps the most interesting results are those from

Table IV

Biflavonoid features which appear in cycads and plant groups other than angiosperms

Plant group	Molecular features
Mosses	1, 2, 3, 4, 5, 6, –, –, –, –, –, –, –, 14, –
Psilotales	1, 2, 3, 4, 5, 6, –, 8, –, –, –, –, –, 14, –
Selaginella	1, 2, 3, 4, 5, 6, –, 8, –, –, 11, –, –, 14, 15
Osmunda	1, 2, –, 4, 5, –, –, 8, 9, –, 11, 12, –, 14, –
Ginkgo	1, 2, –, 4, 5, 6, –, 8, 9, –, –, 12, –, 14, –
Conifers	1, 2, 3, 4, 5, 6, –, 8, 9, –, 11, 12, –, 14, 15
Cycads	1, 2, 3, 4, 5, 6, 7, 8, 9, 10, 11, 12, 13, 14, 15

the genus *Ceratozamia* where members of subgenus *Mexicana* possess the compounds and those of subgenus *Latifolia* lack the compounds. Furthermore, we have detected by thin layer chromatography a compound, possibly propetunidin, which appears to be unique to *Macrozamia* and *Encephalartos*. The aim of our study is to sample as many species as possible to determine the actual distribution of the various compounds.

Phenolic Acids

Wallace (1972) looked at the occurrence of these compounds in 20 species belonging to all 10 genera. The following compounds were found in all the taxa: caffeic; protocatechuic; *p*-coumaric; *p*-hydroxybenzoic; ferulic and vanillic acids. Others were more restricted in distribution: sinapic acid in *Dioon* (one of three species) and *Ceratozamia* (1/1) and possibly in *Encephalartos* (1/7) and *Cycas* (1/4); syringic acid possibly occurs in *Ceratozamia* (1/1); and 2,4-dihydrobenzoic acid occurs in *Encephalartos* (2/7) and *Bowenia* (1/1). Wallace concluded, probably correctly, that no taxonomic conclusions could be drawn from the study.

Lignins

Lignins are complex polymers formed from phenolic alcohols such as coniferyl, sinapyl and *p*-coumaryl alcohols. The actual structure of lignins is probably unknowable, at least with current technology. However, the structure of the compounds can be partially elucidated by indirect methods. One method is alkaline oxidation followed by analysis of the resulting aldehydes: *p*-coumaryl alcohol produces *p*-hydroxybenzaldehyde, coniferyl alcohol vanillin and sinapyl alcohol syringic aldehyde respectively. Towers and Gibbs (1953) showed that *Stangeria* produces all three aldehydes and that syringic aldehyde is rarely produced by conifers; *Ginkgo* was not tested. Recent work by Logan and Thomas (1985) showed that, while syringic aldehyde is actually produced in low amounts by all gymnosperms, the major product from gymnosperms is vanillin. In contrast, the major product from the dicotyledonous angiosperms is syringic aldehyde. As to the cycads, they examined *Cycas, Encephalartos* and *Zamia*, which were all rather similar to one another and were actually the same as *Ginkgo* and the conifers. Logan and Thomas (1987) did some further work on lignin derivatives in fossil plants, including two cycadophytes *Ptilophyllum* and *Pachypteris*. Both plants produced all three aldehydes.

A second way to study lignin structure is to look at the pattern of infrared absorbance of milled material. Sarkanen and Hergert (1971) discuss an earlier report that cycad lignin is in a class of its own due to a strong absorption maximum at 1740 cm^{-1}. They rejected this idea, suggesting instead that cycad lignin is no different from that of conifers. This would appear to be supported by the work of Logan and Thomas (1985). In conclusion, our current knowledge of cycad lignins is of little taxonomic significance.

Other Literature Reports on Cycad Chemistry

Baumann and Yokoyama (1976) published on the evolutionary significance of the seed coat carotenoids of *Dioon, Encephalartos, Macrozamia* and *Zamia*. Obukowicz et al. (1981) reported unidentified phenolic substances detected by ferric chloride in *Cycas revoluta,* postulating that these substances excluded micro-organisms and permitted only the cyanobacterium *Anabaena* to grow in the cycad's coralloid roots. Pena et al. (1985) published a paper on secondary metabolites in *Microcycas, Dioon, Cycas, Zamia* and *Ceratozamia* but the spot tests used are generally rather unspecific and the results are difficult to interpret in a taxonomic sense.

Rates of Evolution in Cycads

All of the compounds that have been discussed have been of little taxonomic significance. The main reason for this is that there is a lack of variation when we compare one genus of cycads with another. Why should there be so little variation? Variation clearly does not reflect the age of the group, for the liverworts are older and they possess a highly variable chemistry. Levin and Wilson (1976) discussed rates of evolution of different groups of seed plants, based on variability in chromosome number within genera. The relative increases in chromosome number diversity per lineage per unit time were as follows: herbaceous angiosperms 100; woody angiosperms 14; conifers 2 and cycads 0. In other words cycads appeared to no longer be evolving. (The cycad figure would now be higher, if recalculated, for cycads are now known to have variable chromosome numbers; see Moretti (this volume) and Norstog (this volume).) Levin and Wilson concluded that the rate of evolution was based on breeding structure. The slowest rates of evolution occurred in groups which were obligate outbreeders, had large effective population sizes and had low dispersability. Such plants have few chances of fixing and dispersing any novel features. Niklas (1982) extrapolated from the results of Levin and Wilson to predict a slow rate of biochemical evolution in the cycads. Indeed, this is what is seen in the known phenolic chemistry of the cycads.

It would be interesting to know if large molecules, such as DNA and proteins, would provide a greater amount of taxonomic information about generic relationships in the cycads. So far data on only one protein is available. The small subunit of ribulose bisphosphate carboxylase of *Macrozamia* has been partially sequenced by Martin and Dowd (1986) and in their study of four gymnosperms and several angiosperms, *Macrozamia* was placed nearer to *Metasequoia* and *Ginkgo* than to any other taxa. Again, the analysis of only one genus does not help us to resolve the relationships of the cycad genera to one another.

Conclusion

A reasonable amount of information is available on the phenolic chemistry of the cycads. However, this knowledge has not helped us significantly in resolving the phylogenetic relationships of the group. Despite earlier claims that chemistry confirms Johnson's (1959) taxonomy, a rigorous analysis of the phenolic chemistry data shows that little can actually be stated about the relationships of the cycad genera. The taxonomic schemes of Johnson and Stevenson are neither supported nor refuted by the known chemical data.

Acknowledgments

The author thanks Dennis Stevenson for awakening his interest in the cycads. Jean-Pierre Sclavo was the catalyst for a memorable meeting in Beaulieu and deserves everyone's thanks. Pat Soteropolous produced most of the preliminary data referred to in the text.

Literature Cited

Bate-Smith, E. C. & N. H. Lerner. 1954. Leuco-anthocyanins. 2. Systematic distribution of leucoanthocyanins in leaves. Biochem. J. **58:** 126–132.

Baumann, A. J. & H. Yokoyama. 1976. Seed coat carotenoids of the cycad genera *Dioon, Encephalartos, Macrozamia* and *Zamia*: Evolutionary significance. Biochem. Syst. Ecol. **4:** 73–74.

Bremer, K. 1985. Summary of green plant phylogeny and classification. Cladistics **1:** 369–385.

Carson, J. L. & J. W. Wallace. 1972. The detection of *C*-glycosylflavones in *Dioon spinulosum*. Phytochemistry **11:** 842–843.

Chatterjee, A., J. Kotoky, K. K. Das & T. Chakraborty. 1984. Abiesin, a biflavonoid of *Abies webbiana*. Phytochemistry **23:** 704–705.

Crane, P. R. 1985. Phylogenetic analysis of seed plants and the origin of angiosperms. Ann. Missouri Bot. Gard. **72:** 716–793.

Dossaji, S. F., E. A. Bell & J. W. Wallace. 1973. Biflavones of *Dioon*. Phytochemistry **12:** 371–373.

———**, T. J. Mabry & E. A. Bell.** 1975. Biflavanoids of the Cycadales. Biochem. Syst. Ecol. **2:** 171–175.

Doyle, J. A. & M. J. Donoghue. 1986. Seed plant phylogeny and the origin of angiosperms: An experimental cladistic approach. Bot. Rev. **52:** 321–431.

Gadek, P. A. 1982. Biflavonoids from the seed testa of Cycadales. Phytochemistry **21:** 889–890.

———**, C. J. Quinn & A. E. Ashford.** 1984. Localization of the biflavonoid fraction in plant leaves, with special reference to *Agathis robusta* (C. Moore ex F. Muell.) F. M. Bail. Austral. J. Bot. **32:** 15–31.

Geiger, H. & W. De G. Pfleiderer. 1971. Über 2,3-dihydrobiflavone in *Cycas revoluta*. Phytochemistry **10:** 1936–1938.

——— **& C. Quinn.** 1975. Biflavonoids. Pages 692–742 *in* J. B. Harborne, T. J. Mabry & H. Mabry (eds.), The flavonoids. Chapman & Hall, London.

——— **&** ———. 1982. Biflavonoids. Pages 505–534 *in* J. B. Harborne & T. J. Mabry (eds.), The flavonoids: Advances in research. Chapman & Hall, London & New York.

Gornall, R. J. & B. A. Bohm. 1978. Angiosperm flavonoid evolution: A reappraisal. Syst. Bot. **3:** 353–368.

Handa, B. K., K. K. Chexal, W. Rahman, M. Okigawa & N. Kawano. 1971. Occurrence of bisflavones in *Zamia*. Phytochemistry **10:** 436–437.

Hegnauer, R. 1986. Chemotaxonomie der pflanzen. Vol. 7. Birkhauser, Basel.

Humphries, C. J. & P. M. Richardson. 1980. Hennig's methods and phytochemistry. Pages 353–378 *in* F. A. Bisby, J. G. Vaughan & C. A. Wright (eds.), Chemosystematics: Principles and practice. Academic Press, London & New York.

Johnson, L. A. S. 1959. The families of cycads and the Zamiaceae of Australia. Proc. Linn. Soc. New South Wales **84:** 64–117.

Kariyone, T. & T. Sawada. 1958. Flavonoids from the leaves of *Cycas revoluta* and *Cryptomeria japonica* var. *araucarioides*. J. Pharm. Soc. Japan **78:** 1013–1015. (Chem. Abstr. **53:** 3203b.)

Kawano, N. & M. Yamada. 1960. Chemical constituents of the plants of Coniferae and allied orders. XLV. Structure of sotetsuflavone. J. Pharm. Soc. Japan **80:** 1576–1578. (Chem. Abstr. **55:** 10424i.)

Lebreton, P. 1980. Les Cycadophytes considérées d'un point de vue chimiotaxinomique. Rev. Gén. Bot. **87:** 133–141.

———, **B. Boutard & S. Thivend.** 1978. Nouvelles données sur la présence de *C*-glycosyl-flavones chez les Conifères. Compt. Rend. Hebd. Séances Acad. Sci. **287:** 1255–1258.

———, **S. Thivend & B. Boutard.** 1980. Distribution des pro-anthocyanidines chez les Gymnospermes. Plant. Méd. Phytothér. **14:** 105–129.

Levin, D. A. & A. C. Wilson. 1976. Rates of evolution in seed plants: Net increase in diversity of chromosome numbers through time. Proc. Natl. Acad. Sci. USA **73:** 2086–2090.

Locksley, H. D. 1973. The chemistry of biflavanoid compounds. Progr. Chem. Org. Nat. Prod. **30:** 207–312.

Logan, K. J. & B. A. Thomas. 1985. Distribution of lignin derivatives in plants. New Phytol. **99:** 571–585.

——— & ———. 1987. The distribution of lignin derivatives in fossil plants. New Phytol. **105:** 157–173.

Martin, P. G. & J. M. Dowd. 1986. A phylogenetic tree for some monocotyledons and gymnosperms derived from protein sequences. Taxon **35:** 469–475.

Mohammad, F., H. M. Taufeeq, M. Ilyas & W. Rahman. 1983. Biflavones from *Encephalartos woodii, E. transvenosus* and *E. villosus*. Lem. (Zamiaceae). Indian J. Chem. **22B:** 184.

Nakajima, M. 1964. Japan survey report for 1957–1961. Pages 208–222 *in* H. R. Arthur (ed.), Symposium on phytochemistry. Hong Kong University Press, Hong Kong.

Niemann, G. J. & H. J. Miller. 1975. C-Glycosylflavonoids in the leaves of gymnosperms. Biochem. Syst. Ecol. **2:** 169–170.

Niklas, K. J. 1982. Chemical diversification and evolution of plants as inferred from paleobiochemical studies. Pages 29–91 *in* M. H. Nitecki (ed.), Biochemical aspects of evolutionary biology. University of Chicago Press, Chicago.

Obukowicz, M., M. Schaller & G. S. Kennedy. 1981. Ultrastructure and phenolic histochemistry of the *Cycas revoluta–Anabaena* symbiosis. New Phytol. **87:** 751–759.

Okuyama, T., Y. Ohta & S. Sabata. 1979. The constituents of *Osmunda* spp. (III). Studies on the sporophyll of *Osmunda japonica*. Shoyakugaku Zasshi **33:** 185–186.

Pena, E., E. Grillo & M. Ruiz. 1985. Metabolitos secundarios en Cycadaceae: 1. Estudios de los tipos de metabolitos secundarios en especies de los generos *Microcycas, Dioon, Cycas, Zamia* y *Ceratozamia*. Revista Jard. Bot. Nac. **6:** 125–133.

Richardson, P. M. 1982. Flavonoids and phylogenetic systematics. Pages 115–123 *in* N. I. Platnick & V. A. Funk (eds.), Advances in cladistics. Vol. 2. Columbia University Press, New York.

Sarkanen, K. V. & H. L. Hergert. 1971. Classification and distribution. Pages 43–94 *in* K. V. Sarkanen & C. H. Ludwig (eds.), Lignins: Occurrence, formation, structure and reactions. Wiley-Interscience, New York.

Sawada, T. 1958. Studies on flavonoids in the leaves of Coniferae and allied plants. V. Relation between the distribution of bisflavonoids and taxonomical position of the plants. J. Pharm. Soc. Japan **78:** 1023–1027.

Stevenson, D. W. 1981. Observations on ptyxis, phenology, and trichomes in the Cycadales and their systematic implications. Amer. J. Bot. **68:** 1104–1114.

Swain, T. 1979. Tannins and lignins. Pages 657–682 *in* G. A. Rosenthal & D. H. Janzen (eds.), Herbivores: Their interaction with secondary plant metabolities. Academic Press, New York.

Towers, G. H. N. & R. D. Gibbs. 1953. Lignin chemistry and the taxonomy of higher plants. Nature **172:** 25–26.

Varshney, A. K., T. Mah, N. U. Khan, W. Rahman, C. W. Hwa, M. Okigawa & N. Kawano. 1973. Biflavones from *Cycas revoluta, C. circinalis* and *C. rumphii*. Indian J. Chem. **11:** 1209–1214.

Wallace, J. W. 1972. A survey for benzoic and cinnamic acids of the Cycadaceae. Amer. J. Bot. **59:** 1–4.

Notes on the Natural History of Two Species of *Encephalartos* from South Africa

YVES DELANGE

Abstract

DELANGE, Y. (Museum National d'Histoire a Paris, Paris Cedex 05, France). Notes on the natural history of two species of *Encephalartos* from South Africa. Memoirs of the New York Botanical Garden 57: 142–143. 1990.—The habitats and ecology of *Encephalartos lanatus* Stapf & Burtt-Davy and *E. transvenosus* Stapf & Burtt-Davy are described.

Key words: *Encephalartos*; Transvaal.

Introduction

In general very little is known about the habitats and ecology of the cycads and the information that is available concerning growing conditions is in the main derived from observations on plants growing in botanical gardens under similar conditions of cultivation. This paper presents information on the habitat and ecology of two South African species of *Encephalartos*.

The cycads of South Africa have a wide distribution from the south to the northeast (Dyer, 1965; Giddy, 1974). They range from biotopes with *Stangeria* in littoral areas of the Transkei to biotopes with *Encephalartos* in the northern Transvaal. This range then encompasses a wide variety of soil composition, climate, flora, and fauna. The present discussion is limited to two species of *Encephalartos* from differing habitats in the Transvaal.

The first species is *E. lanatus* Stapf & Burtt-Davy. This species grows in the Transvaal in an area approximately 130 km east of Pretoria. It is found in the region of the Small Olifants River in the Districts of Witbank and Bronkhorstspruit. It grows in the sheltered part of the valley. The vegetation is arid shrub grassland. The summer temperatures can be very high, reaching 40°C, and the winters may have some frost. The annual rainfall ranges from 550–700 mm. The greatest rainfall is in the summer with an annual distribution as follows: 25 mm or less in winter; 150–200 mm in autumn; 100–150 mm in spring; and 300–400 mm in summer.

The soil is sandy and derived in part from sandstone decomposition. Particle size ranges from fine sand to quite large blocks of decomposing sandstone. This locality, which is on a private estate, is near a locality of similar conditions with large aborescent specimens of *E. eugene-maraisii* Verdoorn. There is only one megasporangiate plant of *E. lanatus* in the proximity of a number of microsporangiate plants. Seed production of this plant is controlled by the local conservation authorities because, although it is known that cycad seeds are highly toxic, especially the Australian cycads (Whiting, 1962), this is apparently not true for this species of *Encephalartos*, and local baboons are very fond of the seeds. Thus, the local authorities protect the megasporangiate cones from baboon predation by enclosing them in grating boxes until the seeds are harvested for planting.

Major flowering plants of the locality include

Protea caffra Meissner (there are at least nine other species of *Protea* in the Transvaal); *Acacia* spp., the Mountain Cabbage Tree or Bergkielpersol, *Cussonia paniculata* Ecklon & Zehyer (Araliaceae); and *Diospyros* spp. and *Euclea* spp. (Ebenaceae).

The second species of *Encephalartos* to be discussed is *E. transvenosus* Stapf & Burtt-Davy. The habitat of this species was made a national monument in 1936 to protect it from eradication. It is found in the Vanda region some 400 km north of Pretoria at the base of Zoutpanberg which is the terminal portion of the Drakensberg Range. One travels some 50 km through extremely dry plains with *Euphorbia ingens* Mayer and *E. cooperi* N. E. Brown until he reaches the cloudy forest region of Modjadji Kraal in the District of Letaba near the Letaba River. This habitat is tropical, whereas that of *Encephalartos lanatus* was south of the Tropic of Capricorn. The annual rainfall is between 1150 mm and 1500 mm. Again the wet period is in the summer, but the winter is less dry. The seasonal rainfall varies as follows: 100–150 mm in winter; 200–250 mm in autumn; 150–200 mm in spring; and 700–900 mm in summer.

The soil is primarily of clay, but there is a well developed forest of natural vegetation and also some introduced taxa such as *Catharanthus roseus* (L.) G. Don, the periwinkle from Madagascar, and more invasively, *Caesalpinia sepiaria* Roxb. of Tropical Asia.

Encephalartos transvenosus is one of the most spectacular species of *Encephalartos* in South Africa. In this area many plants are from 7–8 m in height with 50 or more leaves in a crown. Some plants are up to 13 m tall and on some hills they still form small forests, perhaps giving us a glimpse of their importance at their past apogee. Although this species is widespread in the hills of Zoutpansberg, the species is threatened in some areas as the starch trunks are harvested for porridge by the indigenous population. It is only in the Modjadji area that the taxon seems to be really preserved by the Bantu tribe that is ruled by Rain Queens, who have been governing by hereditary law for many years and who protect this species (Dyer, 1965).

Literature Cited

Dyer, R. A. 1965. South African cycads. Bothalia **8(4):** 400–516.

Giddy, C. 1974. Cycads of South Africa. Purnella, Cape Town.

Whiting, M. 1962. Toxicity of cycads. National Institute of Health, Bethesda, Maryland.

MARCH 1990 MEMOIRS OF THE NEW YORK BOTANICAL GARDEN 57: 144–147

Algunos Datos Etnobotanicos de las Cicadas en Mexico

MARIO VAZQUEZ TORRES

Abstract

VAZQUEZ TORRES, M. (Centro de Investigaciones Biologicas, Universidad Veracruzana, Apartado Postal 294, Xalapa, Veracruz 91000, Mexico). Algunos datos etnobotanicos de las cicadas en Mexico. Memoirs of the New York Botanical Garden **57:** 144–147. 1990.— Although prehispanic and Conquest documents fail to assign any uses for cycads by such ancient cultures as the Aztec, Maya, and Olmec, we nevertheless know from indigenous rural people that four principal classes of use can be ascribed to them. These are: Edibles (*Ceratozamia robusta, Dioon edule,* and *D. spinulosum*); Medicinals (*Zamia loddigesii*); Religious (*Dioon caputoi, D. edule, D. spinulosum,* and *D. tomasellii*); and Ornamentals (*Ceratozamia* spp., *Dioon* spp., and *Zamia* spp.).

Key words: *Ceratozamia*; *Zamia*; *Dioon*; medicinal use; edibility; religious use; ornamental use.

Resumen

En México, en las zonas en las que florecieron las culturas autóctonas como los Olmeca, Maya y Azteca existe al menos una especie de *Ceratozamia, Dioon* o *Zamia.* Los documentos prehispánicos y de la Conquista no consignan el conocimiento y usos de estas plantas por los antiguos mexicanos.

La información actual, recabada con los indígenas y campesinos que habitan las regiones en las que se distribuyen las cícadas, permite establecer su utilización en las siguientes clases principales: Comestibles: *Ceratozamia robusta, Dioon edule* y *D. spinulosum*; Medicinales: *Zamia loddigesii*; Adornos religiosos: *Dioon edule, D. caputoi, D. spinulosum* y *D. tomasellii*; Ornamentales vivas: todas las *Ceratozamia* spp., *Dioon* spp., y *Zamia* spp.

Introduccion

En México se encuentra el 50% de las Cícadas de América (Balduzzi et al, 1982). Si bien es cierto que su distribución geográfica es amplia ocupando las estribaciones de las Sierras Madre Oriental, Occidental y del Sur también es cierto que las poblaciones en general son restringidas o incluso apenas representativas. Por tal motivo, en las áreas en las que florecieron las culturas antiguas como la Olmeca y Maya, y posteriormente los pueblos subsidiarios del Imperio Azteca de las tierras bajas y planicie del Golfo de México, había por los menos una especie de *Ceratozamia, Dioon* o *Zamia.* Sin embargo, la revisión de algunos de los códices, documentos prehispánicos y contemporáneos de la Colonia más importantes, no dan indicios acerca del conocimiento y uso de estas plantas. Este hecho en sí plantea un problema interesante, ¿cómo saber que las utilizaron, y cuál fué su uso?

De no existir información iconográfica ni jeroglífica al respecto, la explicación pudiera ser probablemente porque el conocimiento sobre las plantas medicinales, tóxicas o venenosas fue transmitido generación tras generación de padres a hijos. Esta hipótesis puede sustentarse en el hecho de que actualmente en los núcleos indígenas derivados, representativos de diferentes culturas prehispánicas en México, la posesión y práctica de esos conocimientos se desarrollan de esa manera.

La información obtenida de los indígenas y/o campesinos que reconocen con precisión a las cícadas dentro de su entorno biótico, puede clasificarse por cuanto a sus formas de uso en los siguientes rubros:

I. Comestibles

En esta categoría pueden incluirse las siguientes especies:

i). *Ceratozamia robusta* Miquel ("tushku," vocablo tzeltal) de la que en la zona lacandona de Chiapas es consumido comunmente el megastróbilo una vez maduro, que se hierve o cuece al vapor adicionando al agua "tequixquitl" (vocablo náhuatl; tierra salitrosa con sales de calcio, potasio y sodio principalmente).

Una vez cocido y enfriado, se van desarticulando las megasporófilas con su correspondiente par de semillas; se come la sarcotesta primero se rompe y desecha la testa que por el calor se ha reblandecido y se come la masa amilácea constituida por el megagametofito y el embrión. Nunca se comen las semillas crudas puesto que la experiencia les indica que pueden morirse o por lo menos sufrir artralgias, emaciación y parálisis parcial severas y permanentes, especialmente de los miembros inferiores (conocimiento frecuente de los ancianos en las áreas de distribución de *Dioon* y *Ceratozamia*).

ii). *Dioon edule* var. *angustifolium* Miquel y *Dioon edule* var. *edule* Lindley ("chamal," "tzamal," "quiotamal," tullidora; ¿vocablos náhuatl?). Aún cuando existe el conocimiento por parte de los campesinos de que las semillas de estas especies son consumidas como alimento (incluso el epíteto *edule,* significa *comestible*), ello ha sido practicado en tiempos difíciles solamente (fuerte escasez de alimentos, movimientos revolucionarios de campesinos, etc.) después de procesarlos del siguiente modo:

- Se cosechan las semillas maduras, se secan al sol hasta perder la sarcostesta.
- Se mondan para desprender la testa.
- Se hierven las semillas cambiándoles el agua dos o tres veces.
- Se muelen las semillas y con la masa obtenida se elaboran atoles y tortillas (guisos que implican una cocción adicional).

En el suroeste de Tamaulipas, en el límite con San Luis Potosí, pueden agregarse también en el mismo proceso, semillas de "chamalillo" (*Ceratozamia microstrobila* Vovides & Rees) que vive simpátrica y que tiene una importancia menor, considerando que el volumen de semillas disponibles es bajo.

iii). *Dioon spinulosum* Dyer ("Palma chicalite," ¿voz náhuatl?). Los indígenas y campesinos chinantecos consumen normalmente las semillas de esta cícada (que son bastante voluminosas) y que aún se encuentra representada en abundancia en los cerros calizos de la cuenca media del Río Papaloapan (Oaxaca-Veracruz).

El procedimiento utilizado para la elaboración de "memelitas" (náhuatl: "tlaxcal-mimilli") (cierto tipo compacto de pan) es en esencia el mismo que se utiliza con *Dioon edule.*

iv). *Zamia loddigesi* Miquel ("poua," "cahua," "maíz" de coshca," vocablos zoque-popoluca).

Las semillas una vez maduras se tuestan, se les desprende la testa y se comen.

II. Medicinales

En este renglón lo único que se conoce.

Por medio de frotamiento del vientre de las madres parturientas para acelerar el parto. Una vez ocurrido el nacimiento, se quema el tallo y se ahuma el cordón umbilical (antes de cortarlo). También se asegura que el tallo asado, y aplicado ex ternamente, cura el cáncer dérmico. Estas prácticas se realizan en la Sierra de Santa Marta.

Se sabe también que la ingestión de semillas frescas y crudas de *Ceratozamia mexicana* Brongn., actúan como un potente vermífugo, sin embargo esta información no ha sido corroborada, así como tampoco el poder antineurálgico de las infusiones de semillas de *Dioon edule.*

Zamia spartea A. DC. se dice que es usada como remedio contra mordeduras de serpientes, pero tampoco se ha comprobado.

III. Adornos

Especialmente por motivos de celebraciones religiosas en varias comunidades rurales son utilizadas las frondas de la especie de *Dioon* más cercana al poblado, como motivo estético para resaltar los arreglos florales en las entradas y pórticos de los templos. Por esta razón reciben nombres como "palma de la Virgen" (*Dioon tomasellii* var. *tomasellii* De Luca et al.), "palma de Santa Teresita" (*Dioon edule edule*), o "palma real" (*Dioon caputoi* De Luca et al., *Dioon purpusii* Rose).

El uso prácticamente nulo de especies de *Ceratozamia* o *Zamia* como adornos en festividades religiosas puede deberse a las siguientes causas:

1). Al habitar en ambientes húmedos y sombreados, las hojas están casi siempre cubiertas de epífitas (líquenes, hepáticas y musgos) y su aspecto no es especialmente preferido, además de que los pecíolos y raquis son espinosos.

2). En las áreas en donde se encuentran esas plantas, abundan también varias especies de palmas (*Brahea, Chamaedorea, Orbignya, Paurotis, Sabal, Scheelea,* etc.), cuyas hojas son mucho mas acequibles y apreciadas comparativamente que cualquier especie de cícada, para fines de ornato.

IV. Como Plantas de Ornato Vivas

Este es quizá el uso más importante de las cícadas en México, toda vez que con diferente frecuencia, según la especie de que se trate, casi todas se encuentran como plantas ornamentales en los jardines hogareños, tanto en los pueblos pequeños como en las ciudades cercanas al área de distribución de la especie en particular.

La información registrada en sus correspondientes áreas de distribución, indica que las especies más apreciadas son *Dioon edule, D. merolae, D. spinulosum, D. tomasellii* var. *tomasellii, Ceratozamia mexicana, C. robusta, C. latifolia* Miq., *C. miqueliana* Wendl., *C. kuesteriana* Regel, *C. matudae* Lundell, *Zamia furfuracea* Aiton, *Z. paucijuga* Wieland, *Z. inermis* Vovides et al., *Z. cremnophila* Vovides et al., *Z. splendens* Schutzman. La gran demanda de estas plantas está provocando una fuerte presión sobre sus poblaciones naturales, conduciéndolas al borde de la extinción (Vovides, 1986; Vovides & Peters, 1987).

V. Plantas Venenosas

Las cualidades tóxicas de *Dioon edule* var. *edule* y *D. edule* var. *angustifolium* son ampliamente conocidas por los ganaderos de la zona de la Huasteca, Sierra de San Carlos, Tamaulipas y Sur de Nuevo León toda vez que por motivos de "enchamalamiento" (enfermedad del ganado provocada por la ingestión de hojas de "chamal"), mueren varios animales anualmente lo que representa pérdidas económicas de consideración. Por esta razón en los campos ganaderos es una práctica común la extirpación y destrucción de las plantas de *Dioon.*

VI. Insecticidas

Especialmente para matar moscas. Se usan los conos jóvenes sin importar sexo, para lo cual se cortan en trozos finos, se agregan a un poco de leche o cualquier otra solución dulce y atractiva a estos insectos y minutos después del primer contacto, los insectos van muriendo dentro o alrededor del recipiente.

VII. Misticismo en Torno a las Cicadas

En la Huasteca, de manera especial a las plantas de *Zamia loddigesi* y *Zamia fischeri* Miquel se les deja crecer, se las preserva y cuida dentro de los campos de maíz, cultivo que tiene un alto significado cosmogónico y profundo arraigo en la cultura precortesiana. La explicación de esos cuidados especiales, como plantas sagradas junto al maíz, se desconoce. El nombre que reciben las especies mencionadas es "amigo del maíz."

Conclusiones

Las Cycadales constituyen un elemento florístico cualitativamente importante dentro de las gimnospermas de México.

Entre las distintas Zamiaceae se encuentran especies a las que se les dan diversos usos: comestibles, medicinales, ornamentales y adornos religiosos.

Particularmente por la fuerte presión de extracción y comercialización ejercida sobre las poblaciones naturales, varias especies como *Ceratozamia hildae, C. latifolia, C. miqueliana, C.*

kuesteriana, Dioon tomasellii, Zamia fischeri, Zamia furfuracea, Z. inermis, Z. splendens estan al borde de la extinción.

Por razones de devastación del habitat, *Ceratozamia euryphyllidia, C. norstogii, C. zaragozae, Dioon califanoi, D. merolae, D. rzedowskii, D. spinulosum, Zamia cremnophila, Z. purpurea,* también se encuentran en franco peligro de extinción.

Literatura Citada

Balduzzi, A., S. Sabato & P. De Luca. 1982. (Year of Issue, 1985). A phytogeographical approach to the new world cycads. Delpinoa **23–24:** 185–202.

Vovides, A. 1986. Trade and habitat destruction of threatened Mexican cycads. Traffic **4:** 13.

——— **& C. Peters.** 1987. *Dioon edule*: La planta mas antiqua de Mexico. Ciencia y Desarrollo **13:** 19–24.

MARCH 1990 MEMOIRS OF THE NEW YORK BOTANICAL GARDEN 57: 148–151

Investigation of the Natural Community of *Cycas panzhihuaensis* L. Zhou & S. Y. Yang

ZHOU LIN, YANG SIYUAN, AND ZHOU ZHENG

Abstract

ZHOU, L., S. YANG, and Z. ZHOU (Dukou Horticulture Dept., Dukou City, Sichuan Prov., P. R. China). Investigation of the natural community of *Cycas panzhihuaensis* L. Zhou & S. Y. Yang. Memoirs of the New York Botanical Garden **57**: 148–151. 1990.—A vast stretch of the natural community of *Cycas panzhihuaensis* has been discovered at the hot, dry valley of the Jinsha River, in Sichuan Prov. of China. The *C. panzhihuaensis* community, which is not damaged, is the most extensive community among all known *Cycas* communities. On the basis of an all-around and thorough field investigation, we detail the ecological environment, growth and shape of *C. panzhihuaensis*, and the distribution and synthetic characteristics of the natural community of *C. panzhihuaensis*. We have also provided a sketch of all the species of Chinese cycads which have been formally described for China.

Key words: *Cycas*; Sichuan; China.

Introduction

Cycas is a plant relic which has existed for more than two hundred million years on the earth. If you liken the Chinese giant panda to a living fossil, you must also liken *Cycas* to a living fossil. Nowadays, there are only approximately 120 species of cycads in ten genera, distributed in the tropical and subtropical zones over the entire world. At present, there is found only one genus, *Cycas,* with ten known species, in China. The distribution and ecological environments of these ten species of *Cycas* are described as follows:

1. Cycas revoluta Thunb.

Cycas revoluta is distributed in Fujian, Guandong, Guangxi, Zhejiang, Hunan, Hubei, Sichuan, Guizhou, and Yunnan Provinces. It is planted in gardens in southern China. This species has to winter in greenhouses in northern China with low air temperatures. *Cycas revoluta* prefers a warm, moist and sunny climate and grows well on fertile and deep soil. In southern China, *C. revoluta* blooms and bears seed, but in northern China it does not.

2. Cycas taiwaniana Carruthers

Cycas taiwaniana was first discovered in subtropical jungle along the rivers in Daxi and Qingshui Districts of Beinan County on the island of Taiwan. A vast natural forest of *C. taiwaniana* remains on cliffs or in overhanging rocks in remote mountains of Hongye Village of Taidong County, Taiwan. Later, *C. taiwaniana* was discovered in the Lofu Mountains of Guangdong Province and in some districts of Fujian and Hainan Provinces of China. Now, this species is planted on the coasts of Guangdong and Hainan Provinces. *Cycas taiwaniana* grows very well on any soil where it is hot and the annual rainfall is above 1500 mm, in regions which have frost-free and snow-free winters.

3. **Cycas siamensis** Miquel

Cycas siamensis, which occurs with broad-leaf evergreen trees, has been discovered in Simao, Jinghong, Menghai, Lanchang, and Luxi Districts of Yunnan Province where the annual rainfall is above 2000 mm, and where the annual hot monsoon climate consists of a rainy and a dry season. *Cycas siamensis* also grows in Sichuan, Yunnan, Guizhou, Guangxi, and Guangdong Provinces. It fears fierce cold and sunshine and prefers warm and moist climatic conditions. It is a kind of heliophobic species, and it is not resistant to drought.

4. **Cycas szechuanensis** Cheng & L. K. Fu

Cycas szechuanensis, which is planted artificially, occurs mainly in Ya'an, and Leshan Districts of Szechuan Province and on Mt. Emei. In these areas, there is plenty of rain, fog, and an autumn of long duration. The rain is fine and there is little sunshine. The soil is rather fertile there. *Cycas szechuanensis* also occurs in Sichuan and Yunnan Provinces in areas of warm weather. This species is not drought resistant.

5. **Cycas pectinata** Griffith non Hamilton[1]

Heliophilous *Cycas pectinata* grows in Yunnan and Guizhou Provinces. In Yunnan, it has been found in the Honghe River valley. This drought resistant species, which grows very well on poor soil, is distributed in zones where the annual rainfall is above 1000 mm, where there is plenty of sunshine, warm and dry in the winter, and rainy and hot in the summer. The annual climate of the Honghe River valley consists of a dry season and a rainy season.

6. **Cycas hainanensis** C. J. Chen

Cycas hainanensis, which is originally from the Wanning and Haikou Districts of Hainan Island, is planted in Fujian, Guangdong, and Sichuan Provinces. The annual rainfall in the place of origin of *Cycas hainanensis* is above 2000 mm. The plant, growing separately in tropical jungles, likes a hot, rainy climate, prefers semishady places, and can not tolerate frost or direct strong sunlight.

7. **Cycas rumphii** Miquel

Cycas rumphii grows in all the provinces of southern China where the annual rainfall is greater than 1500 mm. This cycad loves a hot, rainy climate. It is not cold and frost resistant. In recent years, a small area where *Cycas rumphii* is distributed in mountainous regions near Fuzhou City in Fujian has been discovered.

8. **Cycas micholitzii** Dyer

Cycas micholitzii, which is sympatric with *C. pectinata,* grows mainly in Longzhou District of Guanxi Province, but also occurs in the Honghe River Valley. *Cycas micholitzii* prefers a warm, moist climate where the annual rainfall is above 1500 mm. This climate is warm and hot in the summer and warm and frost-free in the winter. This species prefers semishady places in the tropical jungle and is not tolerant of direct, strong sunlight. Individual plants are usually wide spaced.

9. **Cycas guizhouensis** Lan & R. F. Zhou

Cycas guizhouensis is a recently described species of *Cycas* which was discovered in Xinyi District of Guizhou Province. It is distributed mainly in the valleys of the Nanpan and Qingshui Rivers in the southwest part of Quizhou. Plants of *Cycas guizhouensis* grow separately or in clumps on limestone soil and gravel soil above 400–800 m above sea level. *Cycas guizhouensis* grows in places where it is rainy, moist and cool in the summer and moist and frost-free in the winter. The annual climate consists of summer wet and winter dry season. Plants of this species grow quickly and well even on poor soils. *Cycas guizhouensis,* although not frost resistant, is resistant to cold temperatures.

---

[1] The nomenclature of this species needs to be studied. *Cycas pectinata* was first validly published by Hamilton (Mem. Wern. Nat. Hist. Soc. **5(2):** 322–323. 1826). However, even though Hamilton cited a different type specimen, he placed *C. angulata* R. Br. (Prodr. Flor. Nov. Holl. **1:** 348. 1810), a validly published and legitimate name, in full synonymy. Therefore, *C. pectinata* Hamilton is a superfluous name and a *nom. illegit.* Thus, *C. pectinata* Griffith (Notulae ad Plantae asiaticae **4:** 10. Plate 360, fig. 3. 1854) becomes a later homonym because the combination was validly published by Hamilton. Thus, the Griffith combination is also an illegitimate name. A new name is probably needed for the taxon described by Griffith. *Cycas pectinata* has been most commonly used in the sense of Griffith, as it is here.

10. Cycas panzhihuaensis L. Zhou & S. Y. Yang

This species is discussed at length below.

In 1981, in the area of Baguan River of Dukou, Sichuan Province, China, a large natural community of *Cycas panzhihuaensis* of considerable area was found. Soon afterwards, in Ningnan County of Sichuan, and Yuanmou and Huaping counties of Yunnan Province, more natural communities of *C. panzhihuaensis* were discovered, although with smaller areas and numbers of plants. These new discoveries indicate that the distribution of Chinese *Cycas* extends north to latitude 27°11′ in Ningnan County, Sichuan Province. This is the northern boundary of cycad distribution on the mainland of Eurasia, which is of biogeographic significance.

The largest *C. panzhihuaensis* community is distributed along the west slope of the Baguan River to the north bank of the Jinsha River, at latitude 26°36′–26°39′N and longitude 101°33′–101°35′E. Elevation ranges from 1170–2000 m, but most of the population occurs between 1500–1700 m. The average gradient is about 30 degrees. The total area of the community is 310 hectares. There are about 120,000 plants of *C. panzhihuaensis*. Plants range from seedlings to large, reproductively mature, 3 m tall plants.

The habitat area is in a region with a mid-subtropical climatic type, but climatic features of a more southerly subtropical type emerge because of the cut of the Jinsha River Valley with its confined topography and dry and hot climate. For example, Kapok (*Bombax malabaricum*) blossoms and bananas (*Musa* spp.) bear fruits normally. The average annual temperature is 20.3°C. The coldest month is December with an average temperature of 11.8°C. May is the hottest month with an average temperature of 26.3°C. The absolute minimum annual temperature is −1.4°C, and the absolute maximum annual temperature is 40.7°C. The whole accumulated temperature of a year is 7359°C. There are 2709.2 hours of sun per year. The frost-free period is 294 days a year. The first frost occurs on December 6, and the last frost occurs on February 14. Thus, the frost period lasts 66 days in the year.

There is a clear-cut dry and a rainy season each year. The total annual rainfall is 761.6 mm. 89.2% of the total annual precipitation occurs between June and October, the rainy season, whereas only 10.8% of the annual precipitation occurs from November through May. The annual average relative humidity is 61%. The annual average evaporation is 2670 mm, about 3.5 times of annual rainfall.

There are two kinds of soil in the habitat area, namely, a brown soil along the area of the river valley below 1500 m, and mountain loess above 1500 m. The soil layers are comparatively thick. The developing maternal rocks of these two kinds of soil are limestone and sand shale. Soil fertility is relatively low. The physical and chemical properties of the soil along soil profiles are given in Table I.

In its habitat, *C. panzhihuaensis* reproduces by seeds. Its roots are fleshy, and often a main tap root is lacking. Its stem is generally about 50 cm high with the tallest 2.4 m. Stem diameter is normally 15 cm, the thickest being 25 cm. Mature plants of *C. panzhihuaensis* are reproductive every year. Reproductive structures appear in March, and the seeds are mature in September. Observations on natural cycad groves have shown that there are significant differences in seed production from year to year. For example, in 1982, seeds were abundant, totaling about 1500 kg, while in 1983 and 1984 only a few cycads bore seeds, yielding only 50–100 kg a year. Then in 1985, more than 1300 kg of seeds were harvested. In 1977, we transplanted 108 cycad plants to a park. Apparently, because of the better soil and consistent moisture conditions at the park, most of the megasporangiate cycad plants annually produced seeds with minor interruptions.

In general, the germination percentage of the seeds is high. Normally, the seeds are collected in September and sown in the field the next February. They generally germinate within 40 days, with the germination rate generally over 80%.

In habitat, *C. panzhihuaensis* is highly drought resistant, possessing characteristics typical for xerophytes. For example, vegetative leaves are covered by a thick cuticle, and both the large cortex and pith have mucus. Because *C. panzhihuaensis* has many ageotrophic coralloid roots containing nitrogen-fixing *Anabaena,* it can adapt itself to soil poor in nitrogen content. Also, surface fires occur in some cycad groves; but most cycads in these groves can grow normally. Thus, *C. panzhihuaensis* is not only highly drought-resistant, but at the same time is fire-resistant.

In the natural community of *Cycas panzhi-*

Table I
The physical and chemical properties of soil sections

Depth (cm)	Absorbed water (%)	Organic matter content (%)	Total nitrogen (%)	pH	Salt base exchange (ml/100 g)	Property
0–18	6	1.88	0.0824	6.62	34.8	Medium loam
18–55	1.36	0.45	0.0248	6.79	36.3	Heavy loam
55–80	9.83	0.34	0.0178	6.78	40.3	Heavy loam

huaensis, there are 104 species of higher plants. Among them are 7 tree species (6.7% of the total species). The dominant tree species are: *Quercus cocciferoides* and *Rhus succedanea.* An important component of the cycad community are 27 species of shrubs which make up 27% of the total species. Besides the dominance in the number of individual plants of *C. panzhihuaensis,* the main shrubs are *Dodonaea angustifolia, Bauhinia delavayi, Pouzxolzia elegans, Diospyros mooifolia, Clematis delavayi* var. *calvescens, Rhamnus virgata.* There are 56 herbaceous species, amounting to 53.9% of the total species of which grasses are the greatest component contributing to the landscape characteristic of this community. The main herbaceous species are *Themeda hookeri, Eremopogon delavayi, Cymbopogon distans, Heteropogon contortus, Bidens pilosa,* and the fern *Macrothelypteris* sp.

Seven species of lianas and mosses are found, amounting to 6.7% of the total species. The principal lianas are *Stepharia epigaea, Ampelopsis tomentosa,* and *Illigera cordata* var. *mollissima.* Mosses are in a dried-up state during the dry season and recover to grow after the rainy season has commenced. The main moss species are *Holomitrium vaginatum* and *Trachyphyllum inflexum.*

In the whole cycad community, shrubs and grasses constitute the most important components, obviously representing a composition characteristic of a sparse tree-grassland community. We suppose that the sparse tree-grassland here is secondary in development. This is chiefly caused by the occurrence of regular surface fire and human damage, with the cycad community surviving. At present, there are some remaining trees over one hundred years old of *Quercus cocciferoides* growing along the Jinsha River Valley. They show that there was a prior period of heat-loving and drought-resistant oak forest here in the past, instead of the present sparse tree-grassland community.

Acknowledgments

We greatly appreciate the assistance of Mr. Chen Junyu, Professor of Beijing Forestry University and Wuhan City Construction College, and Mr. Guan Zhongtian, Research Fellow of the Sichuan Institute of Forestry Design and Research, in his review of this work.

A New Species of *Encephalartos* (Zamiaceae) from Northeastern Zaire

PAUL BAMPS AND STANISLAS LISOWSKI

Abstract

BAMPS, P. (National Botanical Garden of Belgium, Meise B 1860, Belgium) and S. LISOWSKI (A. Mickiewicz University, Department of Geobotany, PL 61-713 Poznan, Poland). A new species of *Encephalartos* (Zamiaceae) from northeastern Zaire. Memoirs of the New York Botanical Garden 57: 152–155. 1990.—*Encephalartos ituriensis,* a new cycad from northeastern Zaire, is described and illustrated and its relationship with its nearest relatives is presented.

Résumé

Description illustrée d'une nouvelle espèce d'*Encephalartos* en provenance du nord-est du Zaire et discussion de sa position taxonomique vis-à-vis des espèces les plus proches.

Key words: Zamiaceae; *Encephalartos*; Zaire.

Introduction

During a trip to the vicinity of Nduye (01°50′N 28°59′E), about 50 km north of Mambasa, one of the authors (S.L.) discovered some patches of *Encephalartos* on granitic domes which emerge from a rain forest dominated by *Cynometra alexandri* (Caesalpiniaceae) and inhabited by Bambuti pygmies. Provisionally determined as *Encephalartos hildebrandtii* A. Br. & Bouche, it appears upon further examination that it is quite different from that species and from *E. laurentianus* De Wild., which was probably erroneously reported from Ituri on the basis of sterile material, by W. Robyns (1948). It is, in fact, a new species. The differences between this new species, *E. ituriensis,* and *E. hildebrandtii* and *E. laurentianus* are summarized in Table I.

Encephalartos ituriensis Bamps et Lisowski, sp. nov. Type. Zaire. Mt. Mukonza, near Nduye, border of rain forest, alt. 1180 m, Apr 1976, *Lisowski 42909* (holotype: BR; isotype: POZG). Figs. 1–3.

Figs. 1–3 Affinis *E. hildebrandtii* A. Br. & Bouche et *E. laurentiano* De Wild. sed ambobus foliis apice haud dentatis, nervis vix conspicuis, strobilorum ♂ et ♀ squamarum forma mensurisque differt.

Trunk to 6 m high and 50 cm diam., sometimes prostrate. *Leaves* several together at the apex of the trunk, to 2–3 m long and 40 cm wide, the leaflets reduced in size towards the base; median leaflets more or less opposite, sessile, lan-

Table I
Differences between *Encephalartos ituriensis*, *E. hildebrandtii* and *E. laurentianus*

Species	Median leaflets	Male cone	Female cone	Seeds
E. ituriensis	end in one spine; up to 25 cm long and 3 cm wide	26 cm long and 7 cm diam.; bulla 2.5–3 cm wide and 0.6–0.7 cm deep	20 cm long and 12 cm diam.; bulla 5 cm wide and 2 cm deep	2.5–3.5 cm long and 1.5–2.5 cm diam.
E. hildebrandtii	end in 2 or more spiny teeth; up to 35 cm long and 4.5 cm wide	up to 50 cm long and 9 cm diam.; bulla 1.9–2.8 cm wide and 0.9–1.7 cm deep	up to 60 cm long and 25 cm diam.; bulla up to 5 cm wide and 3.3 cm deep	up to 3.8 cm long and 2.6 cm diam.
E. laurentianus	end in 2 or more spiny teeth; 35–50 cm long and 4–7 cm wide	20–35 cm long and 6–10 cm diam.; bulla 1.8 cm wide and 1.2 cm deep	35–45 cm long and about 20 cm diam.; bulla 5–6 cm wide and 3–3.5 cm deep	3–5 cm long and 2–3 cm diam.

ceolate-oblong, shortly narrowed at the base, long attenuate and spinose (one spine) apically, to 25 cm long and 3 cm wide, coriaceous, with 3–9 spinules more or less developed on each side, the longitudinal nerves scarcely visible. *Male cone* fusiform, 26 cm long and 7 cm diam., with a stout peduncle to 16 cm long and 18 mm in diam., microsporophylls deltoid, median ones to 3 cm long and 2.5 cm wide, slightly curved upwards; bulla with a more or less rhomboid terminal facet, 15 mm wide and 6 mm deep. *Ovulate cone* cylindrical-ovoid, 20 cm long and 12 cm diam., somewhat truncate at the base and rounded at the apex with some reduced sterile sporophylls; peduncle short and stout; bulla of median sporophylls rhomboid, to 5 cm wide and 2 cm deep; terminal facet more or less rhomboid, 2 cm wide and 1 cm deep, concave and smooth. *Seeds* ovoid, more or less angular, 2.5–3.5 cm long and 1.5–2.5 diam., smooth, drying reddish brown; micropyle whitish, to 10 × 6 mm.

Paratypes. ZAIRE. Mt. Mukonza, near Nduye, border of rain forest, alt. 1200 m, Apr 1976, *Lisowski 42234* (O BR, POZG); ibid., on rocks, Aug 1975, *Lisowski 41057* (O BR); Mt. Tatotudu, near Nduye, on slope in forest, alt. 1100 m, Apr 1976, *Lisowski 42482* (young plant BR).

→

Fig. 1. *Encephalartos ituriensis* Bamps & Lisowski. **a.** Male cone (*Lisowski 42234*) (× 1/3). **b.** Female cone (*Lisowski 42909*) (× 1/3). **c.** Female plants at the Mukonza Mount, Nduye. Photographs by Lisowski.

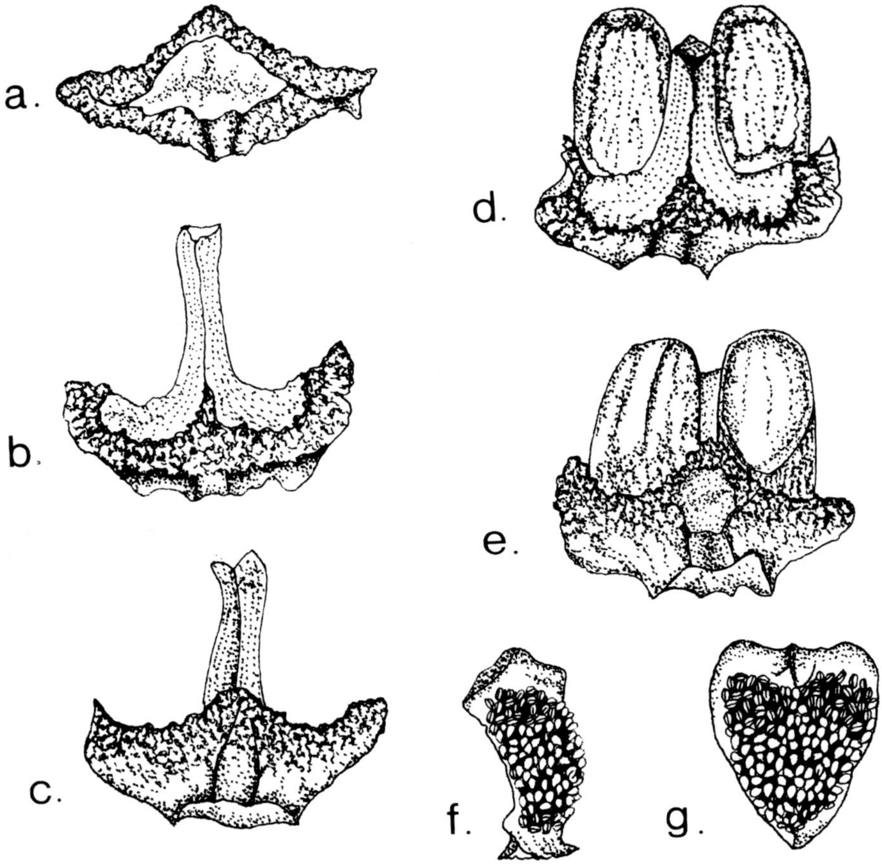

Fig. 2. *Encephalartos ituriensis* Bamps & Lisowski. **a–e.** Sporophylls of a female cone (*Lisowski 42938*). **a.** Apical view (bulla). **b.** Adaxial view. **c.** Abaxial view. **d.** Adaxial view with seeds. **e.** Abaxial view with seeds. **f, g.** Sporophylls of a male cone (*Lisowski 45310*). **f.** Lateral view. **g.** Adaxial view. All natural size.

Discussion

Other specimens, namely *Bequaert 4886* (Irumu, gallery forest in savanna, just one young leaf, Jul 1914), *Christiaensen 1243* (Nyankunde, Bunia Zone, fallow in grass savanna, sterile, Dec 1955), and *Pilette 2* (Ituri, without precise locality or date, mature leaves and photographs of the plant and ovulate cone, about twice as long as that of Lisowski specimens) could belong to *E. ituriensis,* but further investigation and more complete material are needed to confirm this assumption. The specimens collected in Uganda, namely *Lock 68/125* (Mpanga River gorge, Toro, alt. 1230 m, rocky places in grassland, Apr 1968; BR); *Osmaston s.n.* (Toro District, 1956; BR, K); *Osmaston 4055* (Mpanga River falls, on escarpment east of Lake George, short grass on outskirts of gallery forest, Jul 1957; K); and *Bagshawe 1054* (valley of the Mpanga River as cited by Eggerling, 1952), were determined by Melville (1957, 1958) as *Encephalartos hildebrandtii* but are surely different, as noted by Heenan (1977), and could also belong to *E. ituriensis.*

Literature Cited

Eggerling, W. J. 1952. The indigenous trees of the Uganda Protectorate. Ed. 2. The Government Printer, Entebbe. Page 104.

Heenan, D. 1977. Some observations on the cycads of Central Africa. J. Linn. Soc., Bot. **74:** 277–288.

Melville, R. 1957. *Encephalartos* in Central Africa. Kew Bull. **12:** 237–257.

———. 1958. Gymnospermae. Pages 3–10 *in* Flora of tropical East Africa. Crown Agents, London.

Robyns, W. 1948. Cycadaceae. Flore du Congo belge et du Ruanda-Urundi, Spermatophytes **1:** 1–4.

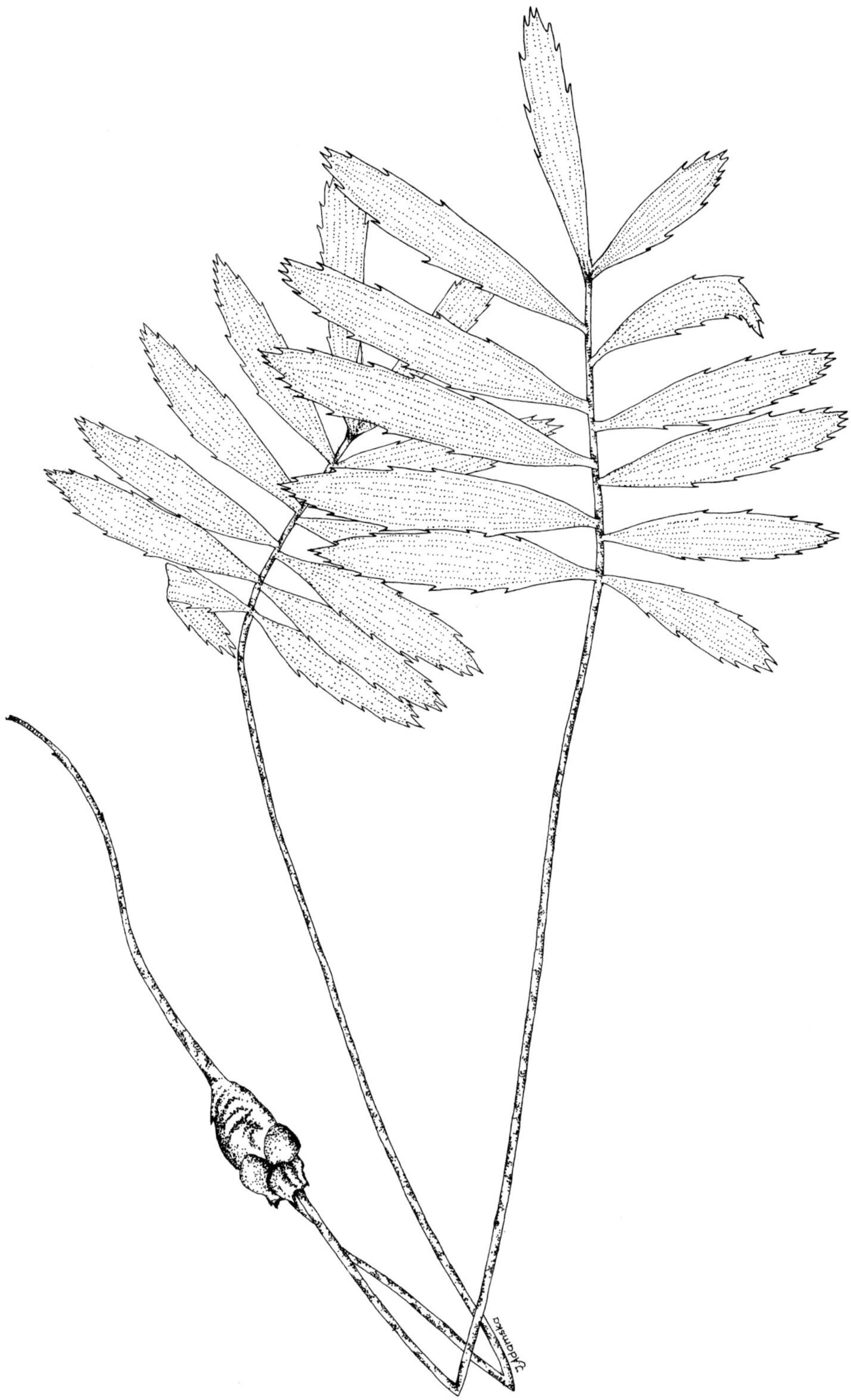

Fig. 3. *Encephalartos ituriensis* Bamps & Lisowski. Seedling (× ½) (*Lisowski 44790*).

MARCH 1990 MEMOIRS OF THE NEW YORK BOTANICAL GARDEN 57: 156–161

A New Species of *Encephalartos* (Zamiaceae) from Tanzania

DENNIS WM. STEVENSON, ALDO MORETTI, AND PAOLO DE LUCA

Abstract

STEVENSON, D. W. (New York Botanical Garden, Bronx, New York 10458, USA), A. MORETTI, and P. DE LUCA (Dipartimento di Biologia Vegetale, via Foria 123, Napoli 80139, Italia). A new species of *Encephalartos* (Zamiaceae) from Tanzania. Memoirs of the New York Botanical Garden **57**: 156–161. 1990.—*Encephalartos sclavoi,* a new species from Central Africa, is a member of a group of species containing *E. bubalinus, E. gratus, E. hildebrandtii, E. tegulaneus,* and *E. voiensis. Encephalartos sclavoi* is most closely related to *E. voiensis* from which it differs in vegetative and reproductive characters and habitat preference.

Key words: *Encephalartos*; Tanzania; leaflet characters.

Introduction

Recently, Moretti et al. (1989) described *Encephalartos voiensis* Moretti, D. Stevenson & Sclavo, which is more closely related to *E. hildebrandtii* A. Br. & Bouche than to other Central African species. *Encephalartos voiensis* differed from the latter principally in vegetative characters. The more important of the leaflet-character differentia include a more coriaceous texture, subfalcate shape, greater leaflet size, spinulose teeth reduced to callose bumps, and pungent apices in *E. voiensis* as compared to the less coriaceous texture, linear-lanceolate shape, smaller leaflet size, spinulose teeth along both margins, and bifurcate apices in *E. hildebrandtii.*

In Tanzania, near the Kenyan border (Fig. 1), we have located another new species of *Encephalartos* which takes the characters of *E. voiensis* to further extremes. The description of this new species is given below and is followed by a discussion of its presumed relationships with other nearby species of *Encephalartos* along with comments on the distribution of these taxa.

Encephalartos sclavoi Moretti, D. Stevenson & De Luca, sp. nov. Type. Tanzania. District Tanga, Gologolo, near Lushoto, 4 Jan 1986, Sclavo 2 (holotype, K; isotypes, MO).

Figs. 2, 3.

Truncus erectus, 1 m altus, 30–35 cm diam. Folia 1.75–2 m longa, oblonga, ad basin gradualiter attenuata, atque spinosa; rachis glabra. Foliola mediana alternata vel subopposite, velala coriacea, intera at revaluta, pungentia, imbricata, adaxiala superne sulcata, 18–28 cm longa, 4 cm lata. Strobilus ♂ subconicus, juvenilis flavavirens, adultus ebracteatus, 2–4 cm longus, 2–3 cm diam. Microsporophylla media ad axin perpendicularia, deltoidea, 2.5 cm longa, 2 cm lata, bulla deflexa praedita. Strobilus ♀ cylindricus vel ovoideo-cylindricus, 30–40 cm longus, 15–20 cm diam., juvenalis viridis vel flevus, adultus flevus vel brunneulus; megasporophylla mediana rhomboidea, pedicullata, 4 cm longa, 6.5 cm lata; pedicellus tribus abaxialilus faciebus praeditus; bullae rhomboidea, 6.5 cm lata, 1.3 cm alta, 3.8–4 cm profunda; bullae terminalis facies hexagona

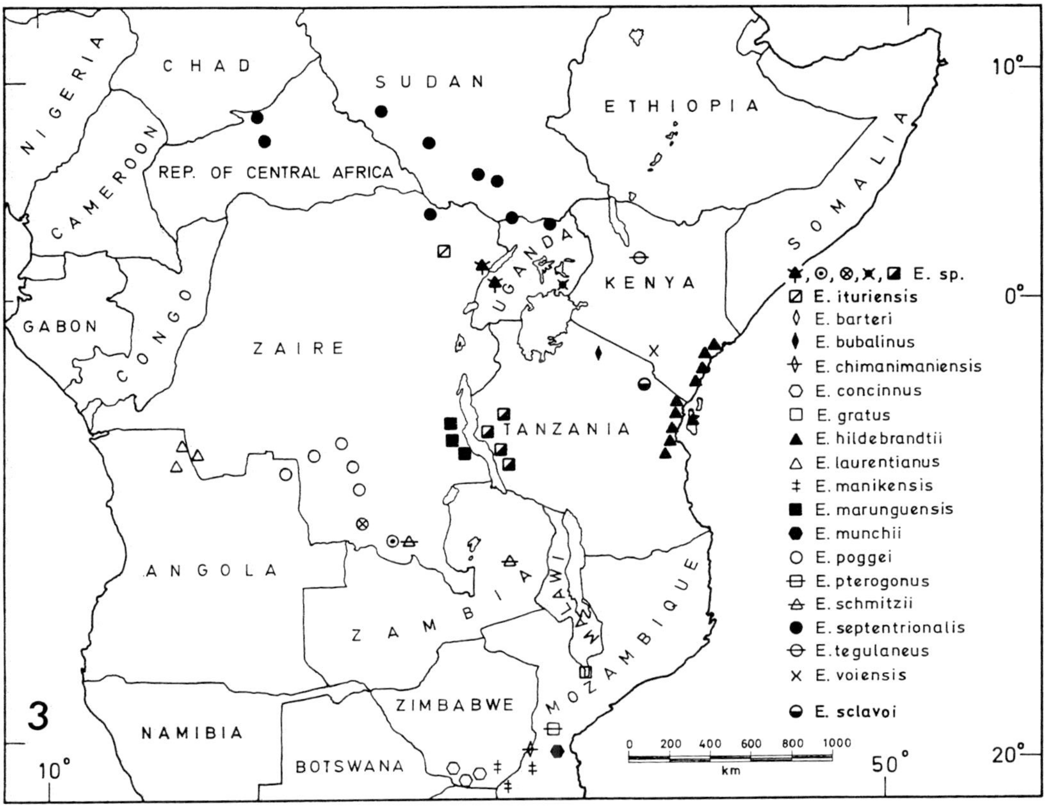

Fig. 1. Distribution map of *Encephalartos* spp. in Central Africa. Derived, in part, from Melville (1957) and Devred (1959).

atque vulticulo praedita; adaxialis facies ex duolus trapezoidibus partibus atque mediano lobo constituto tuberculato margine seminalis fimbria praedita; abaxialis facies ex trapezoidibus partibus atque constituta; mediana partibus cubtata.

Trunk erect to 1 m, 30–35 cm diam., leaf bases persistent. *Leaves* 1.75–2 m long, oblong, gradually tapering basally and becoming spinose at the base. *Petiole* 10–15 cm long, bulbous and tomentose basally. *Rachis* straight, glabrous or only very slightly tomentose and with two shallow longitudinal grooves between leaflet insertions. *Median leaflets* alternate to subopposite, strongly coriaceous, blue-green to dark green and yellow at point of attachment, insertion slightly acute to the rachis and oblique in the ab-adaxial plane, crowded and overlapping, appearing as a half-opened venetian-blind, with median longitudinal groove, oblong to elliptic, slightly inequilateral, 18–28 cm long, 4 cm wide, apex pungent and recurved appearing hook-like, 0–3 short spinose teeth often reduced to callose bumps on basal half of each margin, margins strongly revolute. *Lower leaflets* gradually diminishing in size and then abruptly becoming spinose in the last 3–5 leaflet pairs. *Microsporangiate strobilus* subconical, greenish-yellow when young, yellow when mature, 20–25 cm long, 20 cm diam., peduncle 2–4 cm long, 2–3 cm diam. *Median microsporophylls* deltoid, 2.5 cm long, 2 cm wide, inserted at right angle to axis, bulla deflexed. *Megasporangiate strobilus* cylindric to ovoid-cylindric, 30–40 cm long, 15–20 cm diam., green to yellow when young, yellow to tan when mature; peduncle 3–4 cm long, 2–3 cm diam. *Median megasporophylls* pedicellate with rhomboid bulla, 4 cm long, 6.5 cm wide; pedicel with 3 abaxial facets, the median facet rectangular. *Bulla of megasporophylls* rhomboidal, 6.5 cm wide,

Fig. 2. *Encephalartos sclavoi.* **a.** Adaxial view of a median leaflet with pronounced median groove. **b.** Leaf and leaflets with pronounced overlapping. **c.** Microsporophylls just prior to pollen shed.

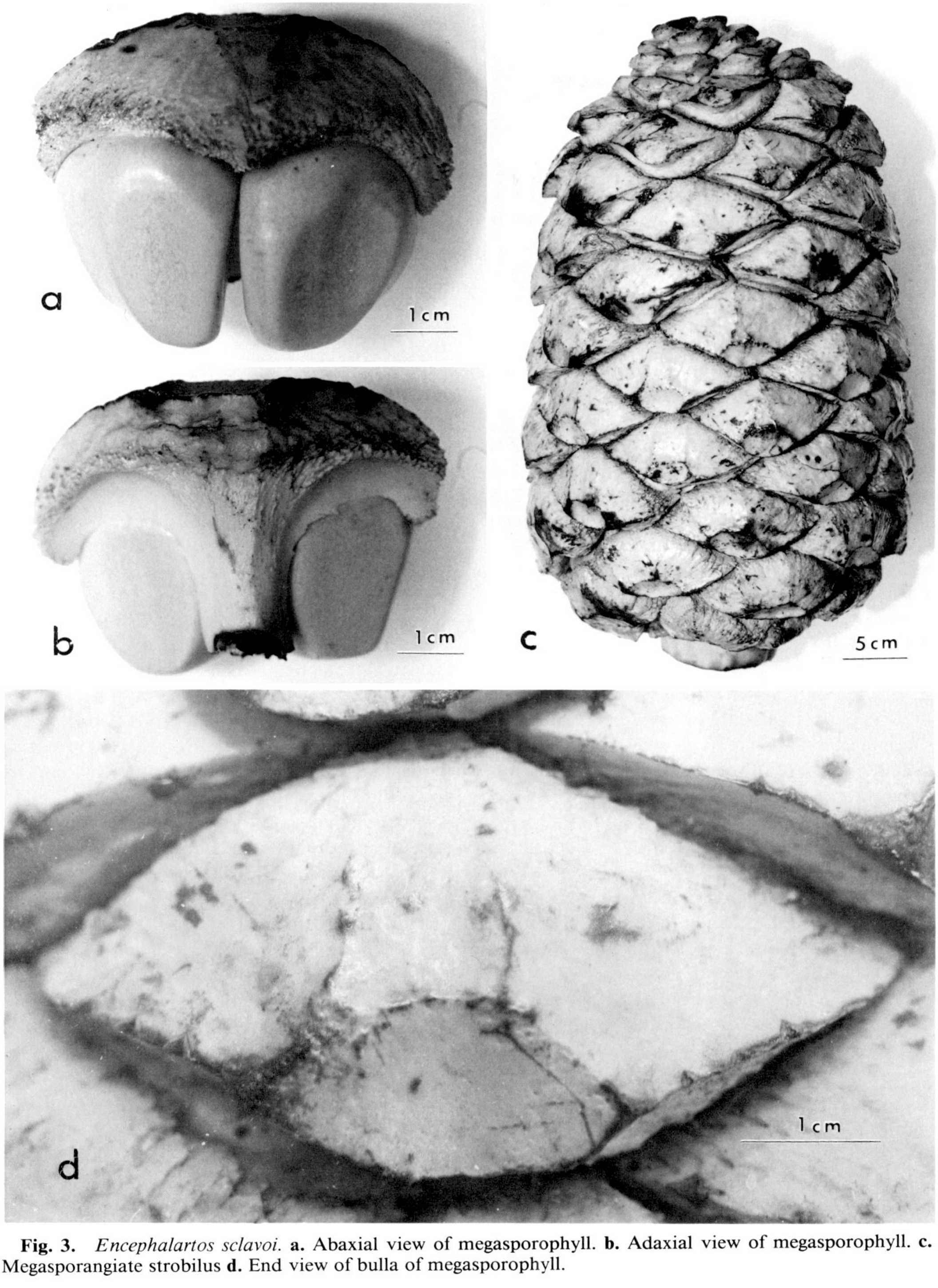

Fig. 3. *Encephalartos sclavoi.* **a.** Abaxial view of megasporophyll. **b.** Adaxial view of megasporophyll. **c.** Megasporangiate strobilus **d.** End view of bulla of megasporophyll.

1.3 cm deep, 3.8–4 cm high; terminal facet slightly concave and hexagonal; adaxial face with two lateral trapezoidal facets, a median trapezoidal facet, a short inconspicuous median lobe that is 2–4 mm wide and long, and with margin and seminal fringe tuberculate; abaxial face with two lateral trapezoidal facets, a median trapezoidal facet with obtuse lateral sagittal ridges, and with the inferior ridge and sagittal crest tuberculate; lateral lobes 5–7 mm long, 2–4 mm wide.

This species is named for Jean-Pierre Sclavo, a dedicated and knowledgeable student of *Encephalartos,* who first discovered and recognized it as new, supplied us specimens, photographs, and generously gave us his extensive field notes.

In our opinion, *E. sclavoi* is most closely related to *E. voiensis,* with which it shares unique vegetative and reproductive characters. In terms of reproductive characters, those of the megasporophylls are the most obvious. That is, both taxa have very short lateral lobes and an inconspicuous median lobe. The former character is not found in the other species (*E. bubalinus* Melville, *E. gratus* Prain, *E. hildebrandtii,* and *E. tegulaneus* Melville) of East Central Africa, which have lateral lobes three times as long as those of *E. sclavoi.* A median lobe is present in *E. gratus,* but in that species it is much larger and very conspicuous. *Encephalartos sclavoi* has mega- and microsporangiate strobili that are one-half the length of those of all the aforementioned species, and, thus, are quite distinct when seen in the field.

With respect to overall vegetative characters and general appearance, *E. sclavoi* is the most distinct species when encountered in the field in East Central Africa, which is here defined in the same sense of Moretti et al. (1989) and encompasses the area between the east coast of Central Africa and a line connecting Lake Victoria, Lake Tanganyika, and Lake Niassa (Fig. 1); i.e., east of the Great Rift Valley. The extremely coriaceous leaves and leaflets with rigid revolute margins combined with the overlapping position of the leaflets produce a half-opened venetian-blind appearance that is unmistakable. In fact, the leaves and leaflets even clatter like venetian blinds in a breeze. This overlapping position of leaflets also occurs in *E. munchii* Dyer & Verdoorn and *E. pterogonous* Dyer & Verdoorn, which are restricted to a small area of Central Mozambique (Fig. 1). These two species differ from *E. sclavoi* in that they have fewer coriaceous leaves and leaflet margins with several distinct spinose teeth. Moreover, they have green strobili and scarlet seeds at maturity, whereas the strobili of *E. sclavoi* are yellow to tan and the seeds are yellow at maturity, as in *E. voiensis* and *E. hildebrandtii.* For this and other reasons discussed below, we are of the opinion that *E. sclavoi* is more closely related to *E. voiensis* and *E. hildebrandtii* than to *E. munchii* and *E. pterogonous.*

Encephalartos sclavoi appears to represent a continuation of morphological changes that are present in *E. voiensis* when the latter is compared to *E. hildebrandtii. Encephalartos voiensis* has leaflets that are more coriaceous and larger in both length and width than those in *E. hildebrandtii.* In addition, the median leaflets in *E. hildebrandtii* have up to nine spinose teeth along each margin whereas the leaflets of *E. voiensis* have smaller spinose teeth only on the basal curve and small, barely discernible callose bumps along the rest of the margin corresponding to the position of the spinose teeth of *E. hildebrandtii.* The leaflets of *E. sclavoi* are even more coriaceous and larger in both length and width than those of *E. voiensis.* Furthermore, the spinose teeth of the leaflets when present in *E. sclavoi* are one-half the length of those of *E. voiensis,* and there are fewer callose bumps. Thus, it appears that there is a trend for increased coriaceousness and decreased spinosity. It should be noted that the habitat of *E. sclavoi* is less shaded and more open than that of *E. voiensis.*

In addition to the morphological evidence given above, *E. sclavoi* is located in the same area of the Tanzania-Kenya border as populations of *E. voiensis* and *E. hildebrandtii* (Fig. 1). In contrast, *E. pterogonous* and *E. munchii* are nearly 2000 km to the south of *E. sclavoi.* Moreover, they are clearly related to *E. manikensis* (Gilliland) Gilliland and its relatives (Dyer & Verdoorn, 1969).

It is interesting to note that *E. sclavoi,* in terms of its overlapping leaflet presentation and extreme coriaceousness, looks remarkably like *E. woodii* Sanders, which is now extinct in the wild. However, this resemblance appears to be homoplasious, as the strobili—in terms of color and sporophyll morphology—are quite different and the last and only known locality of *E. woodii* is some 1500 km distant in the midcoastal region of Natal, South Africa.

As discussed by Moretti et al. (1988), it is not surprising that another very distinct and locally endemic species such as *E. sclavoi* has been found in East Central Africa. Moreover, the discovery of *E. voiensis* and *E. sclavoi* has revealed that the diversity of *Encephalartos* spp. in Central Africa parallels that of Southern Africa and appears also to be related to habitat differences as in Southern Africa (Dyer, 1965; Dyer & Verdoorn, 1969).

Acknowledgments

This research was supported, in part, by National Science Foundation Grants BSR 8607049 and BSR 8796279 to D.W.S. We thank Signore Ortini for drawing Figure 3.

Literature Cited

Devred, R. 1959. Une Cycadacée nouvelle du Congo Belge: *Encephalartos marunguensis* Devred. Bull. Soc. Roy. Bot. Belg. **91:** 103–108.

Dyer, R. 1965. The cycads of Southern Africa. Bothalia **8:** 405–515.

——— **& I. Verdoorn.** 1969. *Encephalartos manikensis* and its near allies. Kirkia **7:** 147–158.

Melville, R. 1957. *Encephalartos* in Central Africa. Kew Bull. **12:** 237–257.

Moretti, A., P. De Luca, J. P. Sclavo & D. Wm. Stevenson. 1989. *Encephalartos voiensis* (Zamiaceae), a new East Central African species in the *E. hildebrandtii* complex. Ann. Mo. Bot. Gard. **77:** (in press).

Zamiaceae, a New Family for Zambia

FRANÇOIS MALAISSE, JEAN-PIERRE SCLAVO,
AND IAN TURNER

Table of Contents

Abstract

MALAISSE, F. (Laboratoire d'Ecologie, Phytogéographie et Phytosociologie, Faculté des Sciences agronomiques, 5800, Gembloux, Belgium), J.-P. SCLAVO (Villa "La Finca," Plateau du Mont Boron, 06300 Nice, France), and I. TURNER (Springs Farm, P.O. Box 2162, Harare, Zimbabwe). Zamiaceae, a new family for Zambia. Memoirs of the New York Botanical Garden **57**: 162–168. 1990.—The presence of the genus *Encephalartos* is reported for Zambia for the first time. The plant agrees with the description of *E. schmitzii*. The phytogeography of the genus *Encephalartos* is stated. Reasons for its scarcity in Central Africa are discussed.

Résumé

La présence du genre *Encephalartos* en Zambie est signalée pour la première fois. La plante correspond à la description de *E. schmitzii*. La phytogéographie du genre *Encephalartos* est précisée. Les raisons de sa rareté en Afrique centrale sont discutées.

Key words: *Encephalartos*; Zambia; Zamiaceae; Africa.

Introduction

Papers dealing with Zamiaceae in south central Africa are scarce. Melville (1957) provided information on Central African *Encephalartos*, introducing a useful terminology. This was further reviewed by Heenan (1977). Observations for the northern part of the Zambezian region were produced by several authors (Devred, 1959; Lisowski & Malaisse, 1971; Malaisse, 1969), while Lewis (1960) reported in volume 1(1) of the *Flora Zambesiaca* the presence of five species of *Encephalartos* Lehm., none of them occurring in Zambia. Dyer's work has also to be quoted (Dyer,

Table I
Previous reports of *Encephalartos* in Central Africa

Name	When and where discovered	Type locality	When described	Complementary description	Type specimen
E. poggei Aschers.	*Pogge s.n.*, 1876 (Luisa-Casserigi)	Between the Luisa and Casserigi Rivers (Zaïre)	P. Ascherson in Verh. Bot. Ver. Brand. **20:** 35 (1878)	–	*Pogge s.n.*
E. laurentianus De Wild.	*Gentil 98*, September 21, 1902 (Kwango River, Kasongo-Lunda)	Kwango River, Kasongo-Lunda (Zaïre)	E. De Wildeman in Ann. Mus. Congo Belge **5(1):** 10 (1903)	J. Lebrun, Rev. Zool. Bot. Afr. **19:** 387 (1930)	*Gentil 98* (BR)
E. marunguensis Devred	*Devred 3924*, July 18, 1957 (Kapona)	Kapona, Muhila plateau, Upper-Shaba distr. (Zaïre)	R. Devred in Bull. Soc. Roy. Bot. Belg. **91:** 104 (1959)	S. Lisowski & F. Malaisse, Bull. Jard. Bot. nat. Belg. **41:** 357 (1971)	*Devred 3924* (BR)
E. schmitzii Malaisse	*Schmitz 5417*, August 1955 (Kundelungu plateau)	Southern part of the Kundelungu plateau, Upper Shaba distr. (Zaïre)	F. Malaisse in Bull. Jard. Bot. Nat. Belg. **39:** 402 (1969)	–	*Malaisse 6286* (BR)
E. ituriensis Bamps & Lisowski	*Bequaert 4886*, July 2, 1914 (Irumu)	Mount Mukondja, Mambasa territory (Zaïre)	P. Bamps & S. Lisowski in this volume	–	*Lisowski 42909* (BR)

Fig. 1. Rocky outcrop with *Encephalartos schmitzii* in a *Brachystegia microphylla* miombo woodland developed on a slope in the vicinity of Mpika (Zambia). Note the female cone in foreground.

1965; Dyer & Verdoorn, 1969). Therefore, the recent discovery of a member of the Zamiaceae in the vicinity of Mpika, Northern Province of Zambia is astonishing. The discovery, presented here, fills a large gap in the geographic distribution of the genus.

Zamiaceae of Zambia

The material collected in Zambia agrees perfectly with the description and material of *Encephalartos schmitzii,* a species only known until now from the type locality on the southern part of the Kundelungu plateau, Shaba Province of Zaïre (Table I). *Encephalartos schmitzii* is a well differentiated species in the *Encephalartos poggei* complex (Malaisse, 1969).

Encephalartos schmitzii Malaisse, Bull. Jard. Bot. Nat. Belg. **39(4):** 402. 1969.

Figs. 1, 2.

Trunk not developed, only the crown above ground. *Leaves* up to 60 cm long, straight; leaflets up to 14 × 1.0 cm, linear or narrowly oblong, ± straight, margin with 0–2 spines on either side, set basically; leaflets reducing in size and becoming oblong towards the base, ultimately simple spines; petiole 1–2 cm long. *Male cones* up to 10 × 3.5 cm, ellipsoid to subcylindric, median cone-scales sessile, terminal facet blunty triangular to rhombic. *Female cone* up to 26 × 12 cm, ovoid-cylindric, green, shortly pedunculate; median cone-scales at right angle to axis, with terminal facet rhombic, 50–60° deflexed, slightly concave. *Seeds* 2.5 cm long, ovoid, orange, brownish with age.

Distribution. ZAMBIA. NORTHERN PROVINCE: Road Mpika–Isoka, km 10 (11°50′S, 31°27′E), *Sclavo & Turner 1* (♀, BR), rocky outcrop with *Xerophyta* in a *Brachystegia microphylla* woodland on a slope.

ZAÏRE. SHABA DISTRICT: Kundelungu plateau; not known elsewhere.

The key presented on page 81 of *Flora Zambesiaca* should be modified as follows:

2. Median leaflets less than 2 cm broad.
   5. Leaflets with well marked spines set apically. . . . . 5. *E. villosus.*
   5. Leaflets with discrete spines set basically or spines absent. . . . . 6. *E. schmitzii.*

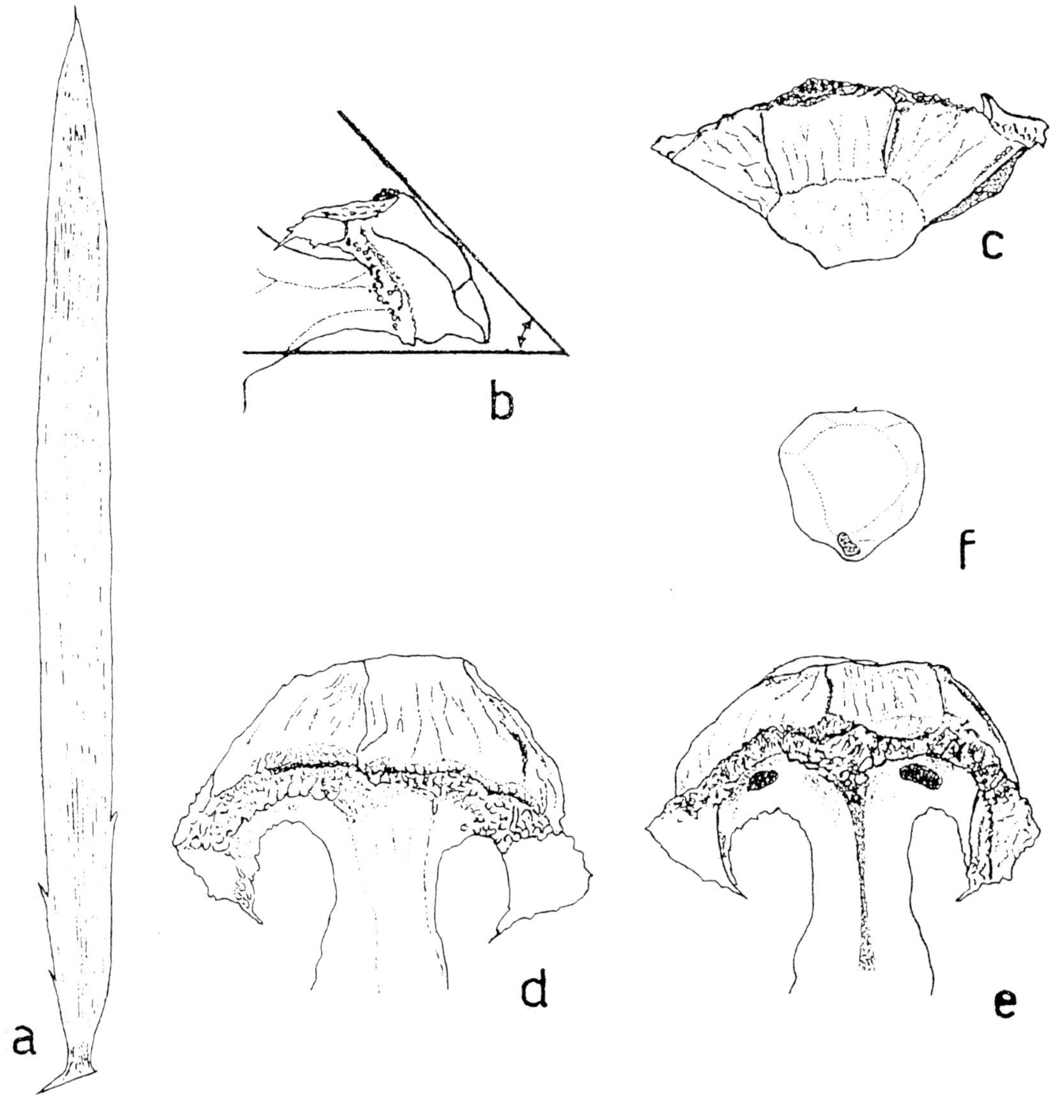

Fig. 2. *Encephalartos schmitzii* Malaisse. **a.** Median leaflet with 3 lateral spines. **b.** Female cone-scale, with terminal facet 50–60° deflexed. **c.** Median female scale, subapical view. **d.** Idem, abaxial view. **e.** Idem, adaxial view. **f.** Seed (after *Malaisse 6286,* from Malaisse, 1969).

Discussion

The presence of *Encephalartos* near Mpika fills a gap of over 1000 km between the Shaban occurrences on the one hand and the Malawian and Zimbabwean localities on the other hand. Moreover, recent prospecting in the neighboring Province of Shaba, in 1988, had led to the discovery of several new localities in the vicinity of Kolwezi, of Fungurume, and also in the southern part of the Marungu plateau between Pepa and Moliro. All these plants belong to the *E. poggei* complex. This led to the conclusion that, even if it is very rare, the genus may be present, as small clumps of about ten individuals in more or less open vegetation, such as the sandy steppish vegetation and the poor miombo of high plateaus or also on rocky outcrops through the Zambezian

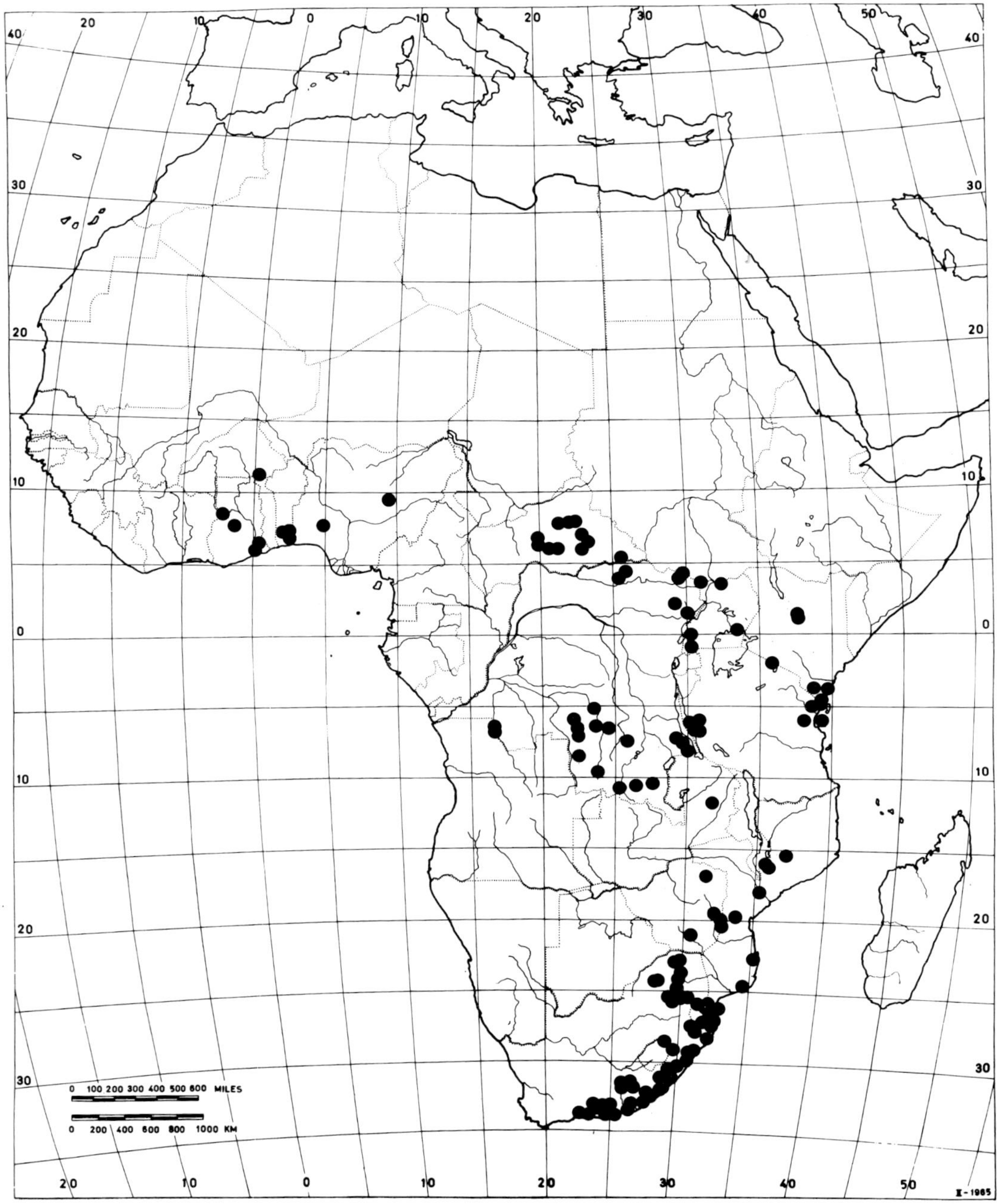

Fig. 3. Distribution of *Encephalartos* Lehm. on continental Africa. The map is based on review of literature (Western, Eastern and Southern Africa) as well as on revision of voucher specimens (Central Africa).

Region. Nevertheless, *Encephalartos* in south central Africa remains relatively scarce.

As far as the African distribution of *Encephalartos* is concerned, recent maps dealing with its distribution in the continental part of Africa generally present it as a large distribution from Senegal to Ethiopia and southwards to South Africa. This, however, is completely incorrect.

The genus *Encephalartos* does not belong to the linked Congo-Guinean–Sudano-Zambezian element, but rather to the latter, just penetrating at the boundaries into the Zaïro-Guinean (=Congo-Guinean) floristic area, namely on the Kwango River and in Ituri (Fig. 3).

Several explanations have been proposed to explain the scarcity of *Encephalartos*. Hall and Jenik (1967), discussing the present distribution of the West African cycad, *E. barteri,* recalled that there have been periods during the last million years when the West African climate has been moister than at present and forest must have covered much of what is now savanna. The authors suppose that *E. barteri* survived in a few places where the shallow, rocky soil prevented the growth of dense forest, but has been unable subsequently to recolonize wide areas of savanna. Both the niche for *E. ituriensis* and the present habitat of *E. schmitzii* are rocky sites. The now prevailing vitality and distribution of *Encephalartos* species in south central Africa reinforce the impression of its inability to recolonize steppe savannas and woodlands. Moreover, the conditions of pollination, fertilization and dissemination have become defective. As Giddy (1984) stated, male cones seldom all mature at the same time, ensuring that pollen is available over a long period of time; but to produce fertile seed in nature a male and a female plant must therefore not only cone at the same time but also grow close enough to one another to effect wind pollination. This becomes rare. On the other hand the importance of the African elephant (*Loxodonta africana*) in dissemination of *Encephalartos* is noted by several collectors. Liben (pers. comm.) has observed, several times, intact seeds in elephant dung. The lack of seedlings in the field may thus result both from reduced chances of fertilization of the female cones as due to the thinning out of natural populations and of absence of endozoochory because of the sharply reduced elephant population.

Human pressure has also to be taken into account. Schweinfurth (1875, vol. **II:** 310) reported that according to the expedition's porters the central part of the stem of *E. septentrionalis,* which contains a large quantity of starch, is used by the Kredi-Ndouggou to prepare a kind of beer, and *E. poggei* has been reported by several Lunda informants to be an important food during famine periods (Malaisse & Parent, 1985). The stem of the latter is used to obtain a flour, being prepared in a manner identical to that for the tuberous roots of cassava. Cut pieces are steeped in running water for about three days, dried in the sun, and then crushed to a fine powder, which is used for the preparation of a pap. The protein content of the stem of *E. poggei* is quite high and also contains a significant amount of lipids (Malaisse & Parent, 1985). However, these uses appear not to be responsible for the present scarcity of *E. schmitzii,* of which paleoendemism remains characteristic. Dyer and Verdoorn (1969) consider that, in modern times, by far the greatest menace to the wild populations of *Encephalartos* is the insatiable demand for sizable cycads to place in cultivation. And they further assert that there is demand for those hardy specimens that have successfully withstood the ravages of fire (namely, by protection of stem-apex by scale-leaves), baboons, and all other hazards. In south central Africa baboons are disappearing, activities of cultivator-collectors have appeared recently and are exceptional, but peasant agriculture, with temporary cassava fields is still operating in the bush where only a few cycads are preserved after transplantation in the vicinity of huts. Because of these conditions there is urgent need for more field exploration and studies.

Acknowledgments

We would like to thank the curator of BR for his assistance, P. Bamps for valuable advice, M. Schaijes for indicating a new *Encephalartos* locality, and K. Kisimba and Y. Muzinga for assistance in the field. We are very grateful to R. Schmid, H. D. Hammond and an anonymous reviewer for valuable comments on a first draft of this paper.

Literature Cited

Devred, R. 1959. Une Cycadacée nouvelle du Congo belge: *Encephalartos marunguensis* Devred. Bull. Soc. Roy. Bot. Belg. **91:** 103–108.

Dyer, R. A. 1965. New species and notes on type specimens of South African *Encephalartos.* J. South African Bot. **31:** 111–121.

——— **& I. Verdoorn.** 1969. *Encephalartos manikensis* and its near allies. Kirkia **7:** 147–158.

Giddy, C. 1984. Cycads of south Africa. Ed. 2. C. Struik Publ., Cape Town.

Hall, J. B. & J. Jenik. 1967. Observations on the West African cycad in Ghana. Niger. Field **32:** 75–81.

Heenan, D. 1977. Some observations on the cycads of Central Africa. Bot. J. Linn. Soc., London **74:** 279–288.

Lewis, J. 1960. Cycadaceae. Fl. Zambesiaca **1:** 79–83.

Lisowski, S. & F. Malaisse. 1971. *Encephalartos marunguensis* Devred, Cycadacée endémique du plateau des Muhila (Katanga, Congo-Kinshasa). Bull. Jard. Bot. Nat. Belg. **41:** 357–361.

Malaisse, F. 1969. *Encephalartos schmitzii* Malaisse, Cycadacée nouvelle du Congo-Kinshasa. Bull. Jard. Bot. Nat. Belg. **39:** 401–406.

——— **& G. Parent.** 1985. Edible wild vegetable products in the Zambezian woodland area: A nutritional and ecological approach. Ecol. Food Nutrition **18:** 43–82.

Melville, R. 1957. *Encephalartos* in Central Africa. Kew Bull. **1957:** 237–257.

Schweinfurth, G. 1875. Au coeur de l'Afrique 1868-1871. Voyages et découvertes dans les régions inexplorées de l'Afrique centrale. Paris, Hachette et Cie, 8° 2 vols. 434 + 508 pp.

MARCH 1990 MEMOIRS OF THE NEW YORK BOTANICAL GARDEN 57: 169–172

Chigua, A New Genus in the Zamiaceae with Comments on Its Biogeographic Significance

DENNIS WM. STEVENSON

Abstract

STEVENSON, D. W. (New York Botanical Garden, Bronx, New York 10458, USA). *Chigua,* a new genus in the Zamiaceae with comments on its biogeographic significance. Memoirs of the New York Botanical Garden **57**: 169–172. 1990.—*Chigua* a new genus in the Zamiaceae from Colombia is described and illustrated. This genus differs from all other cycad genera in having a prominent midrib with longitudinally dichotomous branched lateral veins in its pinnae and hexagonal sporophylls with raised mounds on the peltate face at the angles. Two species are described and illustrated: **C. restrepoi** sp. nov. and **C. bernalii** sp. nov. which differ from each other chiefly in leaflet width, length, and shape. The main biogeographic significance lies in this genus being the only endemic genus in the Cycadales for South America.

Key words: *Chigua*; Zamiaceae; South American endemic.

Introduction

In 1918, Francis Pennell collected a single specimen of a cycad tentatively assigned to *Zamia.* This specimen, which consists of part of an unbranched subterranean caudex and a single leaf, is unusual in that the leaflets have a prominent midrib with lateral veins running longitudinally and branching dichotomously. Because this was a unicate, Sergio Sabato and I thought that perhaps the specimen represented a teratism as we had never seen any other cycad specimens with this leaflet character. Knut Norstog and I recently travelled to Colombia in search of more material that would conform to Pennell's collection. Fortunately, the trip proved successful and more material was collected (along with strobili) that was the same as that collected by Pennell and, moreover, some 50 km distant a second collection was made of plants with quite differently shaped leaflets that were also more numerous per leaf, narrower, and longer. It was also clear that this material represented a new genus in that both the leaflets and the sporophylls were quite unlike those known for any of the currently known cycad genera.

The genera of cycads are easily separable on leaf, leaflet, and sporophyll characters. Thus, for example, *Dioon* Lindl. has leaflets that are laterally decurrent on the rachis and has flat sporophylls with the megasporophylls having two stalked ovules, whereas, in contrast, *Ceratozamia* Brongn. has articulated leaflets and peltate sporophylls with two horns and sessile ovules. A comparison of any other genera would reveal just as significant differences. Because the cycad genera are so distinct from one another in both reproductive and vegetative characters, any new genus should also be unique in these characters. Below is a description of such a new genus and its two species along with a comparison of it to other genera that have some similarity in general in leaflet characters.

Chigua D. Stevenson, gen. nov.

Foliola subopposita vel alternata, con medio nervis praedita, laterale longitudinale e dichotomi; sporophyllus ♀ peltatus, hexagonus, ad quemqua angulum umbunatus; sporophyllus ♂ peltatus, hexagonus, planus.

Leaflets subopposite to alternate, prominent midvein present, lateral veins longitudinal and dichotomously branched; megasporangiate strobili with hexagonal peltate sporophylls; megasporophylls with a conspicuous bump at each angle of the hexagon; microsporangiate strobili with hexagonal peltate sporophylls.

Type Species: *Chigua restrepoi* D. Stevenson.

The genus name is derived from the Spanish transliteration of an Indian common name for cycads in Central and South America.

Chigua restrepoi D. Stevenson. Type[1]. COLUMBIA. Primary rain forest, elev. 150 m, 14 Mar 1987, *D. Stevenson 693* (holotype: HUA; isotypes: FTG, NY). Fig. 1A–H.

Truncus hypogaeus, humilis, elipsoidus, 15 cm diam., 40 cm longus; cataphyllus lanata, triangularis, stipulis; folia pauca, 2–3, glabra; petiolus subteres terves, 60–80 cm longus, validus spinus armatus; rachis semiteres, 60–100 cm longus, in dimidio inferiore pucis spinis armata, supra inermis; foliola inferiore alternata, medio subopposita, supra opposita, 20–30 juga, lanceolata, 3–5 cm lata, 15–25 cm longa, basi cuneata, apicem caudata, margine dentata; strobilus ♂ cylindricus, vel 5 cm longus, 2 cm latus, ferrugineo-tomentosus; microsporophylla peltata, hexagona, plana; strobilus ♀ cylindricus, vel 15 cm longus, 5 cm latus, ferrugineo-tomentosus; megasporophylla peltata, hexagona, ad quemqua angula umbunata, ovulo duo; semina ovidea, rubra.

Stem hypogeous, ellipsoid, 15 cm diam., 40 cm long; *cataphylls* pubescent, triangular, 3–5 cm long, 1–2 cm wide, stipulate; *leaves* 2–3, glabrous, stipulate; *petiole* subterete, 60–80 cm long, armed with prickles; *rachis* semiterete, 60–100 cm long, armed with prickles below, smooth above; *leaflets* alternate below, subopposite in middle, opposite above, 20–30 on each side, lanceolate, 3–5 cm wide, 15–25 cm long, papyraceous, sessile, cuneate at base, caudate at apex, margins with well defined teeth; *microsporangiate strobilus* cylindrical, 5 cm long, 2 cm diam., covered with reddish-brown tomentum; *peduncle* glabrous, 10 cm long, 1 cm diam.; *microsporophylls* peltate, sterile peltate tip flat; *megasporangiate strobilus* cylindrical, 15 cm long, 5 cm diam., covered with reddish-brown tomentum; *peduncle* glabrous, 30 cm long, 1.5 cm diam.; *megasporophylls* peltate with a narrow basal stalk, hexagonal with a conspicuous bump at each angle of the hexagon; *seeds* ovoid, red at maturity.

Paratypes. COLUMBIA. Primary rainforest, 120 m, 24 Jul 1986, *R. Bernal, G. Galeano & D. Restrepo 1138* (COL, FTG); 28 Sep 1986, *I. Turner 3* (FTG); 90–120 m, 5–6 Mar 1918, *F. Pennell 4608* (NY).

This species is named in honor of the late Padre Sergio Restrepo, an excellent local botanist who located the species.

Chigua bernalii D. Stevenson. Type. COLUMBIA. Primary rain forest, 150 m, 27 Jul 1986, *R. Bernal, G. Galeano & D. Restrepo 1189* (holotype: COL; isotype: FTG). Fig. 1I.

Truncus hypogaeus, humilis, ellipsoidus, 10 cm diam., 30 cm longus; cataphyllus lanata, triangularis, stipulis; folia pauca, 2–3, glabra, stipulis; petiolus subteres terves, 1–1.4 m longus, validus spinus armatus; rachis semiteres, 60–160 cm longus, in dimidio inferiore pucis spinis armata, supra inermis; foliola inferiore alternata, medio subopposita, supra opposita, 30–55 juga, lineara vel linear-lanceolata, 1–1.5 cm lata, 30–35 cm longa, basi cuneata, apicem longa caudata, margine dentata e undulata.

Stem hypogeous, ellipsoid, 10 cm diam., 30 cm long; *cataphylls* pubescent, triangular, 3–5 cm long, 1–2 cm wide, stipulate; *leaves* 2–3, glabrous, stipulate; *petiole* subterete, 1–1.4 m long, armed with prickles; *rachis* semiterete, 60–160 cm long, armed with prickles below, smooth above; *leaflets* alternate below, subopposite in middle, opposite above, 30–55 on each side, lin-

[1] As in previous papers, I have purposely omitted precise locality data in an attempt to prevent eradication by commercial collectors. Only a very few plants of this genus have been seen in the wild, and I am convinced that commercial collectors would, either intentionally or unintentionally, cause its extinction.

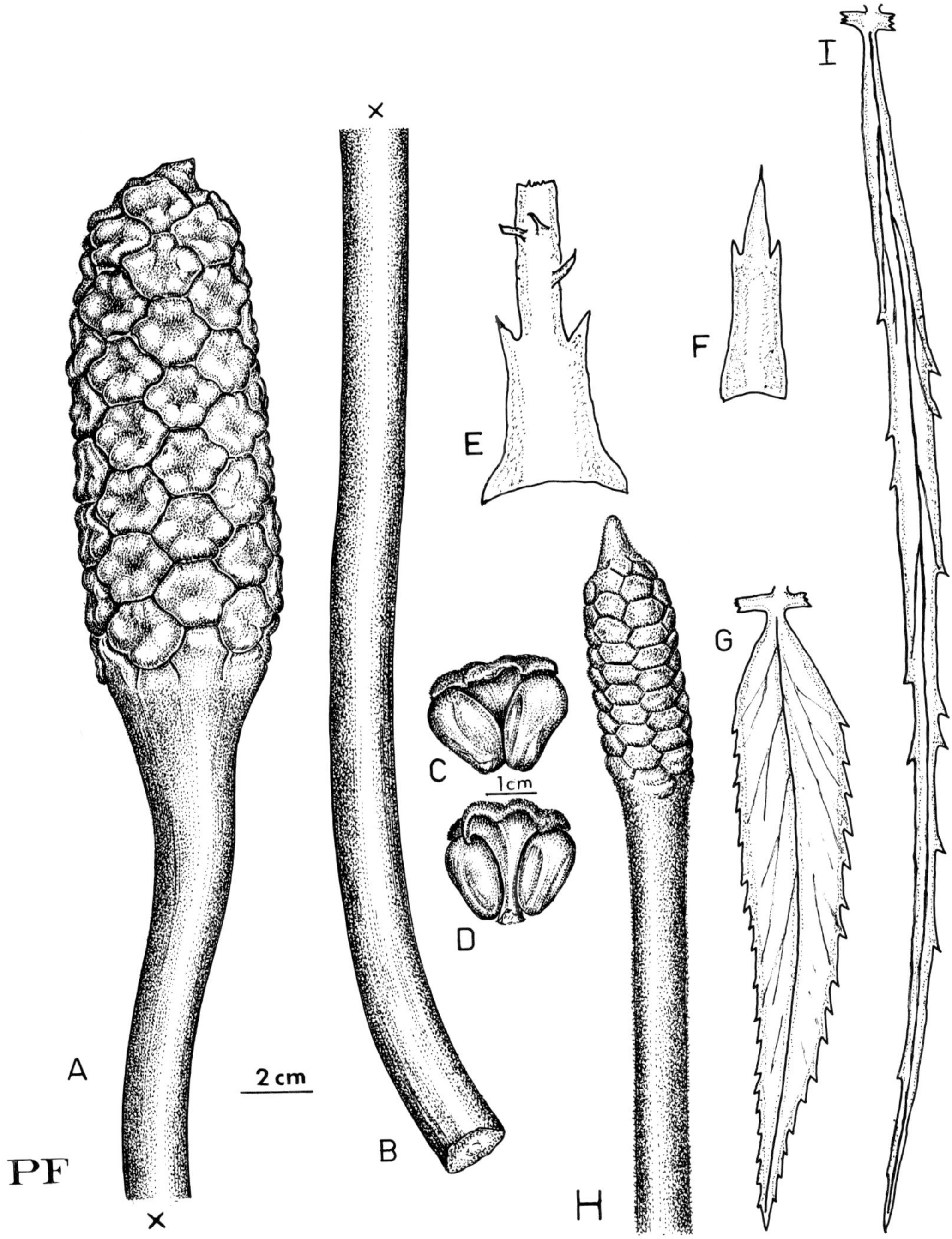

Fig. 1. *Chigua.* **A–H.** *Chigua restrepoi.* **A.** Megasporangiate strobilus. **B.** Peduncle. **C.** Abaxial view of megsporophyll. **D.** Adaxial view of megasporophyll. **E.** Leaf base with stipules. **F.** Cataphyll. **G.** Leaflet. **H.** Microsporangiate strobilus. **I.** Leaflet of *C. bernallii.* (A–D and H by Priscilla Fawcett, E–G and I by D.W.S.)

ear to linear-lanceolate, 1–1.5 cm wide, 30–35 cm long, papyraceous, sessile, cuneate at base, caudate at apex, margins with well defined teeth and undulate. Strobili unknown.

This species is named for Rodrigo Bernal who first discovered it and in recognition of his excellent work in neotropical cycad research.

The major features that distinguish *Chigua* from other cycad genera are the presence of a leaflet midrib with longitudinally dichotomizing lateral veins and the peculiar raised areas that are present at the angles of the hexagonal megasporophyll face. While the raised areas on the sporophyll face is unknown for any other cycad taxa, the midrib is present in two paleotropical genera, *Cycas* L. and *Stangeria* T. Moore, but not in any neotropical genera.

The presence of the midrib in the leaflets of *Cycas* and *Stangeria* needs some discussion. First, the total venation pattern of *Chigua* is quite different from that in the other two genera. In *Cycas,* the midrib is the only vein in the leaflets and it is raised on the abaxial surface with a groove appearing above the midrib on the adaxial surface. In *Stangeria,* the midrib appears raised on both surfaces as in *Chigua,* but the lateral veins depart at a right angle and then dichotomize near the margins, whereas in *Chigua* the lateral veins depart at a quite acute angle and run longitudinally, dichotomizing nearer their points of origin. The leaflets of *Stangeria* are very similar to and easily confused with those of marattiaceous ferns such as *Danaea. Chigua,* on the other hand, bears little or no resemblance to *Danaea.* In addition, *Chigua* has small stipules near the apex of its cataphylls and laterally near the leaf bases, whereas *Cycas* has no stipules and *Stangeria* has no cataphylls and the stipules of its leaves are large and fused into a hooded structure on the adaxial side of the leaf base (Stevenson, 1981). Furthermore, the strobili and sporophylls of *Chigua,* where known, bear no resemblance to either *Cycas* or *Stangeria* and it is chemically distinct from the latter two genera with the mucilage sugar pattern being more similar to that of *Zamia* L. (Stevenson & Siniscalco Gigliano, 1989).

Chigua is being placed in the Zamiaceae on the bases of the near-vertical orthostichies of its sporophylls and their peltate hexagonal morphology, the presence of the small stipules, which also occur in *Ceratozamia* and *Zamia,* and the seedlings of *Microcycas* A. DC., and the previously mentioned pattern of sugars in its mucilage.

In a recent paper on the biogeography of the neotropical cycads, Balduzzi et al. (1982) discussed the lack of a cycad genus endemic to South America. Indeed, we have predicted that an endemic genus either existed and is extinct or had yet to be found. When we first studied the Pennell collection, Sergio Sabato and I realized it represented a potential candidate for that status. Without more material, however, and which, more importantly, possessed distinctively different reproductive characters, the question would remain unresolved.

Cycad genera, in general, show an endemic distributional pattern, with three genera endemic to Australia: *Bowenia* Hooker ex Hooker fil., *Lepidozamia* Regel, and *Macrozamia* Miq.; two endemic to Southern Africa: *Encephalartos* Lehm. and *Stangeria*; one endemic to Cuba: *Microcycas*; and two to Northern Central America: *Ceratozamia* and *Dioon.* The exceptions to this pattern of endemism are *Cycas,* which is distributed widely around the Indian Ocean Basin and in tropical Asia, and *Zamia* which is widespread in the neotropics. What is significant here is that the other pantropical, southern hemisphere continents have endemic genera, so that it is not surprising that there is also one for South America.

Acknowledgments

I wish to thank Rupert Barneby for assistance with the Latin, Sergio Sabato for his stimulating discussions and ideas, Knut Norstog and John Popenoe for field assistance, and Jorge Brand for assistance in the field in Colombia. This work was supported in part by a National Science Foundation Grants BSR 8607049 and 8796279.

Literature Cited

Balduzzi, A., S. Sabato & P. De Luca. 1982 [1985]. Biogeography of the new world cycads. Depinoa **26:** 279–298.

Stevenson, D. 1981. Ptyxis, phenology, and trichomes in the Cycadales and their systematic implications. Amer. J. Bot. **63:** 1034–1046.

——— **& G. Siniscalco Gigliano.** 1989. The systematic value of the monsaccharide composition and distribution pattern of cycad mucilages. Biochem. Syst. Ecol. **17:** 185–190.

MARCH 1990 MEMOIRS OF THE NEW YORK BOTANICAL GARDEN 57: 173–185

West Indian and South American Cycads

SERGIO SABATO

Table of Contents

Abstract

SABATO, S. (Dipartimento di Biologia vegetale, Facoltà di Scienze, Università di Napoli, via Foria 223, 80139 Napoli, Italy). West Indian and South American cycads. Memoirs of the New York Botanical Garden **57**: 173–185. 1990.—The distribution and taxonomy of West Indian and South American cycads are presented. Five species of *Zamia* are recognized for the West Indies. Fourteen species of *Zamia* may be considered as "good species" for South America, where the greatest diversity is found in Colombia. Moreover, a number of *Zamia* species are as yet undescribed for South America. The history of the discovery of the new genus *Chigua* is presented. An illustration of a megasporangiate cone, drawn during the Mutis's 1783 expedition in Colombia, is in some respects unknown for the cycads and can be considered as evidence of the existence of a new genus in Colombia. The causes which determined the present distribution of the New World cycads are discussed.

Resumen

Se discuten las cícadas de las Indias Occidentales y de América del Sur por cuanto a su sistemática y distribución. Se reconocen 5 especies de *Zamia* por las Indias Occidentales y 14 por Sur América. En Colombia las cícadas presenten la variación morfológica mas amplia. Ademas en Sur América unas nuevas especies de *Zamia* son todavia non descritas. Se presenta la historia del descubrimiento del género nuevo *Chigua*. Un dibujo que representa un cono femenino de cícadas, hecho en el curso de la Real Expedición Botánica de Nueva Granada de 1783, es en unos respectos desconocido en las cícadas, y se puede considerar como una prueba de la existencia de un nuevo género de las cícadas de Colombia. Se discuten las causas biogeograficas que determinaron la actual distribución de la cícadas en el Nuevo Mundo.

Key words: Cycads; America; taxonomy; biogeography.

Introduction

Five genera of cycads occur in the New World tropics: *Ceratozamia, Dioon, Microcycas, Zamia,* and *Chigua.*

The genus *Ceratozamia,* with 10 species, occurs in Mexico, Guatemala, and Belize (Stevenson et al., 1986). The genus *Dioon,* with 10 species, occurs in Mexico and Honduras (Sabato & De Luca, 1985).

The genus *Microcycas* is monotypic and endemic to western Cuba, where it occurs in the Cordillera de los Organos. *Microcycas calocoma* (Miquel) A. DC. was originally described by Miquel (1852) as a species of *Zamia.* Some years later, de Candolle (1868) elevated this taxon to generic rank for its arborescent habit, entire leaflets, and microsporophylls with a flattened terminal knob. Apart from the flattened knob, the only other morphological feature that, at present, distinguishes the two genera is the high number of both archegonia and spermatozoids in *Microcycas.* Moreover, *Microcycas* is characterized by a high-numbered and asymmetrical karyotype ($2n = 26$; $2M+2A+22T$), which has recently been demonstrated to be equivalent to that of *Z. roezlii* Linden (Norstog, 1980, 1981) and *Z. paucijuga* Wieland (Moretti & Sabato, 1984). On the other hand, chemical evidence has pointed out that the biflavonoid constituents and the patterns of monosaccharide distribution of the mucilages may separate *Microcycas* from *Zamia* (Dossaji et al., 1975; De Luca et al., 1982).

Zamia, the most widespread of the neotropical genera, is still poorly known. It is distributed between 30°N and 18°S (Fig. 1), occurring in Georgia and Florida, USA, the West Indies (with the exception of the Lesser Antilles), Mexico, Central America, Colombia, Venezuela, Ecuador, Peru, Brazil, and Bolivia. While its distribution is well known in Georgia, Florida, the West Indies, and Mexico, it is poorly known in the rest of Central and in South America, especially in Amazonia.

The taxonomy of *Zamia* is, at best, very confused. Schuster (1932) lists 26 species; Chamberlain (N.d.) includes 29 species in the genus. In my opinion, there are probably 35–40 species of *Zamia,* many of which have yet to be described.

Stevenson and I (Stevenson & Sabato, 1986) have typified all validly published names in *Zamia* and the segregate genus *Aulacophyllum,* exclusive of those for West Indian species which have recently been treated by Eckenwalder (1980a). The genus *Aulacophyllum* was segregated by Regel (1876a) from *Zamia* mainly on the presence of grooves on the adaxial side of the leaflets and for producing leaves simultaneously rather than in succession. Regel included in this genus *Z. lindenii* Regel ex André, *Z. montana* A. Braun, *Z. roezlii, Z. skinneri* Warszewicz, and *Z. wallisii* hort. Veitch ex A. Braun and, moreover, described the new species *Aulacophyllum ortgiesii* Regel, the type of which is identical to the type of *Z. chigua* Seemann. Although some morphological features (i.e., leaflets grooved and sporophylls characterized by the presence of pyramidal apices) support the segregation of *Aulacophyllum* from *Zamia,* the patterns of monosaccharide composition of the mucilages of *Aulacophyllum* and *Zamia* species are very similar and suggest the existence of only one genus (De Luca et al., 1982).

West Indian *Zamia* Species

Eckenwalder (1980a) recognized only one species, *Z. pumila* L., with two subspecies, for the West Indies. Thus, he recognized two taxa instead of the up to 14 taxa that have been recognized by earlier authors. He demonstrated that leaflet width and vein number, the distinguishing characters most frequently used in previous delimitations of West Indian *Zamia,* appear to be closely associated. In fact, the features are structurally related: no matter how wide or narrow the leaflets are, there are always two veins per mm of width. In contrast to leaflet width and vein number, leaflet width vs. leaflet length shows no obvious association and argues for the independence of these features within the entire West Indian area. However, when leaflet width or leaflet length of individuals are plotted in frequency histograms for the West Indies and for each of its subregions, it is apparent that these frequencies follow smooth distributions within areas and sum to a smooth, unimodal distribution for the entire region. These frequency distributions provide little support favoring segregation of taxonomic entities among West Indian *Zamia.* In Cuba, however, Eckenwalder distinguished as *Zamia pumila* subsp. *pygmaea* (Sims) Eckenwalder some populations that contribute to a bi-

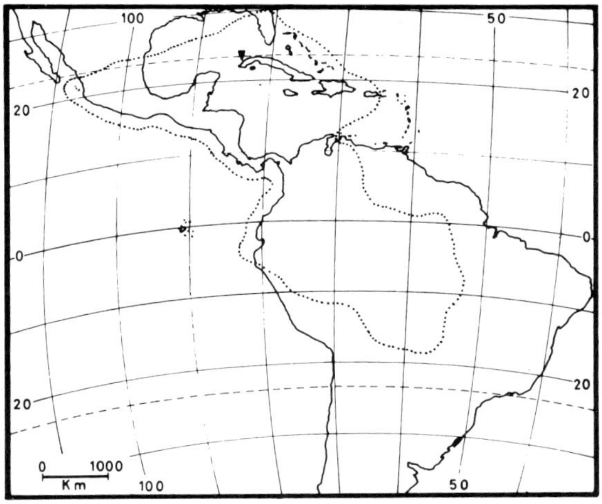

Fig. 1. Distribution of *Microcycas* (triangle) and *Zamia* (dashed line).

modal distribution of leaflet length. In their extreme form in westernmost Cuba, these smaller plants appear strikingly different from individuals of subsp. *pumila* from eastern Cuba.

On the other hand Dennis Stevenson (pers. comm.) took into consideration other characters including overall leaflet shape, denticulation patterns, overall strobilus shapes, and the nature of the sterile tip of the strobili. On the basis of these characters there would appear to be five taxa in this region (Table I). One is a newly described species which is restricted to western Cuba, southern Jamaica and northern Puerto Rico. This new taxon, *Zamia amblyphillida* D. Stevenson, appears to be a giant form of *Zamia pygmaea* Sims.

South American *Zamia* Species

Seventeen species of *Zamia* have been validly published for South America. In my opinion, only 14 may be considered as "good" species. In addition, however, there are a number of *Zamia* species as yet undescribed for South America.

ZAMIA OF COLOMBIA AND VENEZUELA

The greatest diversity is found in Colombia, where *Zamia manicata* Linden ex Regel, *Z. chigua, Z. obliqua* A. Braun, *Z. roezlii, Z. amplifolia* Hort. ex Masters, *Z. wallisii, Z. montana, Z. cupatiensis* Ducke, and *Z. jirijirimensis* R. E. Schultes occur (Fig. 2).

Zamia wallisii was described by Braun (1875) from material collected by Wallis in 1873 in the mountain region of Colombia and cultivated in the garden of James Veitch. The species was later collected in the field by Kalbreyer in 1877 and many plants from his collection were growing at Kew in 1890, according to Hooker (1890) who

Table I
Systematic treatment of West Indian *Zamia* (Stevenson, pers. comm.)

	Leaflet shape	Denticulate pattern	Strobili shape	Locality
Z. pygmaea Sims (dwarf)	obovate	prominent denticulation in the upper third of the leaflet	ovoid with a prominent obtuse to acute sterile tip at maturity	Western Cuba
Z. amblyphillida D. Stevenson (giant)	"	"	"	Western Cuba Southern Jamaica Northern Puerto Rico
Z. integrifolia L. fil.	oblong	very reduced rounded and obscure near the leaflet apex	cylindrical with a prominent acuminate sterile apex	Widespread throughout the region
Z. pumila L. sensu stricto	oblong	prominent in the upper third of the leaflet	cylindrical with a prominent acuminate sterile apex	Hispaniola Southern Puerto Rico Eastern Cuba
Z. angustifolia Jacq.	oblong lanceolate to lanceolate	1-3 apical teeth in the upper third of the leaflet	cylindrical with an acuminate apex	Eleuthera (Bahamas) Oriente Provence (Cuba) Susua Forest (Puerto Rico)

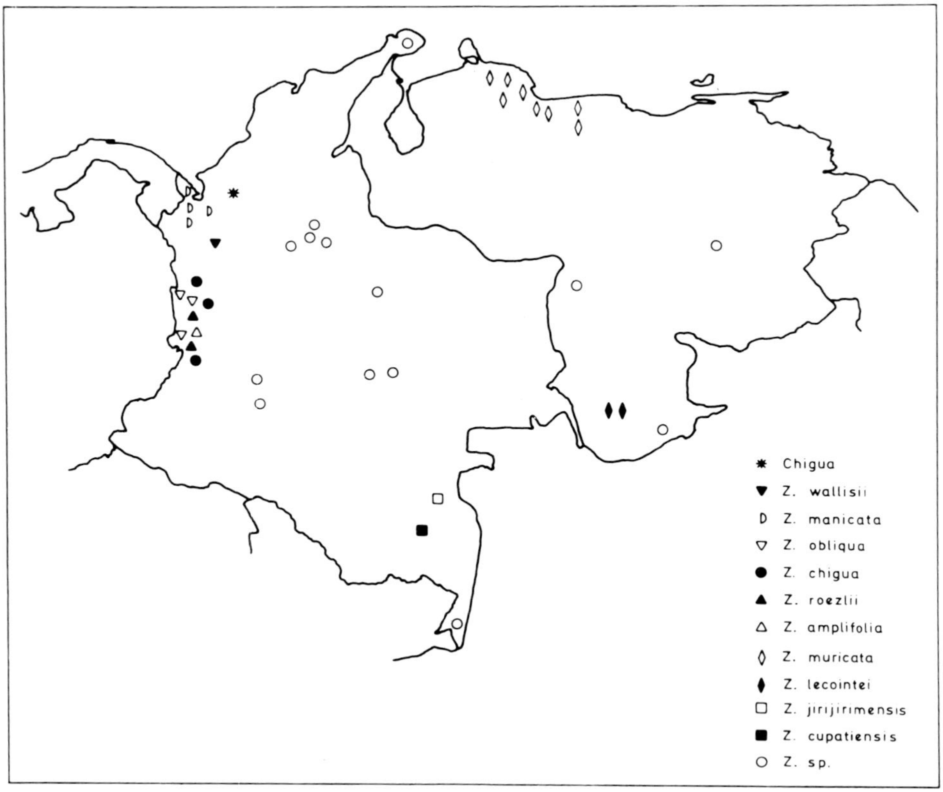

Fig. 2. Distribution of *Chigua* and *Zamia* species in Colombia and Venezuela.

described this plant again. Unfortunately, there are no specimens of the Wallis and Kalbreyer collections in cultivation or in herbaria. Until very recently we had no real concept of the appearance of the plant other than Braun's and Hooker's description, nor even knew whether it still existed. In 1984 the Colombian botanist Rodrigo Bernal rediscovered *Z. wallisii* along the route of Kalbreyer's trip. That same year I had the rare opportunity, together with Knut Norstog, his son Eric, Dennis Stevenson, and Bernal, to recollect this species in the field (Norstog, 1985). *Zamia wallisii* is localized in Antioquia in moist habitats at the border with the Chóco. It is an acaulescent species with leaves more than 2–3 m long, leaflets strongly grooved above, furnished with a very long petiolule, ovate and more than 30–40 cm long.

A petiolule is present also in *Zamia manicata,* an acaulescent species, which occurs in northern Colombia and adjacent southern Panama. Besides the occurrence of a petiolule, each leaflet has an encircling, thickened gland-like structure at the juncture of the base of the leaflet and the petiolule. On the other hand, the seedlings of this species do not exhibit these structures, which characterize only the adult plants. *Zamia manicata* was described by Regel (1876b) from living sterile specimens in the garden of J. Linden, a Belgian horticulturist. Regel based his *Z. manicata* on the presence of a petiolule and of a collar in the basal region of the leaflet. Two years later, Regel (1878) gave an illustration (Fig. 3) of a leaf of this new species and suggested that this species could be the representative of a new genus. In the absence of herbarium material, we (Stevenson & Sabato, 1986) selected this illustration as the neotype. Schuster (1932), in his taxonomic treatment of *Zamia* for Pflanzenreich, inexplicably put this species into synonymy with *Z. obliqua,* whose leaflets appear slightly petiolulate but lack a collar. Consequently, Schultes (1958) many years later described *Z. madida* R. E. Schultes, an independent renaming of *Z. mani-*

Fig. 3. Neotype of *Zamia manicata* Linden ex Regel (from Regel, Gartenflora **27.** Tafel 926, fig. e (at arrow) 1878).

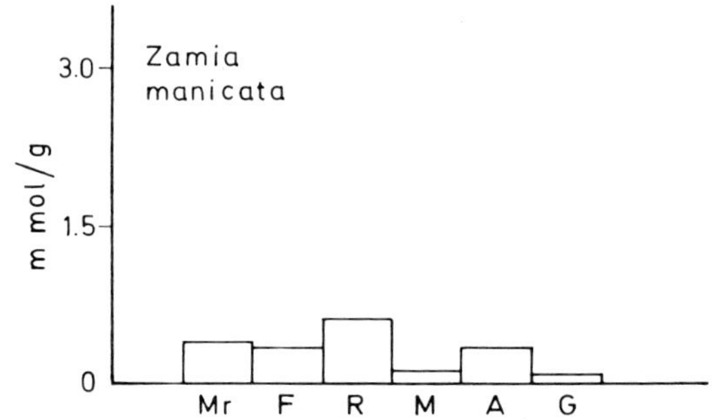

Fig. 4. Monosaccharide distribution histogram of hydrolyzed *Zamia manicata* mucilages. Mr, methylrhamnose; F, fucose; R, rhamnose; M, mannose; A, arabinose; G, galactose. (For methods, see De Luca et al., 1982.)

cata. The micro- and megasporangiate cones of *Z. manicata* are typical of the genus. Moreover, the monosaccharide distribution pattern of the mucilage is typical of *Zamia* (Fig. 4). Thus, both cone morphology and chemical composition of the mucilages of *Z. manicata* indicate clearly that this taxon belongs to the genus *Zamia* and do not support the suggestion of Regel of a possible separation of this species as a new genus.

In the area between Cabo Corrientes and Buenaventura Bay, in Pacific Colombia, *Zamia roezlii, Z. amplifolia, Z. chigua,* and *Z. obliqua* occur.

Zamia roezlii was collected by Roezl along the coast of the Pacific Ocean and described by Linden (1873) from plants cultivated in his garden. In the same year, André gave an excellent color illustration of this species (Fig. 5) which, in the absence of herbarium material, has been selected as the neotype (Stevenson & Sabato, 1986). *Zamia roezlii* is an arborescent plant with a trunk up to 6–7 m high and entire linear lanceolate leaflets, strongly grooved above. Its megasporangiate cones are the largest among the *Zamia* species. Recently, Norstog (1977) observed this plant in the field, where it grows in brackish mud just above the *Rhizophora* zone in mangrove swamps.

Zamia amplifolia was described by Masters (1878) from plants collected in Colombia and cultivated in the garden of Mr. Bull, an English horticulturist. In the absence of herbarium material we (Stevenson & Sabato, 1986) selected *Kiem and Norstog 30* (FTG—3 sheets) as the neotype, because this material fully corresponds to the protologue. The neotype was collected in 1976 near Buenaventura, Dept. Valle del Cauca. *Zamia amplifolia* has deeply veined leaflets about 10–12.5 cm across and somewhat resembles *Z. skinneri,* a Central American species. *Zamia amplifolia* is distinguished by its entire leaflets. In contrast, *Z. skinneri* has leaflets with toothed margins, at least apically.

Zamia chigua was described by Seemann (1854), an English traveller, from plants collected at the mouth of the River San Juan in Darien (Pacific Colombia). Seemann gave a very good illustration (Fig. 6) which, in the absence of herbarium material, has been designated as the lectotype (Stevenson & Sabato, 1986). *Zamia chigua* is an arborescent plant, with overlapping, narrowly lanceolate, often falcate leaflets with long attenuate apices, and petioles and rachis heavily armed. Disjunct populations of *Z. chigua* also occur in central Panama.

Zamia obliqua was described by Braun (1875) from specimens collected by Wallis in 1873 in the warm regions of Colombia. Living plants,

Fig. 5. Neotype of *Zamia roezlii* Linden (from André, Ill. hortic. **20.** Planches 133–134. 1873).

also according to Braun, were in cultivation in the garden of James Veitch. We (Stevenson & Sabato, 1986) were not able to locate any of these specimens and decided to designate as neotype two specimens collected in 1847 by Seemann at Cabo Corrientes and deposited at K, as they conform completely with Braun's description of this species. *Zamia obliqua* is an arborescent plant with a slender trunk up to 2–3 m high. The leaflets are light-green, lanceolate ovate, slightly asymmetrical, with short petiolules and long, sharply denticulate apices. Disjunct populations of *Z. obliqua* occur in central Panama and central Peru.

Zamia montana was described by Braun (1875) from specimens collected by Wallis in 1873 in the high mountain regions of Colombia at 2300–2600 m. Living plants were in cultivation in the garden of James Veitch. It was again met with, in 1880, by Kalbreyer (Thiselton Dyer, 1882). Further attempts to relocate this plant were unsuccessful. Moreover, there are no specimens of these collections in cultivation or herbaria. Stevenson and Sabato (1986) have lectotypified with the description because the description does not match any specimens and the locality is unknown. *Zamia montana* is an arborescent plant 1.20–1.50 m high, with leaflets 30 cm or more long, 5–10 cm wide, strongly grooved above, chartaceous, sublanceolate, apex abruptly acuminate, with usually one prominent and a few obscure teeth. *Zamia montana* resembles *Z. skinneri,* which differs in its leaflets with margins toothed at least in the apical portion.

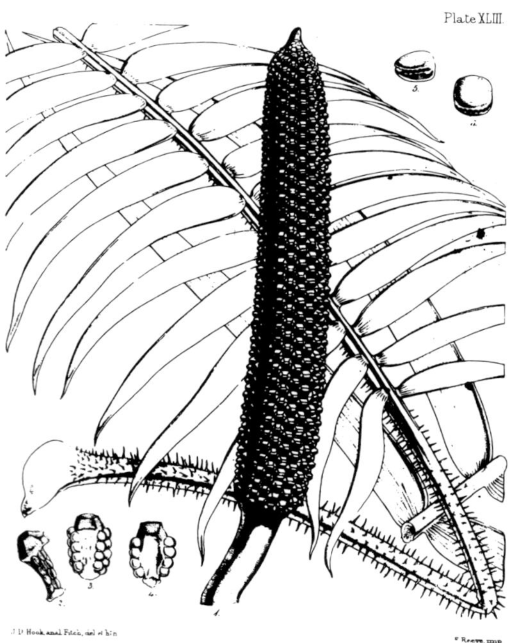

Fig. 6. Lectotype of *Zamia chigua* Seemann (from Seemann, Seem. Bot. Voy. Herald. Tab. 43. 1854).

Zamia cupatiensis, discovered by Ducke (1922) in Amazonian Colombia, has a trunk which is partially subterranean and leaflets ovate lanceolate, apex acute-acuminate, about 21 cm long and 6 cm wide.

In the same region, Schultes (1953) discovered *Zamia jirijirimensis,* which appears to be, in some respects, closely allied to *Z. cupatiensis. Zamia jirijirimensis* has a subterranean stem and very narrow lanceolate leaflets measuring from 20 to 27 cm in length and 1.5 to 2.2 cm in width.

Zamia muricata Willdenow (1806) was described from material collected by Humboldt and Bonpland at Porto Cabello in Atlantic Venezuela. *Zamia muricata* is an acaulescent plant with leaflets oblong, symmetrical, furnished with teeth which usually occur only in the upper half.

Zamia lecointei Ducke, a species described previously from Brazil, probably occurs also in southern Venezuela.

Zamia of Ecuador and Peru

Zamia poeppigiana Martius & Eichler (1863) was described from material collected in 1830 by Poeppig along the Tocaehe River in Amazonian Peru. this species is widespread in Peru and Ecuador. Originally the Ecuadorean material was assigned to *Z. lindenii* Regel ex André (1876). However, Stevenson and I consider *Z. lindenii* and *Z. poeppigiana* conspecific, as the type of *Z. lindenii* and of *Z. poeppigiana* are very similar, as are also the original descriptions. *Zamia poeppigiana* is an arborescent plant with leaflets falcate-lanceolate, sessile, apex acuminate, margins aculeate, 10–20 times as long as wide. See Figure 7.

Some populations of *Zamia* near Pebas (at the border with Brazil), in Amazonian Peru, are of great interest (empty triangle in Fig. 7). These specimens appear very similar to *Z. manicata* in that they have petiolulate leaflets with a prominent glandular structure at the juncture of the blade and the petiolule. In contrast, the Peruvian plants have the gland attached very obliquely. The seedlings, like *Z. manicata,* do not exhibit these structures, which characterize only the adult plants. This new species is acaulescent, with leaves three meters long. The leaflets have margins which are entire or have a few inconspicuous teeth near the apex.

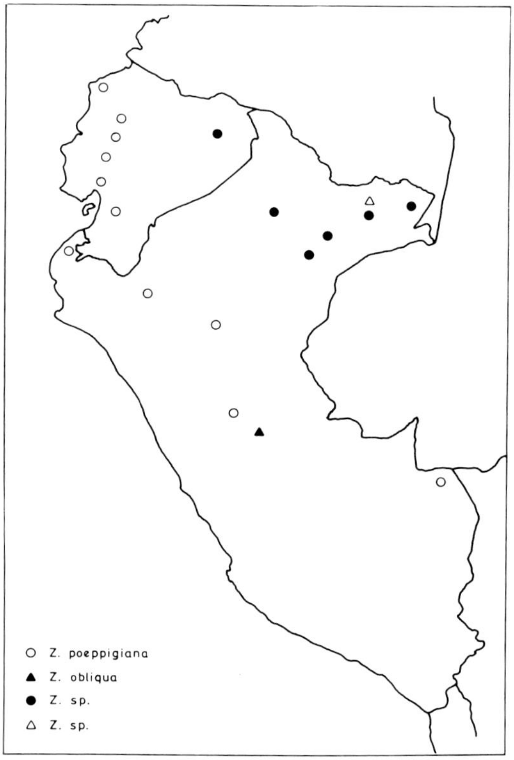

Fig. 7. Distribution of *Zamia* species in Ecuador and Peru.

Zamia obliqua, which is distributed in Colombia and Panama, probably also occurs in central Peru.

Zamia of Brazil and Bolivia

Zamia ulei U. Dammer and *Z. lecointei* are the only two species described for the great territory of Brazil (Fig. 8). *Zamia ulei,* described by Dammer (1907) from material collected in western Brazil, is an acaulescent plant with leaflets obovate to elliptic-obovate, cuneate basally, and with a long denticulate apex. *Zamia lecointei,* described by Ducke (1915) from specimens collected in the region of Rio Erepecuru in northern Brazil, is an acaulescent plant with leaflets linear-lanceolate, long acuminate, entire or bearing occasional inconspicuous teeth. This species occurs also in Amazonian Venezuela.

Zamia boliviana (Brongniart) A. DC. was orig-

Fig. 8. Distribution of *Zamia* species in Bolivia and Brazil.

inally described by Brongniart (1846) as a species of *Ceratozamia* from material collected by d'Orbigny near San Xavier in eastern Bolivia. *Zamia boliviana* is an acaulescent plant with leaflets narrow (0.5–1.5 cm wide), linear-lanceolate, apex attenuate, often with 2–5 inconspicuous teeth near the apex, margins revolute. In some respects, *Z. boliviana* resembles *Z. pumila* from West Indies. It lives in dry habitats of the savanna.

New Genera of American Cycads

The history of the discovery of *Chigua,* the new genus of cycads described in this volume (Stevenson, 1989), began with Francis W. Pennell, who explored Colombia at the beginning of this century. In northern Colombia he collected a cycad specimen which he attributed to the genus *Zamia.* Some years ago, in order to typify all validly published names in *Zamia,* I very carefully examined this specimen at NY. I was impressed by the occurrence of a midrib in the middle of the leaflets, a character which is unknown for the Zamiaceae but occurs in the families Cycadaceae and Stangeriaceae. I showed the specimen to Stevenson, Norstog, and John Popenoe, director of the Fairchild Tropical Garden. We did not know whether this was a teratism or, in fact, a new genus. Personally, I could not understand why South America, unlike other Austral continents, was devoid of endemic genera of cycads. Thus, in 1984 we organized a botanical

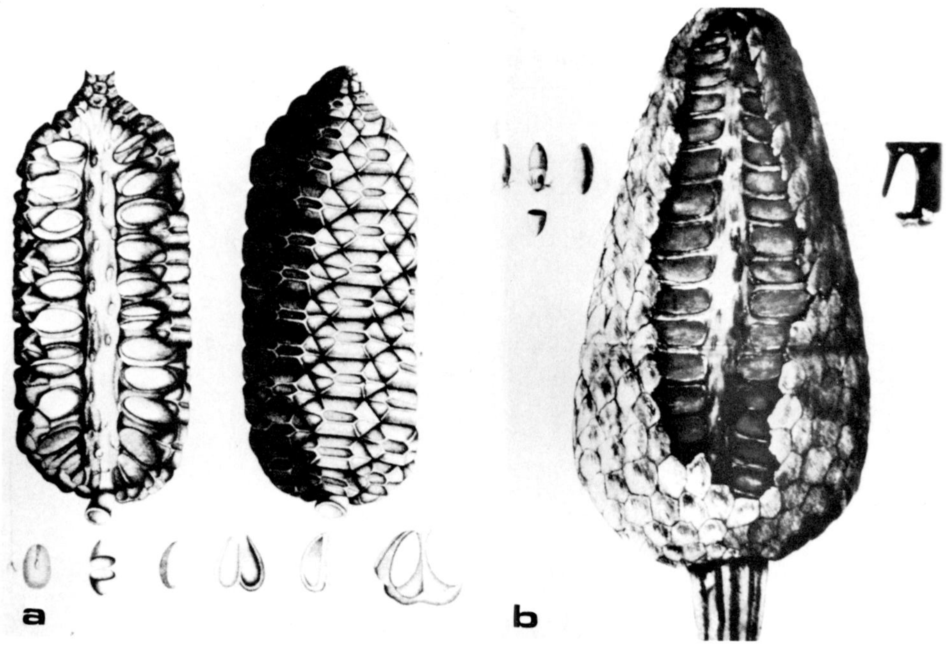

Fig. 9. Illustrations of two megasporangiate cones of cycads, drawn during the Mutis's 1783 expedition in Colombia. **a.** Megasporangiate cone of *Zamia*. **b.** Megasporangiate cone of a new genus.

expedition to Colombia in order to collect this new taxon but unfortunately we could not go to northern Antioquia. In 1987 Norstog and Stevenson were able to collect additional specimens of this plant in Colombia, and we immediately realized that a new genus had been found. We named this new genus *Chigua,* from the vernacular name of this plant.

José Celestino Mutis, a Colombian botanist, organized in 1783 the "Real Expedición Botánica del Nuevo Reyno de Granada." In the course of this expedition more than 15,000 plants were collected and more than 6000 illustrations were drawn. Two illustrations show two megasporangiate cones of cycads. The first (Fig. 9a) belongs, without doubt, to the genus *Zamia.* The second (Fig. 9b) is very significant because it shows a cone which is in some respects unknown for the cycads. It resembles superficially both the megasporangiate cone of *Dioon,* a genus endemic to Mexico and Honduras, and of *Encephalartos,* a genus endemic to Africa. In my opinion, this cone belongs to a new genus of cycads. It is very important to relocate it in the field. This could be the aim of a new botanical expedition to Colombia.

Origin of the New World Cycads

Whereas most genera of cycads are distributed in the southern hemisphere, the American genera of cycads are mainly distributed in the northern hemisphere. In order to discuss the causes which determined the present distribution of the New World cycads, I would like to review our knowledge of the fossil cycads and outline the geographical history of America (Table II).

The fossil record for cycads in America can be traced back to the Permian and possibly the Upper Pennsylvanian. For the Paleozoic Mamay (1976) described two Lower Permian megasporophylls, *Phasmatocycas* and *Archaeocycas,* respectively. Moreover, Mamay summarized his studies, detailing how the cycads could have originated from Late Pennsylvanian and Early Permian taeniopterid pteridosperms. Recently Gillespie and Pfefferkorn (1986) demonstrated that *Phasmatocycas* and *Taeniopteris* were parts of the same plant. These forms also suggest that cycads evolved from taxa with entire leaves, i.e., *Taeniopteris,* rather than from pteridosperms with compound leaves. From the Mesozoic, Delevoryas and Hope (1971) described the new ge-

Table II
Fossil record of American cycads

	Age	Locality
Archaeocycas	Lower Permian	Kansas, Texas
Phasmatocycas	Lower Permian	Kansas, Texas
Mesodoscolea	Lower Cretaceous	southern Argentina
Michelilloa	Triassic	northern Argentina
Chamorgia	Late Triassic	Arizona
Leptocycas	Late Triassic	North Carolina
Lyssoxylon	Late Triassic	Arizona, New Mexico
Bororoa	Paleocene	southern Argentina
Dioon	Paleocene	Alaska
"	Miocene	California
Zamia	Paleocene	Alaska
"	Lower Eocene	Tennessee, Louisiana, Mississippi
"	Upper Eocene or Oligocene	Puerto Rico
"	Oligocene ?	Colombia
"	Oligocene or Lower Miocene	southern Argentina
"	Lower (?) Miocene	Chile
"	Late Tertiary (Pliocene ?)	Brazil
Ceratozamia	Miocene	Mexico
Menucoa	Lower Tertiary (?)	southern Argentina

nus *Leptocycas* from the Late Triassic North Carolina. *Leptocycas* is the only attempt at reconstructing an entire American cycad. Other remains are known from Mesozoic deposits: *Lyssoxylon* (Daugherty, 1941; Gould, 1971), *Michelilloa* (Archangelsky & Brett, 1963), and *Chamorgia* (Ash, 1985). They consist only of silicified stems.

Among the taxa represented by leaves, *Mesodoscolea* from the Lower Cretaceous of southern Argentina appears of great phytogeographical interest because it has been included in the Stangeriaceae (Archangelsky & Petriella, 1971), whose sole living member, *Stangeria,* is endemic to South Africa. On the basis of this fossil, it appears that the Stangeriaceae were already in existence on the American continent during the Lower Cretaceous.

Dioon, Ceratozamia, and *Zamia* are the living cycads represented in the Cenozoic fossil flora of America. *Dioon* fossils consist of leaves found in Paleocene deposits in Alaska (Hollick, 1932) and a trunk found in Miocene deposits in California (Eckenwalder, pers. comm.). *Ceratozamia* fossils consist of pollen grains found in Miocene deposits of Mexico (Rzedowski & Palacios Chavez, 1977). *Zamia* fossils consist of leaves reported from the Paleocene of Alaska, Eocene of Tennessee, Louisiana, Mississippi, Oligocene of Puerto Rico, Colombia, Patagonia, Miocene of Chile, and Late Tertiary of Brazil (Hollick, 1932).

Menucoa (Petriella, 1969) of the Lower Tertiary (?) and *Bororoa* (Petriella, 1972) of the Paleocene are two petrified trunks described from southern Argentina which, because of their polyxylic structure, have been closely related to the Cycadaceae and the Encephalarteae of the Zamiaceae, respectively, rather than to the American Zamiaceae.

From the Permian, when the first Cycadales evolved, until the Early Jurassic, the entire region of North America–South America–Africa was one landmass, part of the supercontinent of Pangaea. During the Mesozoic (by Middle Jurassic time) the continents began to separate, and by the Early Cretaceous North and South America were separated. In the Middle to Upper Cretaceous, the gap between the North and South American subcontinents was beginning to diminish due to the formation of a series of volcanic islands forming an archipelago between Honduras and Colombia (Coney, 1982). Beginning in the Turonian (90 million years BP) parts of these islands were pushed eastwards and became the Greater Antilles, which by Eocene time had achieved their present position (Coney, 1982). It is believed that land never connected North America with the Greater Antilles after their initial separation

(Brown & Gibson, 1983). During Cenozoic time Middle America grew subaerially southwards from North America as a complex volcanic arch-trench system, eventually forming the final land bridge by collision south of Panama in the Pliocene.

In the New World, continental drift not only separated the American subcontinents from the palaeotropics but also separated North and South America, whose floras thus evolved in isolation.

It is most probable that the modern genera of cycads are derived from ancestors which evolved in the Late Mesozoic during the Cretaceous. Unfortunately, the existing genera of cycads appear in the fossil record only at the beginning of the Cenozoic, and there is no earlier information about their evolutionary history.

There is, however, taxonomic and chemical evidence of isolation and early evolutionary divergence of the New World cycads from the African and Australian members of the Zamiaceae. De Luca et al. (1982) have shown that the American genera of the Zamiaceae may be distinguished from other members of the family by the different monosaccharide composition of the mucilages and that the patterns of monosaccharide distribution of the mucilages of the American genera are different from those of the members of Boweniaceae, Cycadaceae, and Stangeriaceae.

Dioon represents the descendant of an ancient evolutionary line characterized by leaf-like megasporophylls, ovules usually borne on a stalk-like outgrowth of the megasporophyll, and pinnae non articulate.

Ceratozamia, Microcycas, and *Zamia* appear closely related and originated from a distinct evolutionary line from which, however, *Ceratozamia* probably diverged earliest. These genera appear morphologically very similar. Their synapomorphies include very reduced sporophylls, ovules sessile, pinnae articulate, and leaves stipulate (in *Microcycas* only in the seedling). Furthermore, the patterns of monosaccharide composition of the mucilages of both South and North American species of *Zamia* are very similar and do not support a distinct origin for the taxa of the two continents and, in fact, support the close relationship of all species of *Zamia* (De Luca et al., 1982).

On the other hand, the new genus *Chigua* appears very distinct, by the fact that its leaflets are furnished with a midrib and by the shape of the female cone, unlike anything in *Zamia.* On the basis of these characters, it could belong to a new family.

When we consider the present distribution of American cycads, the close relationships among *Ceratozamia, Microcycas* and *Zamia,* in conjunction with the fossil record, it all suggests that the geological separation of the two American sub-continents caused the ancestors of *Dioon, Ceratozamia, Microcycas* and *Zamia* to be limited to North America, while the ancestors of the genus *Chigua* were segregated in South America.

Dioon and *Ceratozamia* are distributed in Mexico and northern Central America, and their southern limit coincides with the southern limit of North America (including the Central American Peninsula) during the Cretaceous.

Zamia, as suggested by the fossil record, achieved its modern pan-neotropical distribution at the beginning of the Cenozoic. *Zamia* was probably isolated in the West Indies by the migration eastwards of the Cretaceous chain of volcanic islands, which could have also provided "stepping stones" for dispersal of *Zamia* between the two continents during the Middle and Late Cretaceous. It is probable that hurricanes and strong currents were vectors for interisland transport. They could have permitted seedling establishment from beached rafts of drift material. It is of interest that seeds of *Zamia pumila,* after submergence in seawater for two weeks, showed 92 percent germination (Eckenwalder, 1980b). *Microcycas* was probably isolated in Cuba by the same events that isolated *Zamia* in the West Indies.

The discovery of the new genus *Chigua* in northern Colombia appears of great phytogeographical importance, since hitherto, South America was apparently the only Austral continental mass devoid of endemic genera of cycads. In South America other cycads evolved which became extinct during the Cenozoic. It is probable that the exploration of South America will lead us to the discovery of new and interesting taxa, one of which could be that illustrated during the "Real Expedición Botánica del Nuevo Reyno de Granada of 1783."

Literature Cited

André, E. 1873. *Zamia roezli* Regel. Ill. hortic. **20:** 126–127. Planches 133–134.

———. 1875. *Zamia lindenii.* Pl. 195. Ill. hortic. **22:** 23.

Archangelsky, S. & D. W. Brett. 1963. Studies on Triassic fossil plants from Argentina. III. *Michelilloa waltonii* nov. gen. et spec. from the Ischigualasto Formation. Ann. Bot., N.S., **27(105):** 147–156.

——— **& B. Petriella.** 1971. Notas sobre la flora fósil de la zona de Ticó, Provincia de Santa Cruz. IX. Nuevos datos acerca de la morfología foliar de *Mesodoscolea plicata* Arch. (Cycadales, Stangeriaceae). Bol. Soc. Arg. de Botanica **14(1–2):** 88–94.

Ash, S. 1985. A short thick cycad stem from the Upper Triassic of Petrified Forest National Park, Arizona, and vicinity. Mus. Northern Arizona Bull. **54:** 17–32.

Braun, A. 1875. Die Frage nach der Gymnospermie der Cycadeen. Monatsber. k. Akad. Wiss. Berlin, Apr. **1875:** 376–377.

Brongniart, A. 1846. Note sur un nouveau genre de cycadées du Mexique. Ann. Sci. Nat. Bot., sér. 3, **5:** 5–9.

Brown, J. H. & A. C. Gibson. 1983. Biogeography. The C. V. Mosby Company, St. Louis, Missouri.

Candolle, A. de. 1868. Cycadaceae. Prodromus systematis naturalis **16(2):** 522–548.

Chamberlain, C. J. N.d. A taxonomic monograph of the cycads. Unpubl. Ms., Fairchild Tropical Garden Library, Miami, Florida.

Coney, P. J. 1982. Plate tectonic constraints on the biogeography of Middle America and the Caribbean region. Ann. Missouri Bot. Gard. **69:** 432–443.

Dammer, U. 1907. Cycadaceae. Verh. bot. Ver. Brandenburg **47:** 117–118.

Daugherty, L. H. 1941. The Upper Triassic flora of Arizona. Carnegie Inst. Wash. Pub. **526:** 1–108.

Delevoryas, T. & R. C. Hope. 1971. A new Triassic cycad and its phyletic implications. Postilla **150:** 1–21.

De Luca, P., A. Moretti, S. Sabato & G. Siniscalco Gigliano. 1982. A comparative study of cycad mucilages. Phytochemistry **21(7):** 1609–1611.

Dossaji, S. F., T. J. Mabry & E. A. Bell. 1975. Biflavonoids of the Cycadales. Biochem. Syst. Ecol. **2:** 171–175.

Ducke, A. 1915. Plantes nouvelles ou peu connues de la région amazonienne. Cycadaceae. Arch. Jard. Bot. Rio de Janeiro **1:** 9–10.

———. 1922. Plantes nouvelles ou peu connues de la région amazonienne (II″ partie). Cycadaceae. Arch. Jard. Bot. Rio de Janeiro **3:** 19–20.

Eckenwalder, J. E. 1980a. Taxonomy of the West Indian cycads. J. Arnold Arbor. **61:** 701–722.

———. 1980b. Dispersal of the West Indian cycads, *Zamia pumila* L. Biotropica **12(1):** 79–80.

Gillespie, W. H. & H. W. Pfefferkorn. 1986. Taeniopterid lamina on *Phasmatocycas* megasporophylls (Cycadales) from the Lower Permian of Kansas, U.S.A. Rev. Palaeobot. Palynol. **49:** 99–116.

Gould, R. E. 1971. *Lyssoxylon grigsbyi,* a cycad trunk from the Upper Triassic of Arizona and New Mexico. Amer. J. Bot. **58(3):** 239–248.

Hollick, A. 1932. Descriptions of new species of Tertiary cycads, with a review of those previously recorded. Bull. Torrey Bot. Club **59(4):** 169–189.

Hooker, J. D. 1890. *Zamia wallisii.* Curtis's Bot. Mag. **116:** Tab. 7103.

Linden, J. 1873. Linden Catalogue des plantes nouvelles no. 90. Page 10.

Mamay, S. H. 1976. Paleozoic origin of the cycads. Geol. Surv. Prof. Pap. No. 934, U.S. Gov't. Printing Off., Washington, D.C.

Martius, C. F. P. & A. G. Eichler. 1863. Cycadeae. Fl. Bras. **4(1):** 411–416.

Masters, M. T. 1878. New garden plants. Gard. Chron. **10:** 810.

Miquel, F. A. W. 1852. Sur une espèce nouvelle de *Zamia* des Indes Occidentales, introduite dans l'établissement Van Houtte, à Gand. Flore des serres et des jardins de l'Europe **7:** 141–142.

Moretti, A. & S. Sabato. 1984. Karyotype evolution by centromeric fission in *Zamia.* Pl. Syst. Evol. **146:** 215–223.

Norstog, K. 1977. The spermatozoid of *Zamia chigua* Seem. Bot. Gaz. **138:** 409–412.

———. 1980. Chromosome numbers in *Zamia* (Cycadales). Caryologia **33:** 401–409.

———. 1981. Karyotypes of *Zamia chigua* (Cycadales). Caryologia **34:** 255–260.

———. 1985. Exploring for *Zamia wallisii.* Fairchild Trop. Gard. Bull. **40(1):** 6–17.

Petriella, B. 1969. *Menucoa cazaui* nov. gen. et sp., tronco petrificado de Cycadales, Provincia de Rio Negro, Argentina. Ameghiniana **6(4):** 291–302.

———. 1972. Estudio de maderas pedrificadas del Terciario inferior del área central de Chubot (Cerro Bororó). Rev. Mus. La Plata, N.S., **6:** 159–254.

Regel, E. 1876a. Die Cycadeen, deren Gattungen und Arten. Gartenflora **25:** 140–144.

———. 1876b. Cycadearum generum specierumque revisio. Acta Horti petrop. **4(4):** 275–320.

———. 1878. Die Cycadeen der Garten. Gärtenflora **27:** 3–13.

Rzedowski, J. & R. Palacios Chavez. 1977. El bosque de *Engelhardtia* (*Oreomunnea*) *mexicana* en la region de la Chinantla (Oaxaca, México). Una reliquia del Cenozoico. Bol. Soc. Bot. Mex. **36:** 93–118.

Sabato, S. & P. De Luca. 1985. Evolutionary trends in *Dioon* (Zamiaceae). Amer. J. Bot. **72(9):** 1353–1363.

Schultes, R. E. 1953. Notes on *Zamia* in the Colombian Amazonia. Mutisia **15:** 1–6.

———. 1958. Plantae Austro-Americanae X. Cycadaceae. Bot. Mus. Leafl. Harvard Univ. **18(4):** 114–115.

Schuster, J. 1932. Cycadaceae. *In* Engler, Das Pflanzenreich **99(IV. 1):** 1–168.

Seemann, B. 1854. Cycadeae. Flora of the Isthmus of Panama. Pages 59–254 *in* The botany of the voyage of H.M.S. Herald.

Stevenson, D. W. 1990. *Chigua,* a new genus in the

Zamiaceae with comments on its biogeographic significance. Mem. N.Y.B.G. **57**: 169–172.

——— & **S. Sabato.** 1986. Typification of names in *Zamia* L. and *Aulacophyllum* Regel (Zamiaceae). Taxon **35(1)**: 134–144.

———,———, & **M. Vázquez Torres.** 1986. A new species of *Ceratozamia* (Zamiaceae) from Veracruz, Mexico with comments on species relationships, habitats, and vegetative morphology in *Ceratozamia*. Brittonia **38(1)**: 17–26.

Thiselton Dyer, W. T. 1882. New garden plants. Gard. Chron. **14**: 459–460.

Willdenow, K. L. 1806. Species Plantarum **4(2)**: 847–848.

On the Genus *Glandulataenia*, nov. from the Triassic of Nidhpuri, India

DIVYA DARSHAN PANT

Abstract

PANT, D. D. (Botany Department, University of Allahabad, Allahabad-211002, India). On the genus *Glandulataenia*, nov. from the Triassic of Nidhpuri, India. Memoirs of the New York Botanical Garden **57**: 186–199. 1990.—A new segregate of *Taeniopteris*, *Glandulataenia*, gen. nov., having interstitial glands between veins is described from the Triassic of Nidhpuri, India. Two species of the genus, *G. glandulata* (Srivastava), comb. nov. and *G. triassicus*, sp. nov., are described. The new genus may cast doubts on the reconstruction of *Phasmatocycas* Mamay.

Key words: *Glandulataenia*; interstitial glands; Triassic.

Introduction

The fossil genus *Taeniopteris* Brongniart, which is distinguished by its strong midrib, nonanastomosing and often-forked side veins crossing the lamina at right angles, has been split up from time to time by various authors (Harris, 1932; Florin, 1933; Sahni, 1948; Pant, 1958, and others) on the basis of cuticular and other structural details. However, leaves of the *Taeniopteris* type, showing glands between side veins and sometimes also over the midrib, have continued to be lumped together in the old genus. Obviously, these leaves seem to be distinct from the other nonglandular leaves and they are here included under a new genus *Glandulataenia*.

The genus is based on fossil impressions and compressions of leaves collected from the Triassic *Dicroidium* bed exposed on the bank of the Gopad River cutting through the Sidhi District, on the Nidhpuri village side, Madhya Pradesh, India. The usual methods for investigating compressions, like Schulze's fluid and celloidin pulls, have been used (see Pant, 1958). The measurements generally mention the ranges of size within parentheses and averages between them.

Glandulataenia, gen. nov. Type species: *G. glandulata* (Srivastava), comb. nov.

DIAGNOSIS

Leaves elongated, entire, showing a strong midrib persistent up to apex; lamina arising from the side of the midrib, lateral veins arising at a wide angle or at smaller angles but almost immediately bending outwards and crossing the lamina at a wide angle. Lateral veins occasionally forked, anastomoses between lateral veins extremely rare, interstitial areas showing rows of characteristic rounded dots (glands). Leaves hypostomatic, stomata haplocheilic, usually confined to areas between veins, subsidiary cells irregular, not forming a definite ring.

DISCUSSION

The genus *Glandulataenia* can be distinguished from the older segregates of *Taeniopteris* Brongniart like *Nilssoniopteris vittata, Nilssonia, Doratophyllum* Harris, *Bjuvea* Florin and *Nipaniophyllum* Sahni and *Rhabdotaenia* Pant, by the presence of interstitial rows of glands but it comes closest to *Rhabdotaenia*. Leaves of *Glandulataenia* whose epidermal structure is preserved show that the glands are present in the upper epidermis. The genus can be distinguished from other segregates by the same features as mentioned for *Rhabdotaenia* by Pant (1958) and Pant and Verma (1963). Although Chandra (1974) has stated that "Pant did not give any reasons as to why he instituted a new genus instead of retaining the original genus *Macrotaeniopteris*" and doubted the distinction between *Rhabdotaenia* and *Macrotaeniopteris,* Pant (1958) had clearly pointed out that the type of the ill defined genus *Macrotaeniopteris* had been proved to be bennettitalean.

On the basis of the similarity of the cuticular and epidermal structure of *Rhabdotaenia* with that of *Glossopteris* and their occurrence in Lower Gondwana beds, Pant (1958) and Pant and Verma (1963) had suggested that the two genera were possibly related, but the same cannot be said about *Glandulataenia* and its affinities are for the present uncertain.

Leaves assignable to *Glandulataenia* were earlier described by Srivastava (1971) under the name *Taeniopteris glandulata* and these are here assigned to the new genus on the basis of their characteristic glands and also the cuticular structure of identical leaves in our collection. However, glandless leaves which Srivastava (1976) assigned to *Gopadia* are just like those of *Rhabdotaenia* but distinguished from it merely by the absence of cross connections. They should be transferred to *Rhabdotaenia* since anastomoses between lateral veins are extremely rare even in *Rhabdotaenia* and the genus *Gopadia* should therefore, be regarded as a synonym of *Rhabdotaenia*.

Glandulataenia glandulata (Srivastava), comb. nov. (Pls. I, II, Figs. 1–17; Text-figs. 1A–G; 2A–H)

Taeniopteris glandulata, Srivastava, 1971: 284–290, pls. 6, figs. 1–5, Text-fig. 6A–C.

EMENDED DIAGNOSIS

Leaf large, oblong, oval or spatulate; lamina entire, length unknown (estimated length up to 12 cm), up to 1.2 cm wide in middle base more or less rounded, petiole not seen, midrib up to 1.5 mm wide and with 4–5 longitudinal strands at base, but strands fewer towards apex. Lateral veins arising from midrib at an acute angle of about 70° and almost immediately thereafter arching outwards to run parallel and nearly at right angles to midrib. Lateral veins up to 130 μm thick, generally once forked anywhere in lamina except near the margins. Concentration of veins 15–25 per cm, often lower near midrib; ends of lateral veins curved towards apex, often joined to form a marginal strand, about 70 μm thick; areas between lateral veins showing 4–12 gland-like bodies.

Upper epidermis of lamina nonstomatiferous. Cells in areas between veins polygonoid, (42)63 (–85) μm long × (16–)31(–46) μm wide, sides of cells sinuous, about 3 μm thick, amplitude of sinuosities nearly 6 μm, wave length about 5 μm.

Lower epidermis of lamina stomatiferous, cells between veins polygonoid, arranged in different directions usually (39–)61(–82) μm long × (20–) 33(–46) μm wide. Cell walls more sinuous than those in upper, amplitude of sinuosities about 8 μm, wave length about 7 μm.

Stomata irregularly dispersed and typically confined to areas between veins. Stomata haplocheilic; stomatal frequency 80 per sq mm, index 11, subsidiary cells 4–6, like ordinary epidermal cells but surface papillae directed towards stomatal pore. Polar and lateral subsidiaries not differentiated, guard cells sunken generally 46 μm long × 24 μm wide, outer walls of guard cells thickened, stomatal pore 20 μm long × 6 μm wide.

Cells over veins and strands of midrib in lower and upper epidermis arranged in rows parallel to veins, cells tending to be narrow, rectangular and straight walled, cells over meshes between strands of midrib polygonal and less regularly arranged, sides of cells slightly sinous, stomata rarely present.

Holotype: Specimen No. 33965 of Birbal Sahni Institute of Palaeobotany, Lucknow.

Selected Duplicates: 40436, 51129, 40435, 51128, 51130, 50067, of Divya Darshan Pant

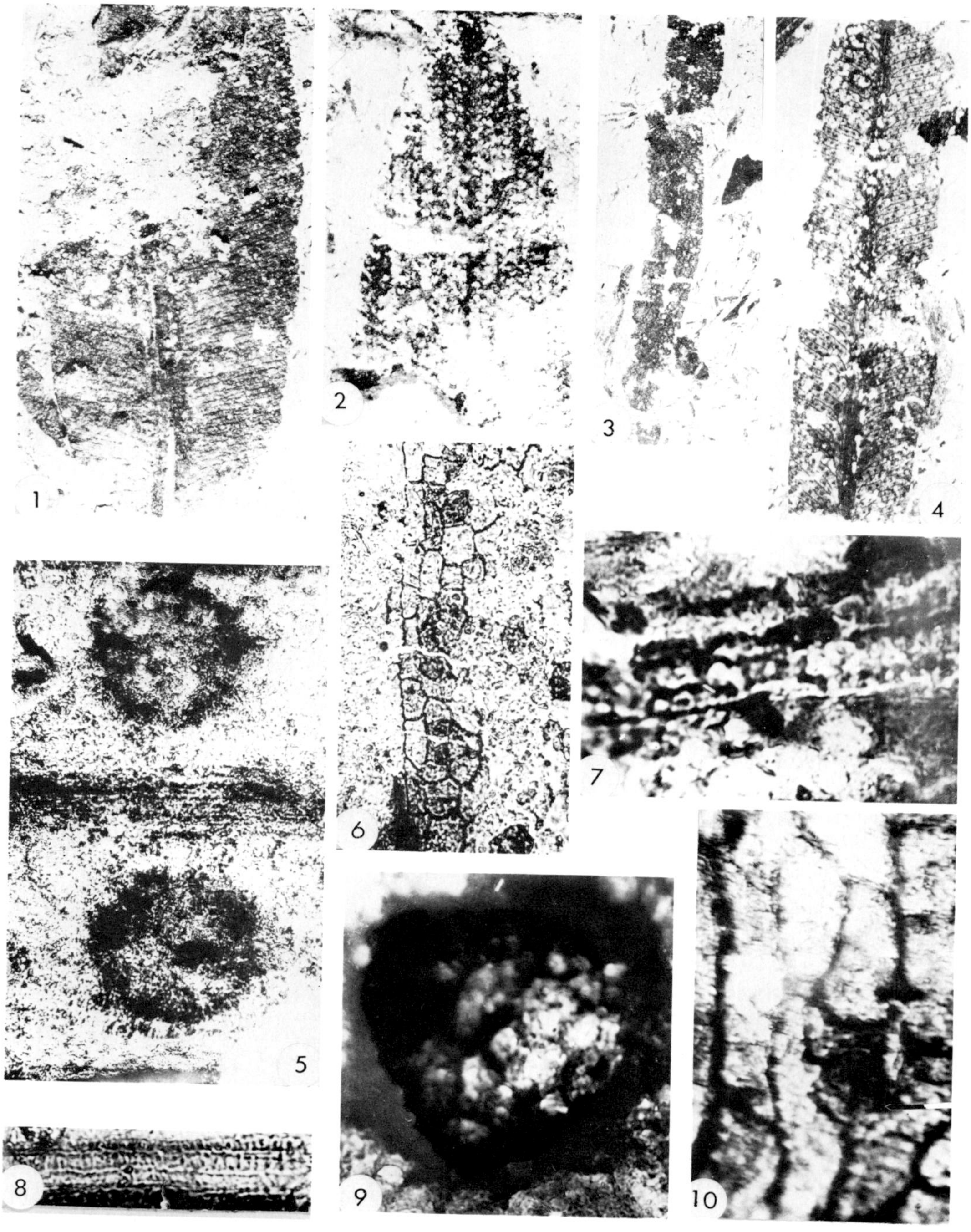

Plate I. *Glandulataenia glandulata* (Srivastava), comb. nov. **1.** Basal fragment of a leaf showing venation. Sp. No. 40436, ×3.5. **2.** Apical fragment of a leaf. Sp. No. 51129, ×3. **3, 4.** Fragments of leaves showing glands between veins. **3.** Sp. No. 51128, ×15. **4.** Sp. 51131, ×3. **5.** Two interstitial glands and an intervening vein. Sl. No. 51172a, ×150. **6.** Pull of epidermis from midrib showing a non-stomatiferous area over a vein and stomatiferous areas on either side. Sl. No. 511716, ×400. **7.** Pull showing midrib tracheids. Sl. No. 51174b, ×500. **8.** Pull showing scalariform tracheids of a side vein. Sl. No. 51175c, ×460. **9.** A partially macerated gland at high magnification to show details. Sl. No. 51170, ×325. **10.** Pull from midrib showing a stoma with remnants of polar lamellae. Sl. No. 51173, ×400.

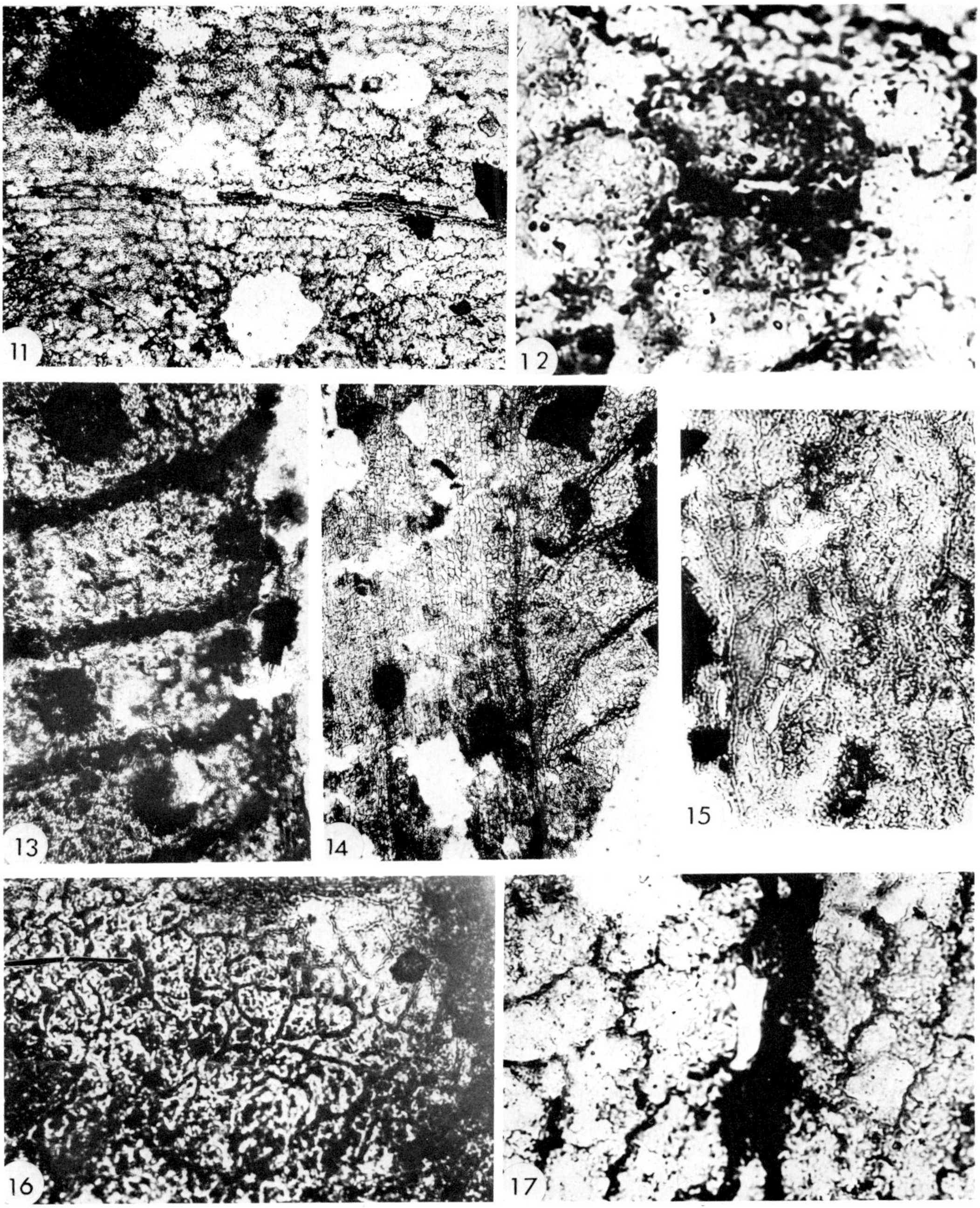

Plate II. *Glandulataenia glandulata* (Srivastava), comb. nov. **11.** Pull showing sinuous-walled cells of upper epidermis of lamina, a vein area and a dark rounded gland and holes left by other glands between the veins. Sl. No. 51170e, ×86. **12.** Pull from lower epidermis showing a single enlarged stoma and outlines of its guard cells. Sl. No. 51121a, ×500. **13.** Magnified pull of leaf margin to show upcurved vein ends and glands. Sl. No. 51138, ×750. **14.** Pull of a leaf showing wide midrib with glands and a portion of side veins on right side. Sl. No. 51137, ×35. **15.** Fragment of lower cuticle of lamina showing cells with rounded papillae and stomata surrounded by dark subsidiaries. Sl. No. 51145, ×200. **16.** Pull from lower epidermis showing a few stomata surrounded by dark subsidiaries. Sl. No. 51133, ×225. **17.** Lower epidermis of leaf showing sinuous-walled epidermal cells and a dumbbell-shaped dark colored protected stoma (s). Sl. No. 51170b, ×520.

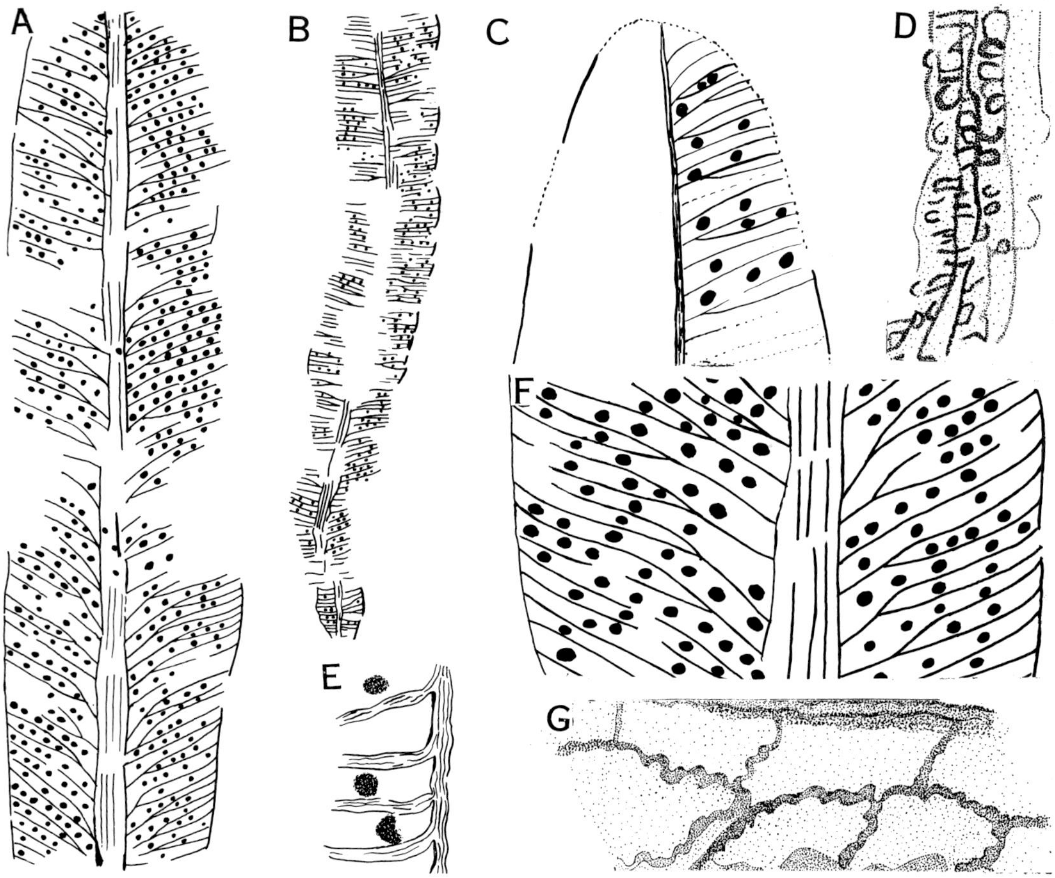

Fig. 1. *Glandulataenia glandulata* (Srivastava), comb. nov. **A.** Fragment of a leaf showing venation and distribution of interstitial glands. Sp. 51131, ×2. **B.** Fragment of a leaf showing venation and distribution of glands. 51128, ×1. **C.** Magnified apical portion of leaf showing veins and interstitial glands. 51129, ×2. **D.** Pull from midrib showing scalariform thickenings. Sl. 51174, ×225. **E.** A portion of leaf margin showing upcurved vein ends seemingly forming a marginal vein. Sl. 40033, ×62. **F.** A portion of **A** further enlarged to show glands and veins. Sp. 51131, ×4.5. **G.** Pull showing sinuous-walled cells from upper epidermis. Sl. 40033, ×150.

Collection, Botany Department, Allahabad University.

Locality and Horizon: Nidhpuri, Sidhi District, Madhya Pradesh, India, Triassic (Middle Gondwana).

Description, Comparison and Discussion

The emended diagnosis of the species is based on the account of the type material given by Srivastava (1971) and ten duplicate fragments of leaves including one base, two apices, and seven pieces of middle regions collected from the type locality and investigated in the present work.

The width of the middle regions ranges from 0.5 to 1.2 cm. Forking of lateral veins may take place anywhere in the width of the lamina. The maximum frequency of forked veins was observed in the regions about halfway between the midrib and margins. No anastomoses between the midrib strands or those of lateral veins could be seen. Although translucent pulls with intact margins also show dark marginal strands joining the ends of the lateral veins, their component cells or their wall thickenings could not be seen. Accordingly, it is not possible to say whether these marginal strands are made up of xylem or fibers.

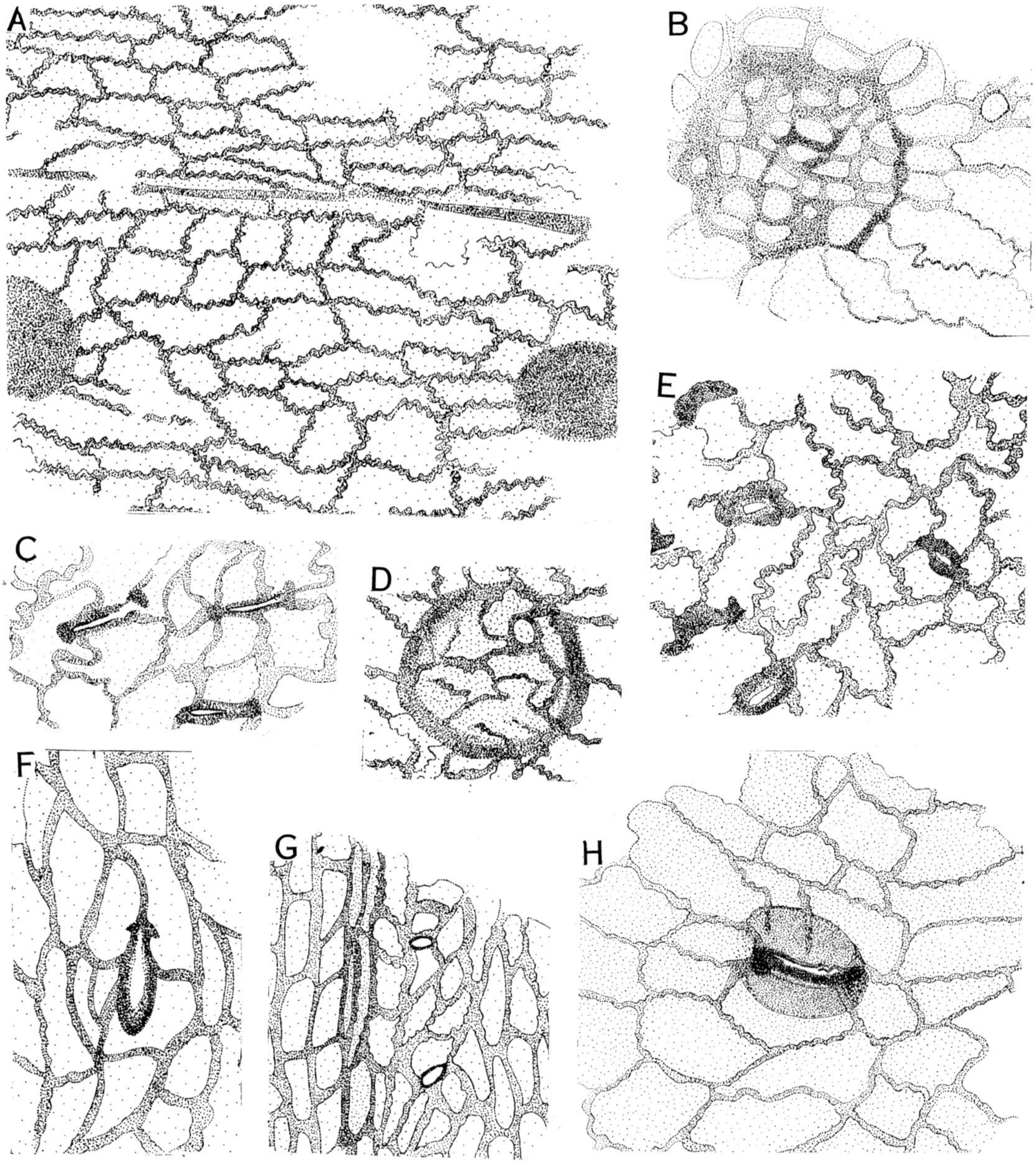

Fig. 2. *Glandulataenia glandulata* (Srivastava), comb. nov. **A.** Pull from upper epidermis of lamina showing two glands. Sl. 51170e, ×75. **B.** Pull from lamina showing cells of gland. Sl. 51172, ×150. **C.** Lower epidermis of lamina showing sinuous-walled cells and three stomata. Sl. 51176, ×150. **D.** A gland from a pull showing a small circular cell (stalk cell). Sl. 51170, ×150. **E.** Lower cuticle of lamina showing stomata and sinuous-walled cells. Sl. 51175, ×150. **F.** Pull from midrib showing a stoma with remnants of polar lamellae. Sl. 51173, ×150. **G.** Pull of epidermis from midrib showing vein area and stomatiferous area. Sl. 51171a, ×75. **H.** Pull from lower epidermis showing a stoma and outline of guard cells. Sl. 51171b, ×150.

In the absence of regular anastomoses between side veins, like those found in *Glossopteris* and *Gangamopteris,* the presence of such marginal strands may account for the rarity of torn laminae.

Pulls from different portions of leaves show that their substance has generally become more or less naturally macerated and the mesophyll has wholly or partially decayed out. However, the xylem of the veins is usually well preserved and often shows scalariform tracheids.

A row of dark oval or rounded patches about 250 μm in diam. are characteristically present in each area between lateral veins. Circumferential walls of the cells at the periphery of the patches are thin. I presume that the patches represent a glandular tissue, since they are filled with dark contents. Below the glands, at a lower focus, the cell outline of the epidermal cells can be seen to be continuous. A small circular area (cell) is also visible under some of the glands, between their glandular head and the underlying epidermal cells. This circular area may be compared with the stalk of the discoid glands of some Convolvulaceae (see Pant & Banerji, 1965) and therefore I believe that the glands were discoid and had a short one celled stalk.

Glandulataenia triassicus, sp. nov. (Pls. III, IV, Figs. 18–30; Text-figs. 3A–F; 4A–G; 5A–I)

DIAGNOSIS

Leaf large, petiolate, elongated, oval; lamina gradually tapering towards apex and base, widest (up to 2.5 cm) in the middle region, length unknown (estimated upper limit about 15 cm) apex obtuse, margin entire, petiole and midrib up to 1.5 mm wide near base, gradually narrowing towards apex. Midrib showing up to seven parallel longitudinal strands at base, strands fewer towards apex. Lateral veins arising from midrib at an acute angle of about 60° and thereafter curving outwards to run almost at right angles to midrib and parallel to each other. Lateral veins up to 118 μm thick at base, generally forked only once anywhere in lamina, but not near margin. Concentration of veins 11–28 per cm, but often less near midrib; ends of lateral veins curved upwards (toward apex) and appearing joined to form a marginal strand, up to 15 gland-like bodies present between laterals.

Upper epidermis of lamina non-stomatiferous, cells in areas between veins, irregularly arranged, polygonoid and elongated in various directions, (60–)73(–95) μm × (35–)40(–45) μm wide; surface of cells appearing finely mottled and side walls slightly uneven due to presence of pits, sides of cells almost straight, about 3 μm thick. Each area between veins showing a row of circular or oval areas (glands) of cells with dark contents. Cells over veins narrower and longer.

Upper epidermal cells over midrib tending to be arranged in longitudinal rows, cells over strands (veins) of midrib narrow and rectanguloid, (75–)87(–100) μm long × (30–)33(–36) μm wide, but epidermal cells in areas between strands of midrib trianguloid to polygonoid, (28–)35 (–48) μm long × (20–)24(–28) μm wide; anticlinal walls of cells straight, periclinal and anticlinal walls of cells thicker than those of cells of lamina, up to 5 μm thick.

Upper cuticle of lamina like upper epidermis, but cell walls over areas between veins often obscure, thinner. Visible portion of cell walls appearing slightly undulate and thinner than those in epidermis, about 203 μm thick, papillae rarely present.

Lower epidermis of lamina stomatiferous, arrangement and form of cells between veins like that in upper epidermis, about (55–)65(–80) μm long × (30–)35(–40) μm wide; cell surface often papillate, anticlinal walls 2–3 μm thick, slightly undulate, amplitude of sinuosities nearly 2 μm, wavelength about 1 μm, but cells near midrib and in angles between forks of side veins tending to be straight. Stomata usually present in areas between veins, irregularly oriented, stomatal frequency about 70 per sq mm, index 2. Stomata haplocheilic, monocyclic, subsidiary cells 4–7, obscurely papillate but otherwise irregular and like ordinary epidermal cells, subsidiaries not forming a ring, polars not differentiated from laterals, guard cell outlines obscure. Stomatal pore 2–6 μm wide; lower epidermal cells over veins narrow and elongated in direction of veins. Glands absent in lower epidermis.

Lower epidermis of midrib showing alternating bands of narrow and wide straight-walled cells, respectively, in areas above and between midrib strands as in upper, cells occasionally papillate, stomata rarely present in areas between strands.

Lower cuticle of lamina like lower epidermis,

Plate III. *Glandulataenia triassicus,* sp. nov. **18.** Leaf with base. Sp. No. 51125, ×1. **19.** Fragment of a leaf showing venation and fine dot-like glands (holotype). Sp. No. 51145. **20.** Portion of lower cuticle of lamina showing papillate cells and stomata. One of them showing outlines of guard cells. Sl. No. 51145b, ×500. **21.** Fragment of leaves with the midvein and faint outlines of side veins and glands. Sp. No. 51148, ×1.5. **22.** Gland showing details. Sl. No. 51138, ×225. **23.** Portion of midrib showing pitted tracheids. Sl. No. 51132, ×900. **24.** Magnified view of a stoma. Sl. No. 51148, ×320.

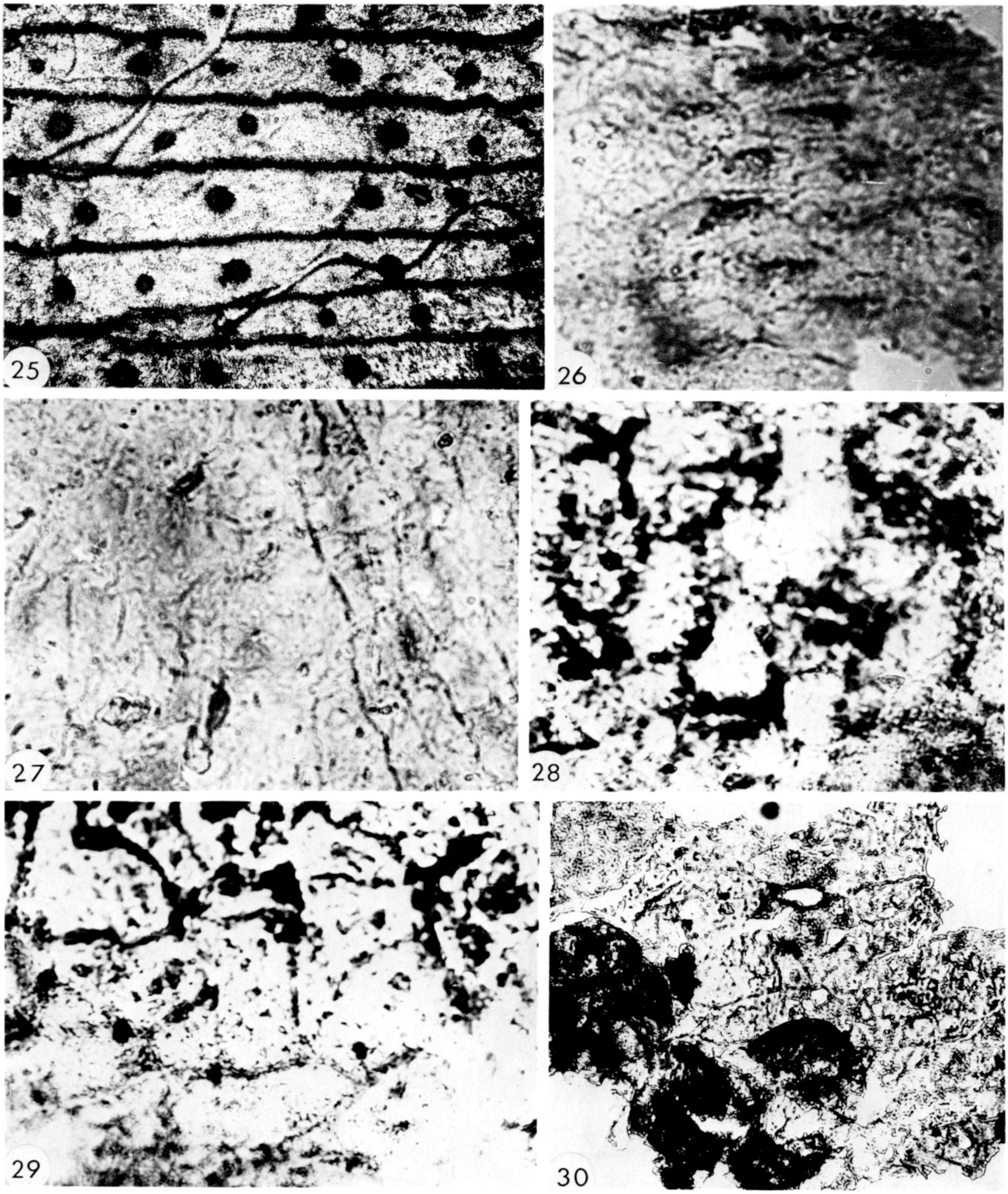

Plate IV. *Glandulataenia triassicus,* sp. nov. **25.** Magnified view of leaf showing distribution of interstitial glands. Sp. No. 51145, ×34. **26, 27.** Portions of the upper cuticle of lamina showing cell outlines and surface papillae. **27** also shows impressions of side veins. Sl. No. 51145, ×225. **28.** Pull from lower epidermis showing stomata with common subsidiary cells. Sl. No. 51133, ×500. **29.** A stoma in Figure 28 more magnified, ×700. **30.** Fragment of lower cuticle of lamina showing papillate cells. Portions of a gland and a stomatal pore. Sl. No. 51145, ×200.

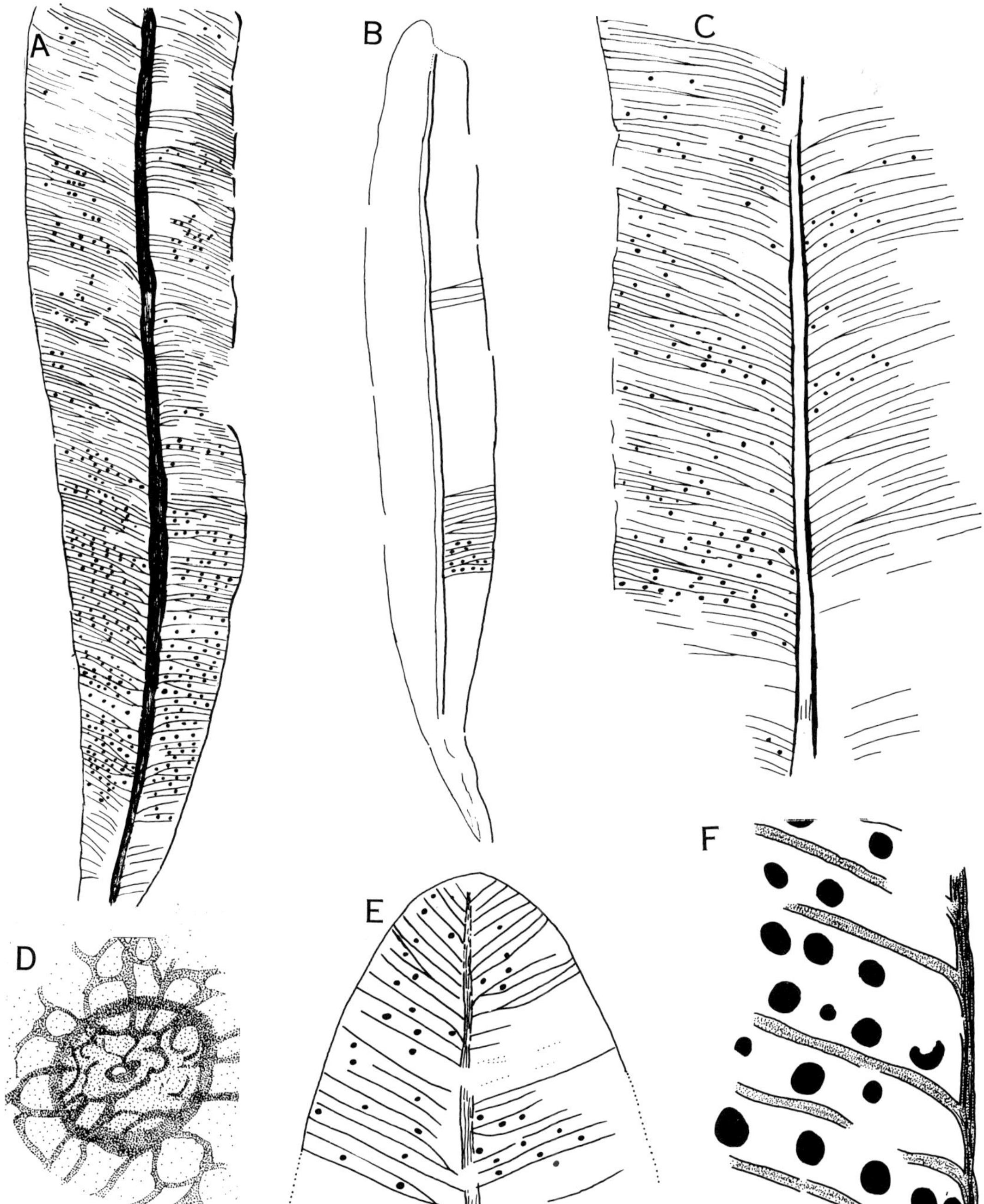

Fig. 3. *Glandulataenia triassicus,* sp. nov. **A.** A fragment of leaf showing interstitial glands and veins. Sp. 51145, ×1.12. **B.** An almost complete leaf with petiolate base. Sp. 51124, ×0.75. **C.** Fragment of a leaf showing veins and gland. Sl. 51128, ×1.12. **D.** Details of a gland. Sp. 51138a, ×75. **E.** Apical portion of a leaf showing venation and glands. Sp. 51135, ×5. **F.** Fragment of a leaf magnified at margin to show details. Sp. 51138, ×0.75.

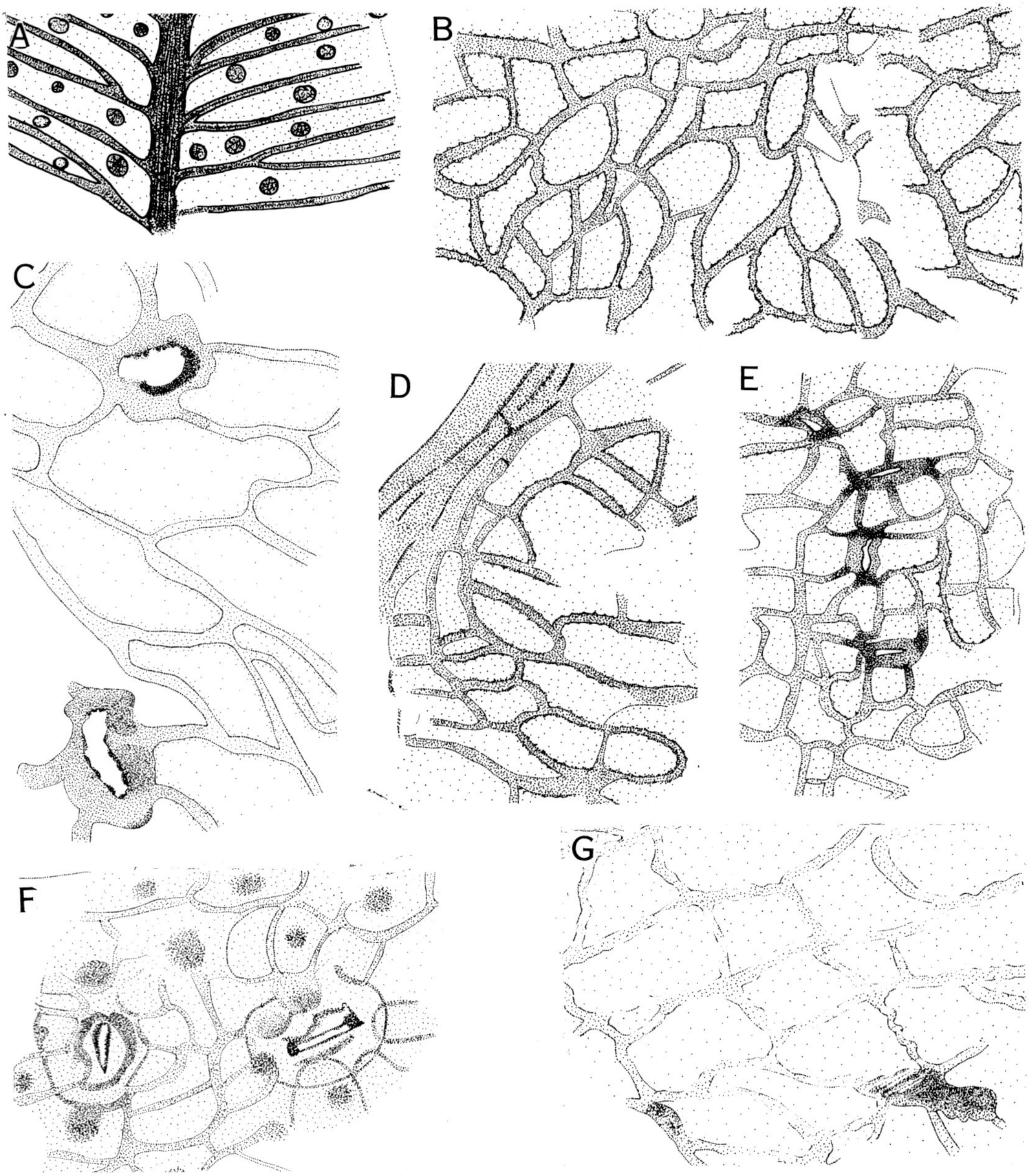

Fig. 4. *Glandulataenia triassicus,* sp. nov. **A.** Fragment of a leaf showing striated midrib, a few forked veins and interstitial glands. 51135, ×7.5. **B.** Upper epidermis showing straight-walled pitted cells. 51133, ×150. **C.** Lower epidermis showing two stomata. 51134c, ×225. **D.** Upper epidermis showing cells over side vein and straight-walled, pitted cells between them. 51133, ×150. **E.** Lower epidermis showing straight-walled cells and irregularly placed stomata in between veins. 51138, ×112. **F.** Lower cuticle of lamina showing papillate cells and stomata, one of the stomata showing outlines of guard cells. 51145b, ×150. **G.** Upper cuticle of lamina. 51145, ×250.

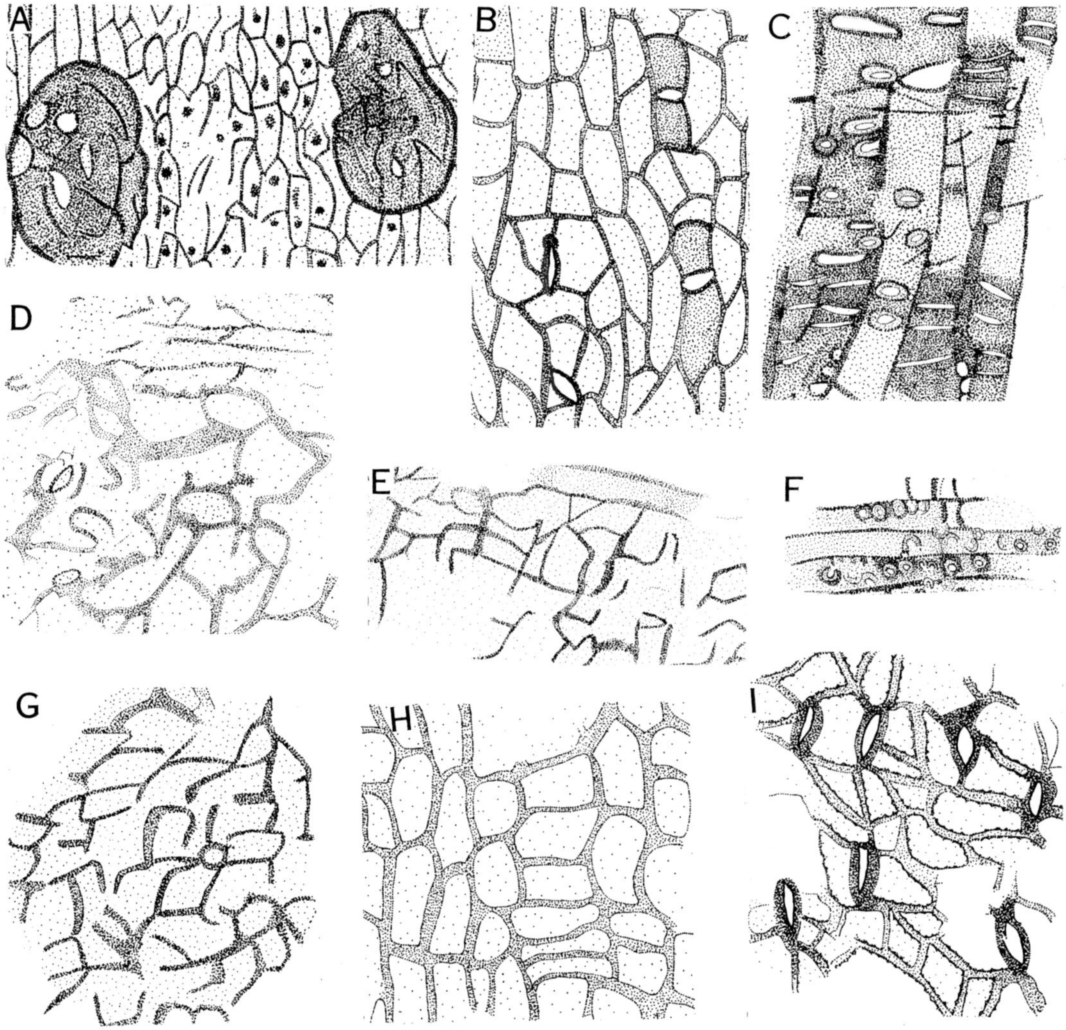

Fig. 5. *Glandulataenia triassicus* sp. nov. **A.** Two glands and straight-walled epidermal cells with median papillae over midrib. Sl. 51137 ×150. **B.** Midrib epidermis with stomata. Sl. 51137 ×225. **C.** Scalariform and pitted tracheids of midrib. Sl. 51132 ×600. **D.** An epidermal portion over a side vein with narrowly elongated cells and elsewhere broken down cells over a gland. Sl. 51138 ×225. **E.** Mesophyll. Sl. 51138 ×125. **F.** Side vein tracheids. Sl. 51141 ×600. **G.** Cells of a gland. Sl. 51138 ×225. **H.** Midrib epidermis with narrow cells over the vein and wide cells over a mesh area. Sl. 51136 ×225. **I.** Lower epidermis of lamina with stomata. Sl. 51133 ×225.

but often showing clear outlines of guard cells and papillate subsidiaries.

Holotype: No. 51145 of Divya Darshan Pant Collection, Botany Department, Allahabad University, India.

Locality and Horizon: Nidhpuri, Sidhi District, Madhya Pradesh, India, Triassic (Middle Gondwana).

Description, Comparison and Discussion

The species is based on six fragments from middle regions of leaves and two apices. In addition there are two almost complete leaves in our collection but a small part of the apex of one of these is broken. The veins fork in the same way as those of *G. glandulata* but forked veins are more frequent on the two sides of midrib, half way between it and margin and they seldom fork near margins. The midrib of *G. triassicus* is also made up of a number of parallel, longitudinal strands like that of *G. glandulata.*

Leaves of *G. triassicus* appear to be very slightly larger than those of *G. glandulata.* The form of the leaf base in *G. triassicus* was always found to be tapering but that of *G. glandulata* was sometimes rounded. The concentration of veins in *G. triassicus* is 11–28 per cm (it is 15–25 per cm in *G. glandulata*).

The structure of glands in *G. triassicus* is just like that of *G. glandulata,* though their number in each vein mesh is slightly higher (up to 15) in *G. triassicus,* whereas they are only up to 12 in *G. glandulata.* Further, the epidermal cells of *G. glandulata* are distinctly sinuous-walled, with the amplitude and wave lengths of their sinuosities being 8 μm and 7 μm, respectively, while those of *G. triassicus* are almost straight-walled and the amplitude and wave lengths of their sinuosities are, respectively, 2 μm and 1 μm. Most of the cells of the lower epidermis of *G. triassicus* have prominent papillae but these are either rare or obscure in *G. glandulata.*

The presence of interstitial glands in *G. triassicus* and *G. glandulata* may also suggest a comparison with other *Taeniopteris* leaves showing similar interstitial dots. Two such leaves were described by Halle (1927), from the Paleozoic of Shanshi, China. Halle (1927, pl. 34, fig. I), also showed the presence of similar bodies in the type specimen of *T. multinervis* (Weiss) from the Permian of Autum, Saxony. Halle thought that the dots were probably glandular. The details of attachment of the lamina to the midrib and epidermal structure of leaves in *Taeniopteris multinervis* is unknown, its glands are described as situated inside the substance of the lamina (see Remy & Remy, 1977).

A middle Triassic leaf, *Taeniopteris dunstanii,* described by Walkom (1917) from the Ipswich Coal Measures, is described as having pits over the midrib. I have examined the type specimen of *T. dunstanii* (No. F.372 in the Geological Survey of Queensland) and found that the leaf is only an impression without any carbon, but it shows pits like those of *Glandulataenia* between side veins and over the midrib. I presume that it could have had similar glands. Obviously, *T. dunstanii* should be remaned as *Glandulataenia dunstanii* comb. nov.

Lately, Mamay (1976) has described, from the Lower Permian rocks of Kansas, some laminar fragments of *Taeniopteris* species which have spherical, intercostal, resinoid, gland-like bodies. These occur in association with another fossil described as *Phasmatocycas kansana* which has an axis with a two ranked arrangement of seeds and similar spherules on the axis. Mamay has joined the two spherulate fossils in his reconstruction of *Phasmatocycas* (see Mamay, 1976: fig. 4) which is regarded as a primitive cycadean megasporophyll. He thinks that the glands on the fossil may have served to attract animals (insects) for pollination of ovules. Even earlier, Halle (1927) and before him Oliver and Scott (1904) had thought that the glands present in Palaeozoic pteridospermous foliage and cupules like those of *Lagenostoma* could have served to attract insects. The glands in the present Triassic leaves could have had a similar function. However, at Nidhpuri they are neither found attached nor associated with ovule-bearing axes of the *Phasmatocycas* type, and this may raise doubts about Mamay's artistically designed reconstruction of the megasporophyll although the recent find of connected portions of taeniopterid laminae and *Spermopteris*-like ovule-bearing fragments attributed to *Phasmatocycas* (see Gillespie & Pfefferkorn, 1986) has been suggested to confirm Mamay's reconstruction of *Phasmatocycas.* However, the specimens described by Gillespie and Pfefferkorn (1986) seem to differ radically from the reconstruction of *Phasmatocycas* made by Mamay (1976)—the taeniopterid lamina of the specimens attributed to *Phasmatocycas* by Gillespie and Pfefferkorn (1986) has its ovules attached to the midrib of the megasporophylls on its abaxial side (cf. Gillespie & Pfefferkorn, 1986: 103, pl. 1, figs. 1, 2 and 4 and reconstructions of megasporophylls in fig. 1), whereas the reconstruction made by Mamay shows ovules attached to the nonlaminar part of the megasporophyll. In fact the attribution of taeniopterid megasporophylls bearing ovules on the abaxial side of their midrib to a cycad becomes doubtful since the megasporophylls of *Cycas* lack laminae in their fertile middle part. Indeed the presently described leaves of *Glandulataenia* seem to indicate that sterile leaves of *Glandulataenia* reported here from the Triassic of India seem to be very different from the two kinds of *Phasmatocycas* described from the Permian of Kansas, one reconstructed by Mamay (1976) and the other by Gillespie and Pfefferkorn (1986). The latter with its midrib attached seeds would seem to suggest, though somewhat remotely, the Glossopteridales instead of the cycads.

A point which may not support the attractive function of glands in the reconstructed megasporophyll is the presence of foliar glands in *Macrozamia* spp., which have so far never been reported to attract insects.

There are also some other taeniopterid leaves with dark dots which are seemingly different. For example, a Lower Cretaceous leaf described by Douglas (1973) from Victoria under the name *Taeniopteris daintreei* also seems to show similar dark bodies between the veins and sometimes above them. The structure of these rather irregularly dispersed dots is unknown but Douglas regards them as fungal bodies. In contrast, the regular form and distribution of the interstitial dots and their structure in my leaves suggest that they are glands on the upper surface.

Presence of the dark interstitial dots in leaves of *Glandulataenia glandulata* and *G. triassicus* may also suggest a comparison with *Nilssonia tenuinervis* Seward, but the dots of this leaf are regarded as resin bodies (see Harris, 1964) and these leaves, in addition, differ from *Glandulataenia* in their generic features.

Acknowledgments

The author is thankful to the Government of India, Department of Science & Technology for financing the project under which this work was completed. He is thankful to Dr. D. D. Nautiyal for useful suggestions. He thanks Dr. D. N. Misra, Dr. D. K. Chauhan and Mr. B. M. Mamgai for technical assistance in preparing slides, photographs and figures and to Dr. Rita Singh for help in checking the manuscript.

Literature Cited

Chandra, S. 1974. *Glossopteris* and allied genera, morphological studies. Pages 126–146 *in* K. R. Surange et al. (eds.), Aspects and appraisal of Indian palaeobotany. Birbal Sahni Institute of Palaeobotany, Lucknow.

Douglas, J. G. 1973. The Mesozoic flora of Victoria. Part 3. Geol. Surv. Victoria Mem. **29:** 18–23.

Florin, R. 1933. Studien über die Cycadales des Mesozoikums. Kungl. Svenska Ventenskapsakad. Handl. III **12:** 1–134.

Gillespie, W. H. & H. W. Pfefferkorn. 1986. Taeniopterid lamina on *Phasmatocycas* megasporophylls (Cycadales) from the Lower Permian of Kansas, U.S.A. Rev. Palaeobot. Palynol. **49:** 99–116.

Halle, T. G. 1927. Palaeozoic plants from Central Shanshi. Palaeontol. Sinica, Ser. A **2(i):** 5–217.

Harris, T. M. 1932. The fossil flora of Scoresby Sound, East Greenland. Part 2. Description of seed plant *insertae sedis* together with a discussion of certain cycadophyte cuticles. Meddel. Grønland **85(3):** 1–112.

———. 1964. The Yorkshire Jurassic flora II, Caytoniales, Cycadales and Pteridosperms. British Museum (Nat. Hist.) London.

Mamay, S. H. 1976. Palaeozoic origin of cycads. U.S. Geol. Survey Prof. Paper No. 934.

Oliver, F. W. & D. H. Scott. 1904. On the structure of the Paleozoic seed, *Lagenostoma lomaxi,* with a statement of the evidence upon which it is referred to *Lyginodendron* (abstract). Ann. Bot. **18:** 321–322.

Pant, D. D. 1958. The structure of some leaves and fructifications of the Glossopteris flora from Tanganyika. Bull. Brit. Mus. (Nat. Hist.) London Geol. **3(4):** 127–175.

——— **& R. Banerji.** 1965. Epidermal structure and development of stomata in some Convolvulaceae. Senckenberg. biol. **46(2):** 155–173.

——— **& B. K. Verma.** 1963. On the structure of leaves of *Rhabdotaenia* Pant from the Raniganj coal field, India. Palaeontology, London **6(2):** 301–314.

Remy, W. & R. Remy. 1977. Die Floren des Erdalterums. Einführung in Morphologie, Anatomie, Geobotanik und Biostratigraphie der Pflanzen des Paläophytikums, Gluckauf GmbH, Essen.

Sahni, B. 1948. The Pentoxyleae: A new group of Jurassic gymnosperms from the Rajmahal Hills of India. Bot. Gaz. **110:** 47–80.

Srivastava, S. C. 1971. Some gymnospermic remains from the Triassic of Nidpur, Sidhi District, Madhya Pradesh. Palaeobotanist **18(3):** 280–296.

———. 1976. Some macro-plant fossils from the Triassic rocks of Nidpur, India. Palaeobotanist **23(1):** 44–48.

Walkom, A. B. 1917. Mesozoic flora of Queensland. Part I (concluded) (d) Ginkgoales (e) Cycadophyta (f) Coniferales. Publ. Geol. Surv. Queensland **259:** 1–30.

A World List of Cycads

DENNIS WM. STEVENSON, ROY OSBORNE,
AND JOHN HENDRICKS

Abstract

STEVENSON, D. W. (New York Botanical Garden, Bronx, New York 10458, USA), R. OSBORNE (University of Natal, King George V Avenue, Durban 4001, South Africa), and J. HENDRICKS (110 Brookmeade Avenue, Statesville, North Carolina 28677, USA). A world list of cycads. Memoirs of the New York Botanical Garden **57**: 200–206. 1990.—A list of cycad genera, species, and infraspecific taxa is given together with a general indication of their distribution. Brief comments are given about the status of the taxonomy of each genus. Those genera in the most need of taxonomic revision are indicated.

Introduction

Taxonomy is that discipline of botany which provides a rational basis for the naming of plants. It is an objective science in that set procedures work toward developing a meaningful systematic genealogy. However, problems relating to the selection of criteria of rank result in subjective judgment and opinion; this is especially true in the order Cycadales.

Most cycad species names originated from studies of isolated plants or habitats. When descriptions of these species were later compared on a broader regional basis, many appeared to be synonymous or less differentiated than others, thus names and classifications had to be reassigned. This process may be repeated as often as justified.

Apart from the as yet unresolved confusion of the early literature over the identification of some taxa, there are also problems of judgment and of opinion. In addition, there are known but undescribed new species and there are difficulties arising from situations where botanically invalid names have been applied to taxa by collectors that differ from or have not been synonymized with valid names of taxa. Another important factor is that some cycad habitats have yet to be fully explored and the plants properly classified. Fortunately, *Dioon, Encephalartos,* and *Macrozamia* are largely defined, although some changes can be anticipated. Only *Bowenia, Ceratozamia, Lepidozamia, Microcycas,* and *Stangeria* are considered taxonomically complete at this time.

Major advances have recently been made in *Zamia* taxonomy by information from chromosome karyotyping (Moretti & Sabato, 1984; Norstog, 1980). Biochemical studies also are being used to define genera and give promise of differentiation at lower taxonomic rank (Stevenson & Gigliano, 1989). Other modern analytical techniques may have application in future years. Taxonomy in its purest form reflects evolution; the best advance will be made by combining the knowledge from the many relevant interacting scientific disciplines.

Our list of validly named cycad species was first published in 1985 (Osborne & Hendricks, 1985), and a supplement followed in 1986 (Osborne & Hendricks, 1986). The following list is a further update, in which several changes have been made on the basis of recent information. It itemizes the currently valid taxonomic ranking at or below the species level to the best of our existing knowledge. It is not a listing of all validly published species, nor does it represent our opinion as to 'good' species or suitable classification in all cases. Indeed, our opinions and those of many others are at considerable variance from this list, particularly in assignments in the genera *Cycas* and *Zamia*.

Work is proceeding toward an improved cycad taxonomy, though perhaps neither of the scope nor at the rate which might be desired. In the interim, it is recommended that labelling of specimen plants be supplemented with any additional appropriate information.

World List of Genera and Species of Cycads

BOWENIA Hook. ex Hook. f. (2 species)	Australia
B. serrulata (W. Bull) Chamberlain	Queensland
B. spectabilis Hook. ex Hook. f.	Queensland
CERATOZAMIA Brongn. (10 species)	Mexico; Guatemala; Belize
C. euryphyllidia V-Torres, Sabato & D. Stevenson	Veracruz, Mexico
C. hildae G. Landry & M. Wilson	San Luis Potosí & Querétaro, Mexico
C. kuesteriana Regel	Tamaulipas, Mexico
C. latifolia Miq.	Hidalgo, San Luis Potosí, & Querétaro, Mexico
C. matudae Lundell	Chiapas, Mexico; N. Guatemala
C. mexicana Brongn.	Pueblo & Veracruz, Mexico
C. miqueliana H. Wendl.	Veracruz, Mexico
C. norstogii D. Stevenson	Chiapas, Mexico
C. robusta Miq.	S. Mexico; Guatemala; Belize
C. zaragozae Medellin-Leal	San Luis Potosí, Mexico

No changes are expected in *Ceratozamia*. For a recent treatment on the species concepts, ecology, and distribution of the genus see Stevenson et al. (1986).

CHIGUA D. Stevenson (2 species)	South America
C. bernalii D. Stevenson	Colombia
C. restrepoi D. Stevenson	Colombia
CYCAS L. (17 species)	
C. baguanheensis L. K. Fu & S. Z. Cheng	China
C. cairnsiana F. Muell.	Queensland, Australia
C. calcicola J. R. Maconochie	N. Territory, Australia
C. chevalieri Leandri	Vietnam
C. circinalis L.	Asia, Africa
subsp. *circinalis*	
forma *circinalis*	
forma *glauca* (Miq.) Schuster	
forma *gothanii* Schuster	
forma *undulata* (Desf.) Schuster	
subsp. *madagascariensis* (Miq.) Schuster	Madagascar
forma *madagascariensis*	
forma *trigonocarpoides* Schuster	
subsp. *papuana* (F. Muell) Schuster	New Guinea

subsp. *ruiminiana* (Porte) Schuster	Philippines
var. *ruiminiana*	
var. *curranii* Schuster	
forma *curranii*	
forma *apertorum* Schuster	
forma *chamberlainii* W. H. Brown & Keinholz	
forma *maritima* Schuster	
var. *scratchleyana* (F. Muell.) Schuster	New Guinea
subsp. *vera* Schuster	
var. *vera*	
var. *beddomei* Dyer	India
var. *pectinata* (W. Griff.) Schuster	S.E. Asia
C. guizhouensis K. Lan & R. Zou	China
C. hainanensis C. J. Chen & C. Y. Chen	Hainan Island, China
C. media R. Br.	
var. *media*	N. Territory, Australia
var. *basaltica* (C. A. Gardn.) Schuster	Western Australia
var. *furfuracea* (W. V. Fitzg.) Schuster	Western Australia
var. *lane-poolei* (C. A. Gardn.) Schuster	Western Australia
C. micholitzii Dyer	
var. *micholitzii*	Vietnam; China
var. *simplicipinna* Smitinand	S.E. Asia
C. nathorstii Schuster	Sri Lanka
C. panzhihuaensis L. Zhou & S. Y. Yang	China
C. pruinosa J. R. Maconochie	Western Australia
C. revoluta Thunb.	China; Japan; Ryukyu Is.
var. *revoluta*	
var. *robusta*	
var. *taiwaniana* (Carruth.) Schuster	China; Taiwan
C. rumphii Miq.	S.E. Asia; Pacific Is.
subsp. *rumphii*	
var. *rumphii*	
var. *subinclusa* Schuster	
forma *subinclusa*	
forma *palauica* Kaneh	Palau
forma *papuana* (F. Muell.) Kaneh	New Guinea
forma *seemannii* (A. Braun) Kaneh	Fiji
subsp. *normanbyana* (F. Muell.) Schuster	Queensland, Australia
subsp. *zeylanica* Schuster	Sri Lanka
C. szechuanensis C. Chen & L. K. Fu	China
C. siamensis Miq.	S.E. Asia
subsp. *siamensis*	
subsp. *balansae* (Warburg) Schuster	Vietnam
C. wadei Merrill	Philippines

The taxonomy of *Cycas* needs extensive revision, including assignment of species, subspecies, and varietal status. Several deletions and new insertions are to be expected. It should be noted that the nomenclature of *Cycas* is a complete mess and recent work on typification and valid publication of names has yet to be completed although Sabato and Stevenson are currently working on this aspect and expect to complete their work in the near future. The list presented here for *Cycas* is largely derived from Schuster (1932). However, Schuster's taxonomic concepts, for the most part, are vague and ill-defined at best. His concept of "forma" would appear to be applicable to cultivated plants, but this is by no means clear. The phytogeography of taxa within *Cycas* is very poorly known, thus we can not give the distribution for the lower ranks. One should not assume that these taxa are local endemics because in many cases they probably represent the morphological extremes. The paucity of collections in herbaria do not alleviate this gap in our knowledge, which will only be enhanced by extensive field work.

Taxon	Distribution
DIOON Lindley (10 species)	Mexico; Honduras; Nicaragua
D. califanoi De Luca & Sabato	Oaxaca, Mexico
D. caputoi De Luca, Sabato, & V-Torres	Puebla, Mexico
D. edule Lindley	
var. *edule*	Veracruz & San Luis Potosí, Mexico
var. *angustifolium* (Miq.) Miq.	Nuevo León, San Luis Potosí, & Tamaulipas, Mexico
D. holmgrenii De Luca, Sabato, & V-Torres	Oaxaca, Mexico
D. mejiae Standley & L. O. Williams	N. Honduras; Nicaragua
D. merolae De Luca, Sabato, & V-Torres	Chiapas, Mexico
D. purpusii Rose	Oaxaca, Mexico
D. rzedowskii De Luca, Moretti, Sabato, & V-Torres	Oaxaca, Mexico
D. spinulosum Dyer	Veracruz & Oaxaca, Mexico
D. tomasellii De Luca, Sabato, & V-Torres	
var. *tomaselli*	S.W. Coast, Mexico
var. *sonorense* De Luca, Sabato, & V-Torres	N.W. Coast, Mexico

The taxonomy of *Dioon* is fairly complete and no significant changes are anticipated. For a thorough treatment of the species and their ecology and distribution see Sabato and De Luca (1985). It should be noted that the spelling *Dioon* was recently conserved over that of *Dion* by the nomenclature committee at the XIV International Botanical Congress held in Berlin. Thus, the spelling of the generic name is now mandatory and stable.

Taxon	Distribution
ENCEPHALARTOS Lehmann (46 species)	Africa
E. altensteinii Lehm.	E. Cape & Transkei, South Africa
E. arenarius R. A. Dyer	E. Cape, South Africa
E. barteri Carruth. ex Miq.	
subsp. *barteri*	Benin; Ghana; Nigeria; Sudan; Togo
subsp. *allochrous* L. E. Newton	Nigeria
E. bubalinus Melville	Tanzania, Kenya
E. caffer (Thunb.) Lehm.	E. Cape, South Africa
E. cerinus Lavranos & Goode	Natal, South Africa
E. chimanimaniensis R. A. Dyer & Verdoorn	Mozambique; Zimbabwe
E. concinnus R. A. Dyer & Verdoon	Zimbabwe
E. cupidus R. A. Dyer	Transvaal, South Africa
E. cycadifolius (Jacq.) Lehm.	E. Cape, South Africa
E. dolomiticus Lavranos & Goode	E. Transvaal, South Africa
E. dyerianus Lavranos & Goode	E. Transvaal, South Africa
E. eugene-maraisii Verdoorn	
subsp. *eugene-maraisii*	W. Transvaal, South Africa
E. ferox Bertol f.	Zululand and N. Natal, South Africa; Mozambique

E. friderici-guilielmi Lehm.	E. Cape & Transkei, South Africa
E. ghellinckii Lehm.	Transkei and Natal, South Africa
E. gratus Prain	Malawi; Mozambique
E. heenanii R. A. Dyer	Swaziland & S.E. Transvaal, South Africa
E. hildebrandtii A. Braun & Bouche	
var. *hildebrandtii*	E. Africa
var. *dentatus* Melville	Tanzania; Uganda
E. horridus (Jacq.) Lehm.	E. Cape, South Africa
E. humilis Verdoorn	E. Transvaal, South Africa
E. inopinus R. A. Dyer	Transvaal, South Africa
E. ituriensis Bamps & Lisowski	Zaire
E. laevifolius Stapf & Burtt-Davy	E. Transvaal & Swaziland, South Africa
E. lanatus Stapf & Burtt-Davy	Transvaal, South Africa
E. latifrons Lehm.	E. Cape, South Africa
E. laurentianus De Wild.	Angola; Zaire
E. lebomboensis Verdoorn	N. Natal, S.E. Transvaal, Swaziland, South Africa; Mozambique
E. lehmannii Ecklon & Zehyer ex Lehm.	E. Cape, South Africa
E. longifolius (Jacq.) Lehm.	E. Cape, South Africa
E. manikensis Gilliland	Mozambique; Zimbabwe
E. marunguensis Devred	Zaire
E. middelburgensis Vorster et al.	Transvaal, South Africa
E. munchii R. A. Dyer & Verdoorn	Mozambique
E. natalensis R. A. Dyer & Verdoorn	Natal, South Africa
E. ngoyanus Verdoorn	Zululand, N. Natal, S.E. Transvaal, & Swaziland, South Africa
E. paucidentatus Stapf & Burtt-Davy	E. Transvaal & Swaziland, South Africa
E. poggei Ascherson	Angola; Zaire
E. princeps R. A. Dyer	E. Cape, South Africa; Zaire
E. pterogonus R. A. Dyer & Verdoorn	Mozambique
E. schmitzii Malaisse	Zaire; Zambia
E. sclavoi D. Stevenson, Moretti, & De Luca	Kenya
E. septentrionalis Schweinfurth	Zaire; Sudan; Uganda; Central Afr. Rep.
E. tegulaneus Melville	Kenya
E. transvenosus Stapf & Burtt-Davy	N. Transvaal, South Africa
E. trispinosus (Hook.) R. A. Dyer	E. Cape, South Africa
E. turneri Lavranos & Goode	Mozambique
E. umbeluziensis R. A. Dyer	Swaziland, South Africa; Mozambique
E. villosus Lem.	E. Cape, Natal, S.E. Transvaal, Transkei, Zululand, & Swaziland, South Africa; Mozambique
E. voiensis Sclavo, Moretti & D. Stevenson	Kenya
E. woodii Sander	Natal (Extinct in nature)

While little in the way of revision is expected, additions from Central Africa are certain.

LEPIDOZAMIA Regel (2 species)	Australia
L. hopei Regel	Queensland
L. peroffskyana Regel	New South Wales & Queensland
MACROZAMIA Miquel (14 species)	Australia
M. communis L. A. S. Johnson	New South Wales
M. diplomera (F. Muell.) L. A. S. Johnson	N.S.W.

M. fawcettii C. Moore	N.S.W.
M. heteromera C. Moore	N.S.W.
M. lucida L. A. S. Johnson	Queensland & N.S.W.
M. macdonnellii (F. Muell. ex Miq.) A. DC.	Central Australia
M. miquelii (F. Muell.) A. DC.	Queensland & N.S.W.
M. moorei F. Muell.	Queensland & N.S.W.
M. pauli-guilielmi W. Hill & F. Muell.	
subsp. *pauli-guilielmi*	Queensland
subsp. *flexuosa* (C. Moore) L. A. S. Johnson	N.S.W.
subsp. *plurinervia* L. A. S. Johnson	Queensland
M. riedlei (Gaud.) C. A. Gardn.	S.W. Australia
M. secunda C. Moore	N.S.W.
M. spiralis (Salisb.) Miq.	N.S.W.
M. stenomera L. A. S. Johnson	N.S.W.

Some minor revision and additions are to be expected. However, no basic departures from the revision of Johnson (1959) are expected.

MICROCYCAS A. DC. (1 species)	Cuba
M. calocoma (Miq.) A. DC.	W. Cuba
STANGERIA Th. Moore (1 species)	South Africa
S. eriopus (Kunze) Baillon	E. Cape, Natal, Transkei, & Zululand
ZAMIA L. (43 species)	Neotropics
Z. acuminata Oersted ex Dyer	Nicaragua; Costa Rica; N. Panama
Z. amblyphyllidia D. Stevenson	Puerto Rico; Cuba; Jamaica
Z. amplifolia Hort. ex Masters	Chóco, Colombia
Z. angustifolia Jacq.	Bahamas; Cuba
Z. boliviana (Brongn.) A. DC.	N. Bolivia
Z. chigua Seemann	Chóco, Colombia; Panama
Z. cremnophila Schutzman, Vovides & Dehgan	Tabasco, Mexico
Z. cupatiensis Ducke	Colombia; Brazil
Z. fairchildiana L. D. Gomez	Costa Rica; Panama
Z. fischeri Miq.	Querétaro, San Luis Potosí, & Veracruz, Mexico
Z. furfuracea L. fil. *in* Aiton	Veracruz, Mexico; Belize
Z. herrerae Calderon & Standley	El Salvador; Honduras; Guatemala; Chiapas, Mexico
Z. inermis Vovides, Rees & V-Torres	Veracruz, Mexico
Z. integrifolia L. fil. *in* Aiton	Florida & Georgia, N. America; Bahamas; Cuba; Caiman Is.
Z. jirijirmensis R. E. Schultes	E. Colombia
Z. lawsoniana Dyer	Oaxaca, Mexico
Z. lecointei Ducke	Amazonian Brazil & Venezuela
Z. lindenii Regel ex Andre	Ecuador
Z. lindleyi Warsz. ex A. Dietrich	Panama
Z. loddigesii Miq.	Mexico
Z. manicata Linden ex Regel	N. Colombia; S. Panama
Z. montana A. Br.	N. Colombia
Z. monticola Chamberlain	Veracruz, Mexico
Z. muricata Willd.	N. Venezuela
Z. obidensis Ducke	Brazil

Z. obliqua A. Br.	Chóco, Colombia; S. Panama
Z. paucijuga Wieland	W. Mexico
Z. picta Dyer	W. Guatemala; Chiappas, Mexico
Z. peoppigiana Martius & Eichler	Peru
Z. portoricensis Urban	Puerto Rico
Z. pseudomonticola L. D. Gomez	Costa Rica
Z. pseudoparasitica Yates *in* Seemann	N. Panama
Z. pumila L.	Dominican Republic; Puerto Rico; Cuba
Z. purpurea Vovides, Rees & V-Torres	Veracruz, Mexico
Z. pygmaea Sims	W. Cuba & Isla de Piños (Isla de Juventud)
Z. roezlii Linden	Chóco, Colombia
Z. skinneri Warsz. ex A. Dietrich	Nicaragua; Costa Rica; N. Panama
Z. soconuscensis Schutzman, Vovides & Dehgan	Chiapas, Mexico
Z. spartea A. DC.	Oaxaca, Mexico
Z. splendens Schuztman	Chiapas, Mexico
Z. sylvatica Chamberlain	Oaxaca, Mexico
Z. tuerckheimii J. Donnell Smith	Guatemala
Z. ulei Dammer	W. Brazil, E. Peru
Z. variegata Warsz.	Guatemala
Z. verschaffeltii Miq.	Mexico
Z. wallisii A. Br.	Colombia

The taxonomy of *Zamia* is in need of substantial revision with input from further field and analytical studies. All of the names given here have been validly and legitimately published (Eckenwalder, 1980; Stevenson & Sabato, 1986). However, many of them are undoubtedly synonyms and at least 12 species of *Zamia* have yet to be described, which will appear in a Flora Neotropica monograph on neotropical cycads in the near future.

Acknowledgments

This work was supported, in part, by National Science Foundation Grants BSR 8607049 and BSR 8796279 to D.W.S.

Literature Cited

Eckenwalder, J. 1980. Taxonomy of the West Indian cycads. J. Arnold Arb. **61:** 701–722.

Johnson, L. A. S. 1959. The families of cycads and the Zamiaceae of Australia. Proc. Linn. Soc. New South Wales **84:** 64–117.

Moretti, A. & S. Sabato. 1984. Karyotype evolution by centromeric fission in *Zamia* (Cycadales). Pl. Syst. Evol. **146:** 215–233.

Norstog, K. 1980. Chromosome numbers in *Zamia* (Cycadales). Caryologia **33:** 419–428.

Osborne, R. & J. Hendricks. 1985. A world list of cycads. Encephalartos **3:** 13–17.

——— **&** ———. 1986. A world list of cycads—Supplement. Encephalartos **5:** 27.

Sabato, S. & P. De Luca. 1985. Evolutionary trends in *Dion* (Zamiaceae). Amer. J. Bot. **72:** 1353–1363.

Schuster, J. 1932. Cycadaceae. *In* A. Engler (ed.), Das Pflanzenreich **99(IV. 1):** 1–168.

Stevenson, D. & G. Siniscalco Gigliano. 1989. The systematic value of the monosaccharide composition and distribution pattern of cycad mucilages. Biochem. Syst. Ecol. **17:** 185–190.

——— **& S. Sabato.** 1986. Typification of names in *Zamia* L. and *Aulacophyllum* Regel (Zamiaceae). Taxon **35:** 134–144.

———, ——— **& M. Vazquez Torres.** 1986. A new species of *Ceratozamia* (Zamiaceae) from Mexico, with comments on vegetative morphology, habitats, and species distribution in *Ceratozamia*. Brittonia **38:** 17–26.

Index to Common Names

Index to Scientific Names

New genera and species are in **boldface**. Page numbers followed by an asterisk (*) indicate illustrations or maps.